Exploration seismology
Volume 1
*History, theory, and
data acquisition*

Exploration
SEISMOLOGY
Volume 1

History, theory,
& data acquisition

R. E. SHERIFF
Professor of Geophysics,
University of Houston

L. P. GELDART
Coordinator, Canadian International
Development Agency Program for Brazil

Cambridge University Press
Cambridge
London   New York   New Rochelle
Melbourne   Sydney

Published by the Press Syndicate of the University of Cambridge
The Pitt Building, Trumpington Street, Cambridge CB2 1RP
32 East 57th Street, New York, NY 10022, USA
10 Stamford Road, Oakleigh, Melbourne 3166, Australia

First published 1982
Reprinted with corrections 1985

Printed in Great Britain at the University Press, Cambridge

Library of Congress catalogue card number: 81–18176

British Library Cataloguing in Publication Data

Sheriff, R. E.
Exploration seismology
Vol. 1: History, theory, and data acquisition
1. Seismology
I. Title   II. Geldart, L. P.
551.2′2   QE534.2

ISBN 0 521 24373 4

Contents

Contents of Volume 2

Preface

Chapter 4 of *Applied Geophysics* (Telford, Geldart, Sheriff and Keys, CUP, 1976) was the starting point for this book. In writing that chapter, the present authors were restricted by lack of space from including much material they would have liked to have included. Moreover, the comprehensive nature of *Applied Geophysics* restricted the exposition to the basic principles at a fairly elementary level and precluded going into details or advanced topics. We have been encouraged by colleagues to expand this chapter into a stand-alone book because many important aspects of exploration seismology are not systematically described in the literature. Among such are the early history of seismic exploration, the vector wave equation, transversely isotropic media, the Kirchhoff formula, diffraction solutions, various types of surface waves, both the Knott and Zoeppritz approaches to energy partition at interfaces, normal-mode propagation, reflector curvature effects, resolution considerations, three-dimensional methods, vertical profiling, extended resolution, S-wave studies, etc.

We found so many topics that needed to be included that we have had to divide our work into two volumes, volume 1 dealing with *theory and data acquisition* and volume 2 with *data processing and interpretation*. With only a few exceptions, the basic division came naturally. Many university post-graduate courses divide on similar bases, so that volume 1 should be sufficient for some courses and volume 2 for others. Occasional references to chapters 7 to 10 refer to volume 2.

This book is intended both as a text for students in advanced courses and also as a reference book for persons engaged in seismic exploration. The theoretician and student need to appreciate what is practical, and practitioners need to understand theory and the possibilities it may suggest. We have tried to strike a balance between theoretical aspects and practical details, that is, to include sufficient theory to enable the reader to follow most of the current literature and enough practical details to make the book of value to the practicing seismologist.

This book is structured to be as consistent and systematic as possible. Effort has been made to be consistent with mathematical conventions, definitions and symbols. It has been necessary, however, to use certain symbols for several purposes, mainly to conform to accepted practices. A listing of the mathematical symbols is given immediately following this preface.

We give the systematic derivation of relationships from first principles of physics, except for a few cases where the derivations are excessively lengthy or involve higher mathematics. At the same time we want to make our work intelligible to geologists and geophysicists who do not feel comfortable with mathematics. A reader willing to take the mathematics on faith should be able to jump over the equations and still see the implications of the mathematical conclusions. We have endeavored to explain in words the significance of equations rather than merely let the equations speak for themselves. We have been encouraged by Sir Harold Jeffreys' preface to *The Earth* (1924):

> If the geologist cannot follow a part of the book, I hope he will omit it and go on ... Mathematically trained readers, with few exceptions, will do just the same.

Terms are *italicized* where they are defined. Precision of terminology is often one of the clearest indications of a person's degree of understanding and we have made special effort to define and use the specialized vocabulary of seismology precisely.

Each chapter begins with an overview so that the reader may see clearly the aim of the various sections. Each chapter (except for chapter 1) ends with problems which endeavor to illucidate aspects not treated in the text as well as develop proofs and additional relationships.

We acknowledge the assistance we have had from many people in the preparation of this book and express our thanks to them. We particularly relied on many people in preparing the history section, and especially thank Messrs Harry Mayne, Dan Skelton, Bill Laing, O. Leenhardt and Bruce Frizelle for their help. We especially cite Margaret Sheriff and Leslie Denham for their help.

R. E. Sheriff
L. P. Geldart   1981

Mathematical conventions and symbols

(a) General rules and definitions

$\leftrightarrow$ denotes corresponding functions in different domains, the arguments indicating the type of transform

$\mathbf{A}$ vector quantity, magnitude is $|\mathbf{A}|$

∇ del, the vector operator $\mathbf{i}(\partial/\partial x) + \mathbf{j}(\partial/\partial y) + \mathbf{k}(\partial/\partial z)$

∇^2 Laplacian operator, $\partial^2/\partial x^2 + \partial^2/\partial y^2 + \partial^2/\partial z^2$

$\sum_k g_k$ sum of g_k over appropriate values of k

$\exp x$ e^x

$\mathrm{sinc}\,(t)$ $(1/t)\sin t$

$\mathrm{step}\,(t)$ unit step function, $\mathrm{step}\,(t) = 0$ for $t < 0$, $+1$ for $t > 0$.

$\delta(t)$ unit impulse at $t = 0$.

(b) Latin symbols

a rate of increase of velocity with depth, element spacing

a, b, c constants

a_i, b_i angles of incidence

A, B, C amplitudes of waves or of displacement potential functions

$\mathscr{A}, \mathscr{B}$ amplitudes of displacements

$c_{\mathscr{V}}, c_{\mathscr{P}}$ specific heat at constant volume, pressure

$D_\mathrm{s}, D_\mathrm{w}$ depth of shot, weathering

E Young's modulus, energy, energy density, energy ratio

$E_\mathrm{s}, E_\mathrm{d}$ elevation of surface, datum

$E_\mathrm{R}, E_\mathrm{T}$ fraction of energy reflected, transmitted

F array response, magnitude of a force

$\mathbf{F}$ force per unit mass

h distance to reflector or refractor, thickness, depth, damping factor

$h(t)$ output

H magnetic field strength

i current

i_0 angle of raypath at shotpoint

i, j, k   unit vectors in x-, y-, z-directions

I intensity

k constant, bulk modulus

K force on geophone coil per unit current, effective elastic modulus

l, m, n direction cosines relative to x-, y-, z-directions

L self inductance, kinetic energy per unit volume

m mass

m, n constants, integers, exponential decay parameters

N noise

p raypath parameter

$\mathscr{P}$ pressure

P power

Q quality factor

r integer, distance, radius, radial coordinate

R reflection coefficient, radius, resistance

s Laplace transform parameter

S spring constant, signal

$\mathscr{S}$ area, surface

t time, traveltime

t_0 traveltime at shotpoint

t_i, t_{iu}, t_{id} refraction intercepts

t_{uh} uphole time

$\Delta t_0, \Delta t_s, \Delta t_g$ near-surface corrections

Δt_c differential weathering correction

Δt_n normal moveout (NMO)

$\Delta t_d / \Delta x$ dip moveout

T period, transmission coefficient

u, v, w displacements in x-, y-, z-directions

$\dot{u}, \ddot{u}$ time derivitives of u, $\partial u / \partial t$, $\partial^2 u / \partial t^2$

U group velocity

v amplitude of geophone velocity

V velocity of a wave, phase velocity, interval velocity

$\overline{V}$ equivalent average velocity

$\overline{\overline{V}}$ rms velocity

V_a apparent velocity

V_R, V_L, V_T velocity of Rayleigh, Love waves, tube waves

V_u, V_d apparent velocity in updip, downdip directions

V_H, V_W velocity below, in the low-velocity layer (LVL)

$\mathscr{V}$ volume

W S-wave acoustic impedance

x offset, displacement, distance

x', x_c critical distance, crossover distance for a refraction

z depth

Z P-wave acoustic impedance

(c) Greek symbols

α P-wave velocity, angle, angle of approach

β S-wave velocity, angle

γ phase or phase difference, specific heat ratio

Γ geophone sensitivity

δ logarithmic decrement, delay time, angle

δ_s, δ_g delay time associated with shotpoint, geophone

$\delta(t)$ unit impulse at $t = 0$

Δ dilatation, sampling interval

ε eccentricity, phase shift upon reflection

$\varepsilon_{xx}, \varepsilon_{xy}$ normal, shearing strains

ζ vector displacement $= u\mathbf{i} + v\mathbf{j} + w\mathbf{k}$

η absorption coefficient

$\boldsymbol{\eta}$ outward-drawn unit normal

θ angle, spherical coordinate (colatitude), polar coordinate

θ_x angle of rotation about x-axis

Θ critical angle

$\boldsymbol{\Theta}$ vector rotation $= \theta_x\mathbf{i} + \theta_y\mathbf{j} + \theta_z\mathbf{k}$

κ $2\pi(\text{wavenumber}) = 2\pi/\lambda$

λ wavelength, Lamé constant

λ_a apparent wavelength

$\lambda_{\parallel}, \lambda_{\perp}$ Lamé constants for transversely isotropic medium

μ rigidity (shear) modulus, a Lamé constant

$\mu_{\parallel}, \mu_{\perp}, \mu^*$ rigidity moduli for transversely isotropic medium

v frequency $= \omega/2\pi = 1/T$

v_0 natural frequency

ξ dip, distance from origin to moving point on a curve

Ξ strike

ρ density, radius (of curvature)

σ Poisson's ratio, strength per unit length, standard deviation

σ_{xy} stress in x-direction on surface perpendicular to y-axis

τ damping factor

γ potential function, source density

ϕ spherical coordinate (longitude), porosity, angle, P-wave displacement potential function, magnetic flux, loss angle

Φ transform of P-wave displacement potential function

$\chi, \boldsymbol{\chi}$ S-wave displacement potential function

ψ wavefunction, disturbance

ω angular frequency $= 2\pi v$

ω_0 natural frequency

$\boldsymbol{\Omega}$ vector potential function for rotation

1
Introduction

Overview

Exploration seismology deals with the use of artifically-generated elastic waves to locate mineral deposits (including hydrocarbons, ores, water, geothermal reservoirs, etc.), archaeological sites, and to obtain geological information for engineering. Exploration seismology provides data which, when used in conjunction with other geophysical, borehole and geological data and with concepts of physics and geology, can provide information about the structure and distribution of rock types. Usually seismic exploration is part of a commercial venture and hence economics is an ever-present concern. Seismic methods alone cannot determine many of the features which make for a profitable venture and, even when supplemented by other data, a unique interpretation is rarely evident. Seismic exploration usually stops long before unambiguous answers are obtained and before all has been learned which might possibly be learned, because in someone's judgement further information is better obtained in some other way, such as by drilling a well. Seismic methods are in continual economic competition with other methods.

The importance of seismic work in the exploration for petroleum is evidenced by its extensive application. Almost all oil companies rely on seismic interpretation for selecting the sites for exploratory oil wells. Despite the indirectness of the method – most seismic work results in the mapping of geological structure rather than finding petroleum directly – the likelihood of a successful venture is improved more than enough to pay for the seismic work. Likewise, seismic methods are important in groundwater searches and in civil engineering, especially to measure the depth to bedrock in connection with the construction of large buildings, dams, highways and harbor surveys, and to determine whether blasting will be required in road cuts, if potential hazards such as limestone caves or forgotten mine workings underlie building sites, if tunnels or mine drifts are likely to encounter water-filled zones, or if faults are present which might be hazards to a nuclear power plant. On the other hand, seismic techniques have found little application in direct exploration for minerals because they do not produce good definition where interfaces between different rock types are highly irregular. However, they are useful in locating features such as buried channels in which heavy minerals may be accumulated.

Exploration seismology is an offspring of earthquake seismology. When an earthquake occurs, the earth is fractured and the rocks on opposite sides of the fracture move relative to one another. Such a rupture generates seismic waves which travel outward from the fracture surface. These waves are recorded at various sites using seismographs. Seismologists use the data to deduce information about the nature of the rocks through which the earthquake waves traveled.

Exploration seismic methods involve basically the same type of measurements as earthquake seismology. However, the energy sources are controlled and movable and the distances between the source and the recording points are relatively small. Much seismic work consists of *continuous coverage* where the response of successive portions of earth is sampled along lines of profile. Explosives and other energy sources are used to generate the seismic waves and arrays of seismometers or geophones are used to detect the resulting motion of the earth. The data are usually recorded in digital form on magnetic tape so that computer processing can be used to enhance the signals with respect to the noise, extract the significant information and display the data in such a form that a geological interpretation can be carried out readily.

The basic technique of seismic exploration consists of generating seismic waves and measuring the time required for the waves to travel from the source to a series of geophones, usually disposed along a straight line directed towards the source. From a knowledge of traveltimes to the various geophones and the velocity of the waves one attempts to reconstruct the paths of the seismic waves. Structural information is derived principally from paths which fall into two main categories: *head-wave* or *refracted* paths in which the principal portion of the path

is along the interface between two rock layers and hence is approximately horizontal, and *reflected* paths in which the wave travels downward initially and at some point is reflected back to the surface, the overall path being essentially vertical. For both types of path, the traveltimes depend upon the physical properties of the rocks and the attitudes of the beds. The objective of seismic exploration is to deduce information about the rocks, especially about the attitudes of the beds, from the observed arrival times and (to a limited extent) from variations in amplitude, frequency and waveform.

A brief outline of the seismic reflection and refraction methods is given first (§1.1); this explanation ignores complications and variations which are the subjects of future chapters.

Exploration seismology is a fairly young activity, having begun only about 1923. The early history of seismic exploration is summarized in §1.2. The seismic method is by far the most important geophysical technique in terms of capital expenditure (§1.3) and number of geophysicists involved. The predominance of the seismic method over other geophysical methods is due to various factors, the most important of which are the high accuracy, high resolution and great penetration of which the method is capable. Seismic literature is discussed in §1.4.

1.1   Outline of seismic methods
1.1.1   *Seismic reflection method*

Seismic techniques have changed considerably within recent years and many variations exist. The technique described below provides a background to the understanding of subsequent discussions; the reasons for various steps and various modifications of techniques will be described in subsequent chapters.

Assume a land crew using an explosive charge as the energy source. The first step after determining proper locations is the drilling of a vertical hole in the earth at the *shotpoint*, the hole diameter being perhaps 10 or 12 cm and the depth usually between 6 and 30 m. A *charge* of 1 to 25 kg of explosive is armed with an electric blasting *cap* and then placed near the bottom of the hole. Two wires extend from the cap to the surface where they are connected to a *blaster* which is used to send an electrical current through the wires to the cap which then explodes, initiating the explosion of the dynamite (the *shot*).

Two *cables* 2 to 4 km long are laid out in a straight line extending each way from the hole about to be fired. The cables contain many pairs of electrical conductors, each pair terminating in an electrical connector at both ends of the cable. In addition, each pair of wires is connected to one of several outlets spaced at intervals of 25 to 100 m along the cable. Several *geophones* (*seismometers*) are connected to each of these outlets so that each pair of wires in the cable carries the output energy of a *group* of geophones back to the recording instruments. Because of the small spacing between the geophones in the group attached to one pair of wires, the whole group is approximately equivalent to a single fictitious geophone located at the center of the group. Usually 48 or more geophone groups are located at equal intervals along the cable. When the dynamite charge is exploded, each geophone group generates a signal which depends upon the motion of the ground in the vicinity of the group. The net result is the generation of signals furnishing information about the ground motion at a number of regularly-spaced points (the *group centers*) along a straight line passing through the shotpoint.

The electrical signals from the geophone groups go to an equal number of amplifiers. These amplifiers increase the overall signal strength and partially eliminate (*filter out*) parts of the input deemed to be undesirable. The outputs from the amplifiers along with accurate timing signals are recorded on magnetic tape and on paper records. Thus the recorded data consist of several *traces*, each trace showing how the motion of one geophone group varies with time after the shot.

The data are usually processed to attenuate noise *vis-a-vis* reflected energy based on characteristics which distinguish them from each other, and the data are displayed in a form suitable for interpretation.

Events, that is, arrivals of energy which vary systematically from trace to trace and which are believed to represent reflected energy, are identified on the records. The *arrival times* (the interval between the shot instant and the arrival of the energy at a geophone group, also known as the *traveltime*) of these events are measured for various geophone groups. The location and attitude of the interface which gave rise to each reflection event are then calculated from the arrival times. Seismic velocity enters into the calculation of the location and attitude of the interfaces. The results are combined into cross-sections and contour maps which represent the structure of the geological interfaces responsible for the events. Patterns in the seismic data are sometimes interpreted in terms of stratigraphic features or as indicators of hydrocarbons. However, the presence or absence of hydrocarbons or other minerals is usually inferred from the structural information.

We have introduced above a number of terms used in a specialized sense in seismic work (indicated by *italics*), for example, shotpoint, group, trace, events, arrival time. Exploration seismology abounds in such technical terms.

We shall henceforth use italics to indicate that we are defining a term; we shall follow the definitions given in the *Encyclopedic Dictionary of Exploration Geophysics* (Sheriff, 1973) for seismic terms and the *Glossary of Geology* (Bates and Jackson, 1980) for geologic terms.

1.1.2 *Seismic refraction method*

The principal difference between reflection and refraction methods is that for refraction the distance between shotpoint and geophones is large relative to the depths of the interfaces being mapped whereas it is small or comparable to the depths for reflection. Consequently the travel paths in refraction work are predominantly horizontal whereas for reflection work they are predominantly vertical. Head waves or refractions (see §2.4.7) enter and leave a high-velocity bed at the critical angle and only a bed with velocity significantly higher than any bed above it can be mapped. Consequently the applications of refraction methods are more restricted than those of reflection. (It should be noted that refraction is used in two different senses in seismology, to refer to the bending of raypaths due to changes in velocity and in the present sense of involving head waves. The classical mapping of high-velocity masses such as saltdomes is also classed as a refraction method, although refraction at the critical angle is not necessarily involved; see §6.1.2.)

Since refraction exploration generally involves greater distances than reflection work, stronger sources are required. Because distributed in-line geophones would attenuate the head waves which have appreciable horizontal component of motion, geophones are either bunched together or distributed perpendicular to the shot–geophone line. Otherwise, however, the same equipment can often be used.

1.2   History of seismic exploration

This account is based on articles by Barton (1929), Heiland (1929a, b), Mintrop (1931), Shaw *et al.* (1931), Rosaire and Lester (1932), DeGolyer (1935), Rosaire (1935), Leet (1938), Weatherby (1940), Schriever (1952), Born (1960), McGee and Palmer (1967), Elkins (1970), Laing and Searcy (1975), Owen (1975), Petty (1976), Sweet (1978), and Green (1979), supplemented by conversations with individuals who were personally involved in early geophysical work.

1.2.1 *Preliminary events*

Geophysical exploration for oil began with the torsion balance, which was developed by Baron Roland von Eötvös about 1888. While gravity surveys with the torsion balance were made in Europe on a limited scale, beginning about 1900, to map geologic structures, the first extensive surveys for petroleum objectives were in the United States and Mexico in the 1920s. In December 1922 a survey of the known Spindletop saltdome in Texas gave a gravity anomaly but subsequent surveys were disappointing until 1924 when the Nash Dome was discovered. This resulted in the first geophysical oil discovery in January 1926. Through 1929 sixteen salt-domes found by torsion balance surveys subsequently resulted in hydrocarbon discoveries (Sweet, 1978).

The theory of seismic waves might be dated from Robert Hooke's law enunciated in 1678, but most of the theory of elasticity was not developed until the 1800s. Baron Cauchy's memoir on wave propagation won the Grand Prix of the French Institute in 1818 and S. D. Poisson showed theoretically the separate existence of P- and S-waves around 1828. C. G. Knott (1899) presented a paper on the propagation of seismic waves and their reflection and refraction, and Emil Wiechert and Karl Zoeppritz (1907) published their work on seismic waves. Lord Rayleigh (1885), A. E. H. Love in 1911 (see Love 1927) and R. Stoneley (1924) developed the theories of the surface waves which bear their names.

Robert Mallet (1848, 1851) began experimental seismology by measuring the speed of seismic waves using black powder as the energy source and a disturbance of the surface of a bowl of mercury as the detector. Mallet obtained very low velocities; probably low sensitivity allowed him to see only the later cycles of Rayleigh waves, then unknown. H. L. Abbot (1878) measured P-wave velocities using essentially the same type of detectors but a very large explosion. John Milne (1885) and T. Gray used a falling weight as a source (as well as explosives) in a series of seismic wave studies using two seismographs in line, probably the first seismic spread. Otto Hecker (1900) used nine mechanical horizontal seismographs in line to record both P- and S-waves.

The possibility of employing the seismograph to define subsurface conditions was first put forward by Milne in 1898 (Shaw *et al.*, 1931):

> As an earthquake wave travels from strata to strata, if we study its reflection and changing velocity in transit, we may often be led to the discovery of certain rocky structures buried deep beneath our view, about which without the help of such waves it would be hopeless ever to attain any knowledge … Earthquakes are gigantic experiments which tell the elastic moduli of rocks as they exist in nature, and when properly interpreted may lead to the proper comprehension of many ill-understood phenomena.

L. P. Garret in 1905 suggested the use of seismic refraction to find saltdomes but suitable instruments had not yet been developed (DeGolyer, 1935).

1.2.2 *Early applications to petroleum exploration*

After the sinking of the Titanic by an iceberg in 1912, Reginald A. Fessenden worked on inventions for iceberg detection. Among the methods was the use of acoustic waves in water, and an outcome of this was the first (US) patent (fig. 1.1) on the application of seismic waves to exploration, applied for in 1914 and issued in 1917, entitled 'Method and apparatus for locating ore bodies'. Fessenden's patent said

> The invention described herein relates to methods and apparatus whereby, being given or having ascertained two or more of the following quantities, i.e., time, distance, intensity and medium, one or more of the remaining quantities may be determined.

He proposed using sources and detectors in water-filled holes, and locating ore bodies by both the use of reflections from them and by variations they introduce in traveltime measurements between holes. His patent was subsequently challenged (unsuccessfully) by (among others) Mintrop (1931) because Fessenden used 'acoustic' waves rather than 'seismic' waves and because his use of boreholes for sources and detectors did not accord with subsequent practice.

Ludger Mintrop in Germany in 1914 devised a seismograph with which he could make observations of explosion-generated waves with sufficient accuracy to make exploration feasible.

The Germans and Allies both experimented during World War I with the use of three or more mechanical seismographs to locate enemy artillery, but airwaves generally proved more satisfactory than seismic waves for this purpose. Among those involved in these experiments were Mintrop and the Americans, Fessenden, E. A. Eckhardt, W. P. Haseman, J. C. Karcher and Burton McCollum. These six were predominant in the development of commercial application of seismic waves after the war. McCollum attributed the idea of applying seismic methods to petroleum exploration to Haseman (unpublished 'Recollections re McCollum' by R. L. Palmer). Mintrop's work was clearly independent and Fessenden was apparently only brought into application efforts about 1925.

Mintrop in 1919 applied for a German patent on 'Method for the determination of rock structures' which was issued in 1926. Mintrop's patent said

Fig.1.1. First page of Fessenden's patent.

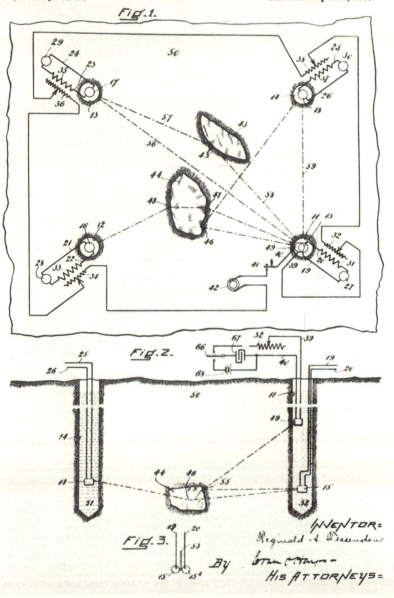

R. A. FESSENDEN.
METHOD AND APPARATUS FOR LOCATING ORE BODIES.
APPLICATION FILED JAN. 15, 1917.

1,240,328.

Patented Sept. 18, 1917.

Fig. 1.

Fig. 2.

Fig. 3.

INVENTOR:
Reginald A. Fessenden
BY
HIS ATTORNEYS:

Where the problem is to obtain . . . the approximate composition of the strata, the divining rod has been used as is well known. However, . . . it has not yet been possible to ascertain a connection of unique meaning between the indication of the divining rod and the geologic particularities of the subsoil . . . According to my invention . . . the connection of mechanical waves with the characteristic properties of the strata is much more immediate . . . mechanical waves are artificially generated . . . by detonating a certain amount of explosives, their elastic propagation through the various formations is recorded by a seismometer located at a suitable distance . . . from the records of the latter the velocities of the various waves and the depth to which they penetrated can be determined, which allows conclusions as to the succession, thickness, density as well as the direction of the strike and dip of rock formations.

John William Evans and Willis B. Whitney in 1920 applied for a British patent on 'Improvements in and relating to means for investigating the interior of the Earth's crust' which was issued in 1922. Their patent said

The present invention . . . is characterized in that the sound waves . . . are received simultaneously or approximately so at a plurality (at least two . . .) of receiving stations . . . for the following reasons: Even in the simplest case when it is known that the stratum to be examined is horizontal there are two unknown quantities namely (1) the average velocity of the reflected wave . . . and (2) the depth of the reflecting stratum and therefore two equations . . .

and two observations are consequently necessary. Despite their rather complete grasp of reflection seismology, this patent does not figure prominently in subsequent developments which concentrated on refraction.

Udden (1920) wrote in the *Bulletin of the American Association of Petroleum Geologists* (AAPG) (and illustrated it with fig. 1.2):

. . . it ought to be possible, with present refinements in physical apparatus and their use, to construct an instrument that would' record the reflections of earth waves started at the surface, as they encounter such a well-marked plane of difference in hardness and elasticity as that separating the Bend and Ellenberger formations (in North-central Texas) . . . A seismic wave might be started by an explosion at the surface of the Earth, and a record of the emerged reflection of this wave . . . might be registered on an instrument placed at some distance from the point of explosion . . . It ought to be possible to notice the point at which the first reflection from the Ellenberger appears . . . With a map of the surface of the Ellenberger, it seems to me that millions of dollars worth of drilling could be eliminated.

In 1920 the Geological Engineering Company was founded by Haseman, Karcher, Eckhardt and McCollum to apply seismic exploration to finding petroleum. Karcher had recorded a seismic reflection from waves generated by artillery at the Indian Head test range in Maryland in 1917 and in a quarry (fig. 1.15a) in Washington, D.C., in 1919 (Karcher 1974). They converted an oscillograph into a 3-trace recorder and constructed electrodynamic geophones from radiotelephone receivers. In June, 1921,

Fig.1.2. Reflection expected from the contact between the Bend formation and the underlying Ellenburger limestone. (From Udden, 1920.)

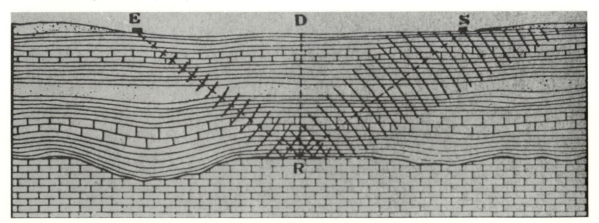

Karcher, Haseman, I. Perrine and W. C. Kite at Belle
Isle (Oklahoma City) obtained a clear reflection from the
contact between the Sylvan shale and the Viola limestone
(fig. 1.3). About five months of reflection and refraction
experimentation were carried out. One experiment involved dropping dynamite from an airplane in an attempt
to obtain more nearly plane waves (Karcher had tried
using aerial fireworks as a source in 1919). The company
ran out of funds when a surplus of oil forced the price
down to 15¢/barrel. The principals returned to their
former jobs, except for McCollum. McCollum agreed to
settle with the company's creditors in return for the
company's patents and equipment.

During 1920–1 Mintrop shot refraction lines across
two known saltdomes in northern Germany and discovered another, the Meissendorf dome, although it had
no commercial significance. In 1921 he founded Seismos
to do geophysical exploration and subsequently wrote a
number of pamphlets promoting refraction exploration.
In 1922 Seismos tried seismic methods in Sweden for
mining objectives and in Holland for coal mapping.

Everett Lee DeGolyer wrote on 3 October 1922 to
J. B. Body in London (the following three extracts are
from DeGolyer's papers in the library of Southern
Methodist University, Dallas, Texas):

> You will remember that during the past summer
> Dr Barton, of the Amerada Petroleum Corpora-
> tion's staff, spent some time in Europe receiving
> instruction in the use of Eötvös torsion balances,
> and while there made several visits to Germany to
> investigate other physical methods of approach to
> geologic problems.
>
> One of the methods which interested him
> very much, and which he seemed to think had
> considerable possibility, was the Seismic method . . .
> I should like to suggest that it be called to Dr Erb's
> [Shell's] attention with the recommendation that
> he consider its availability for use in the Mecatepec–
> Papantla District [Mexico] . . .

Body wrote to DeGolyer on 14 December 1922:

> You will have seen my cable No. 88 . . . 'Negotiating
> with Seismos from Hannover for using Mexico their
> method measuring with seismograph transmission
> waves caused by explosions thereby determining
> depth positions subterranean Tamasopo also out-
> line saltdomes. Method gave satisfactory results
> central Europe and are assured can be used Mexican
> conditions. Our intention is send out party . . .'

Seismos party 1 began work in the Golden Lane
area of Mexico for Mexican Eagle (Shell) in 1923. The
contract for this work provided:

> Seismos bind themselves to organize an expedition
> in order to carry out the . . . investigations . . . This
> expedition shall consist of 2 seismologists and
> 1 mechanic, all of them experts with thoroughly
> up-to-date technical knowledge and possessing such
> zeal and sense of duty as is requisite for the success
> of their work. [They were] to be equipped with two
> complete seismic field-stations with the necessary
> instruments . . . to carry out in Mexico during 25
> days effectual observations . . . in a geologically
> known territory . . . Upon arrival the expedition
> shall confer with the local manager . . . who shall
> decide where and when their operations shall be
> carried out and to the solutions of which geological
> problems same shall be applied, on the under-
> standing that as far as purely scientific questions
> are concerned . . . they shall use their own dis-
> cretion . . .

Compensation was to be $600 American for the two
seismologists together and $150 for the mechanic during
the time the expedition was in Mexico and $500 for the
instruments. If Mexican Eagle so elected, the contract
could be replaced 'by a new contract concluded for an
indefinite period and for observations in Mexican regions
geologically unknown'. In this case, monthly compensation was to be increased to $800 and $250 for the men and
$1000 for the instruments. The geological problem in-
volved finding high-velocity limestone reefs under a shale
cover, a situation for which refraction appeared to be
ideal.

Seismos party 2 began work in Oklahoma and Texas
for Marland Oil Company (a predecessor of Conoco),
also in 1923. Seismos party 1 moved to Texas to work
for Gulf Oil in 1924 and in June discovered the Orchard
Dome southwest of Houston, which is usually considered
to be the first seismic (refraction) hydrocarbon discovery,
a claim disputed by McCollum (see below). Early refraction records are shown in fig. 1.4a, b.

The Seismos crews used a mechanical seismograph
(fig. 1.5) consisting of a mass suspended by a horizontal
leaf spring with a natural frequency of about 10 Hz. The
only amplification was mechanical and optical, and
recording was done by directing a beam of light onto a
mirror connected to the mass by a hair so that the mirror
rotated when the mass moved, and then onto a strip of
photographic paper moved by a hand crank turned by
the observer. Shotpoint-to-seismograph distance was
surveyed and a blastphone (fig. 1.6) was used to record
the airwave to find the shot instant. (Subsequently radio
was used to determine the shot instant and the airwave
arrival to find the shot-to-detector distance.) The overall

Fig.1.3. First application of the reflection seismograph to exploration. (From Schriever, 1952.) (*a*) Two reflection records made in September, 1921; E marks the explosion time, R the reflection from the Viola lime-stone, BP the airwave (blastphone). (*b*) First depth section, at Vines Branch, Oklahoma, 9 August, 1921. (*c*) First seismic structure map, near Ponca City, Oklahoma, September, 1921.

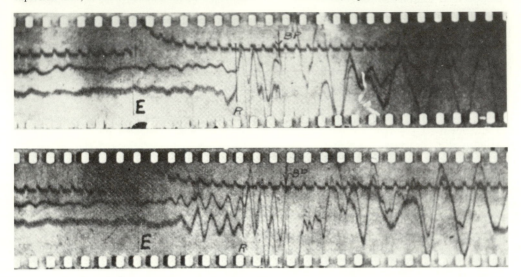

(*a*)

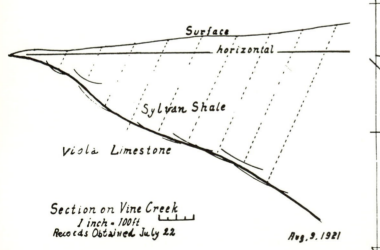

(*b*)

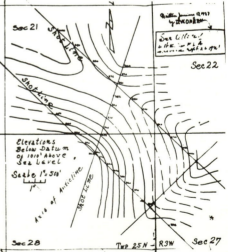

(*c*)

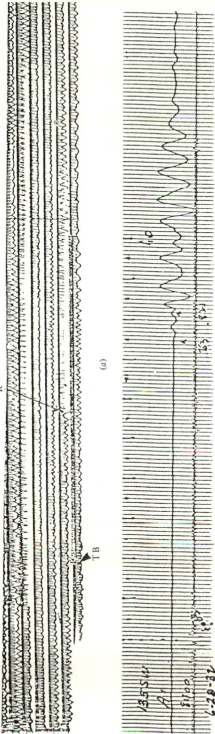

(a)

(b)

Fig.1.4. Early refraction records. (Courtesy Conoco.) (a) Record obtained with mechanical seismograph, 1924 or 1925; recording was helical about a drum so that traces at the right-end continue again at the left-end; TB = shot instant obtained by radio, R = refraction arrival. (b) Refraction record, Texas, June 1932.

sensitivity and precision were low and profiles were only $3\frac{1}{2}$ miles long, which gave limited penetration so that Seismos crews missed a number of domes at moderate depths. L. P. Garrett of Gulf Oil, for whom Seismos was working, developed the fan-shooting method (§6.1.2) about 1925, which increased the effectiveness in locating saltdomes. By 1929 the refraction method had found fifty saltdomes which resulted in hydrocarbon discoveries (Sweet, 1978). During the same period 'geology and accident' discovered one dome (Barton, 1929, p. 616).

Following the failure of the Geological Engineering Company in 1922, McCollum obtained the backing of Atlantic Refining and formed McCollum Geological Exploration to carry out refraction work. New equipment was built and in 1924 both reflection and refraction was used in the Tampico area of Mexico. The first well drilled

on a seismic location, La Gatero No. 4, was dry although the seismic prediction was correct. In May, 1924 the Zacamixtle 199 well in the Golden Lane area succeeded in finding oil, to dispute the claim of the Orchard dome in Texas as the first seismic discovery. However, the Mexican well was non-commercial at the time because of its remote location (Owen, 1975). In 1928 the Atlantic–McCollum joint venture was dissolved; McCollum and Atlantic divided the four sets of instruments and McCollum formed McCollum Exploration Company.

The Marland Oil Company had supported two months of the 1921 Geological Engineering Company reflection experimentation (which was unsuccessful) and had brought Seismos party 2 to Texas in 1923. The Seismos party failed to find any saltdomes for Marland. In 1925 Marland hired Haseman, Eckhardt, Eugene

Fig.1.5. Mintrop's mechanical seismograph. Movement of the case with respect to the inertial mass tilted the aluminum cone, pulling on the hair and rotating the mirror. (From Malamphy, 1929.)

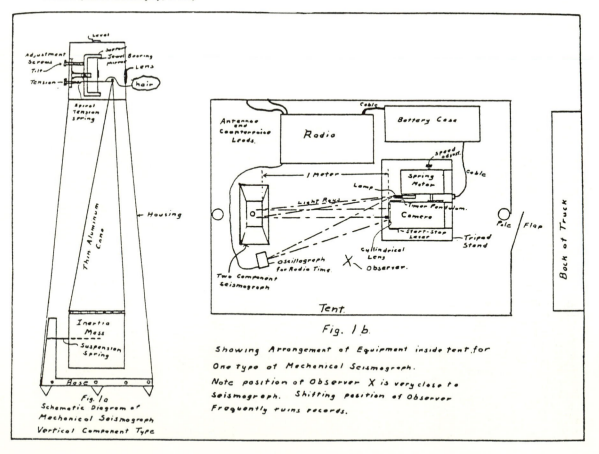

McDermott and others to develop a more sensitive electrical seismograph. The Marland field party began exploration in 1926, replacing the Seismos party. The equipment worked well but Marland never recorded a saltdome discovery.

1.2.3 *The Geophysical Research Corporation*

DeGolyer was at first disappointed with the refraction method, but success by Seismos crews working for Gulf changed his mind and he began to search for personnel to develop seismic methods. He learned of Karcher's 1921 experiments and in May 1925, Amerada, Rycade (an Amerada subsidiary) and Karcher formed the Geophysical Research Corporation (GRC). They acquired Fessenden's patent and his services as consultant.

GRC built an electrical seismograph which was much more sensitive than the Seismos mechanical seismograph. The detector was a variable-reluctance type and the amplifier was resistance-coupled using a vacuum tube.

The oscillograph used two galvanometers and the recording film was hand-cranked. Timing lines were obtained by shining a light through slits attached to the prongs of a 50 Hz tuning fork. The time-break (shot instant) was transmitted by interrupting a CW (continuous wave) transmitter.

GRC fielded seven field parties in 1926 and refraction exploration greatly expanded. A GRC crew under E. E. Rosaire was forbidden to shoot profiles more than $3\frac{1}{2}$ miles in length, which had become the standard distance since it had been successful for Mintrop, but the observers 'got lost' and discovered the Port Barre saltdome (Sweet, 1978). Thereafter the standard distance became 6 miles. Refraction at the time was used as a reconnaissance method and was usually followed by detailing with a torsion balance (and later gravimeter) survey.

GRC party 6, an experimental crew, tried reflection work in Kansas in 1926. They soon moved to Texas and obtained usable reflection records from the caprock of the

Fig.1.6. Blastphone used to detect airwave for determining shot–detector distance. The diaphragm d is a pie tin, the transducer a carbon granule microphone.

(Photographed at Museum of the Geophysical Society of Houston.)

Nash saltdome. Other GRC parties also experimented at recording reflections. In 1927 party 6 moved to the Seminole Basin of Oklahoma, an area ideally suited to reflection work, where they soon found a structure which became the first discovery by the reflection method, the Maud Field (in 1928). This success was quickly followed by others and by 1930 the reflection method began to take over from the refraction method. Early reflection records are shown in fig. 1.15.

1.2.4 *Other activities in the 1920s*

Humble Oil Company, at the instigation of Wallace E. Pratt, established a geophysical department in 1924 under Dr N. H. Ricker and the following year fielded two refraction crews using mechanical seismographs designed by O. H. Truman (Carlton, 1946). These crews began using a telephone line to carry the time-break, but before the end of 1925 they used radio for both communications and transmission of the time-break.

Frank Rieber in 1924 obtained funding (from General Petroleum, Standard Oil of California, Associated and Shell) for a refraction survey in the San Joaquin valley of California. This survey was unsuccessful in obtaining deep information. Rieber carried out other surveys in California in 1927–8 but his company failed in 1930. In 1932 he began work on reflection instruments and in subsequent years introduced a number of instrumental innovations.

In 1925 the Petty Geophysical Engineering Company was formed by Dabney E. Petty and Olive Scott Petty (and other family members). They felt they could easily improve on Mintrop's mechanical seismograph and in 1926 fielded a crew equipped with condenser-type geophones (fig. 1.7) and vacuum-tube amplifiers. They used string galvanometers (fig. 1.8) and a camera with photographic paper pulled along by a spring motor; shadows cast by the moving strings on the paper were recorded. The Pettys did appreciable experimentation to

Fig.1.7. The Petty prototype geophone. The 'steady mass' m is on a long beam hinged (h) at the left and supported by a strong spring s. At the right end the beam is attached to one plate of an air-gap condenser c. Movement of the case with respect to the steady mass changes the separation of the condenser plates, the change in capacity being proportional to the displacement. The dimensions are $48 \times 32 \times 15$ cm. (Photographed at Museum of the Geophysical Society of Houston.)

find a quicker, easier way to locate saltdomes. They discovered that a salt forerunner (fig. 1.9), which could be distinguished by its amplitude, could be used to tell if a saltdome had been encountered even without knowing the shot-to-detector distance. They also found that the Rayleigh-wave pattern changed when a saltdome intervened and used this fact when they could not get a readable P-wave. The increased sensitivity of their equipment plus their interpretational ingenuity allowed them to survey with smaller shots than others used.

In 1927 the first well velocity measurements were made. A geophone lowered 5000 ft (1500 m) into a Gulf well in Kansas recorded the traveltime from a shot at the surface. Also in 1927, C. A. Heiland established the first course in exploration geophysics at the Colorado School of Mines.

McCollum successfully mapped the Barbers Hill Dome by reflection in 1928 using 100 detectors at each station spaced to attenuate horizontal waves, but the use of multiple detectors was too cumbersome and so was abandoned; it was revived about the mid 1930s with 4–6 detectors per group.

In the late 1920s seismic exploration began to move abroad; to Persia (Iran) and Venezuela in 1927, to Australia in 1929, to the Netherlands East Indies in 1930.

Donald C. Barton (1929), who subsequently became first president of the Society of Exploration Geophysicists (SEG), described early methods:

For work with the mirage [refraction] method, a troop rather commonly consists of one firing unit, two, three or four receiving units, a squad of hole diggers, a chief of party, a 'landman,' a calculator, and in some cases a crew of surveyors, and in some a hole-filling crew . . . The firing unit . . . is equipped with the necessary apparatus to fire the charge, . . . with meteorological apparatus and with a sending and receiving wireless set, which is used to communicate with the receiving units and to send out the instant of the explosion . . . The receiving unit . . . is equipped with a seismograph, . . . a wireless sending and receiving set, and meteorological apparatus . . . To set up a station, a 3 inch hole is dug to a depth of 3 ft, the geophone cable is reeled out and the geophone is dropped down the hole . . . Each receiving unit . . . signals that tentatively it is ready to receive . . . When all . . . have signaled their o.k. for firing, the firing master sets his wireless sending out a continuous wave note, . . . waits a standard short interval and then fires the charge. The wireless key is held down by a circuit which goes around the dynamite. The explosion instantaneously breaks the circuit, causes the release of the wireless key and the instantaneous [*sic*] cutting off of the wireless note . . . The average charges used . . . range from 40 to 250 lbs. For the same shots, Seismos Gesellschaft would use two to three times as large a charge . . .

Fig.1.8. String-galvanometer harp. Currents pass through fine wires (some are broken on this harp) held taut by small springs, causing the wires to be deflected in a magnetic field. Shadows of the wires were focused onto photographic paper. A string-galvanometer record is shown in fig.1.15*e*. (Photographed at Museum of the Geophysical Society of Houston.)

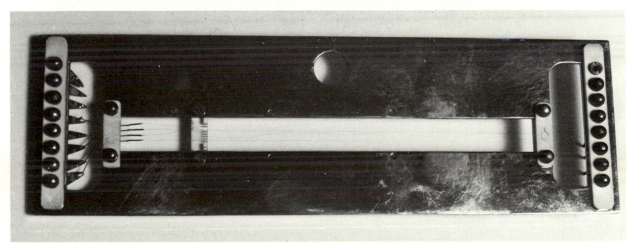

Fig.1.9. Refraction recording, 1930. (From Petty, 1976.)
(*a*) Normal (no salt) record; (*b*) record showing salt forerunner; (*c*) recording truck in use about this time (Courtesy GSI).

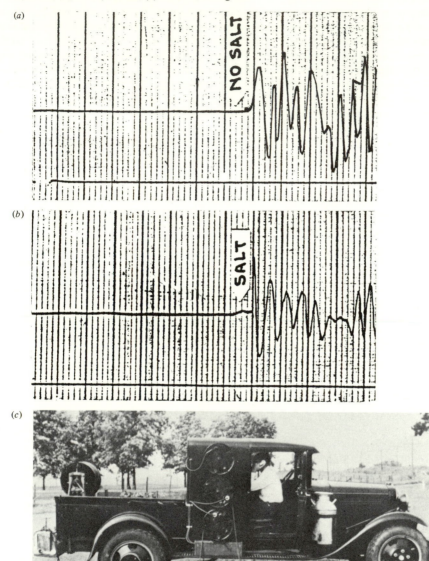

In the reflection type of shooting, the charge is very much smaller ... The practice ... [is to place the] main charge 17 to 25 ft down a 6 inch hole and the auxiliary charge at the surface. The latter is used to produce an air wave. The holes are dug with hand augers ... The distance between the firing point and the seismograph ... ranges from 1.2 to 1.8 times the depth of the formation which it is desired to map ...

A scout from a rival company not uncommonly is set to watch the troop and to report their activity to his company and especially to report anything to indicate that possibly they may have picked up a salt dome. He often gets to be on good terms with the troop, but at critical moments they go to all sorts of strategy to outwit him.

1.2.5 *Development of the geophysical contracting industry*

Burton McCollum in 1922 applied for a patent on 'Method and apparatus for determining the contour of subterranean strata' which was issued in 1928 along with two other patents on variations of seismic methods. McCollum sold two of his patents to the Texas Company in 1928 and seven more between 1929 and 1935. These and other patents were transferred to The Texas Development Company who tried to collect royalties from others but were mainly unsuccessful. In 1934 they sued Sun Oil Company for patent infringement. Almost the entire petroleum industry joined with Sun in the defense; the matter was settled out of court in 1937. The settlement involved companies forming a Seismic Immunities Group and granting each other royalty-free licenses of their patents

Fig.1.10. The 'Amerada Tree', a diagram drawn in 1950 to show geophysical contracting companies formed by people who left GRC. (W. J. Zwart and K. M. Lawrence located this historic document.)

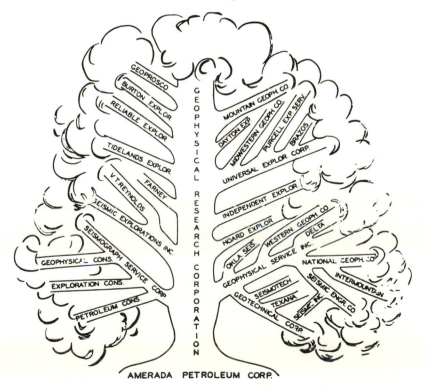

and of patents for which they might file within a year of withdrawal from the group. Initially 64 patents were involved, including 2 of Mintrop's, 10 of McCollum's, 32 of Harvey C. Hayes', 8 of Fessenden's and 2 of Karcher's. Several payment schemes could be elected; one involved a lifetime payment of $10 000 per party, a party being defined as either (*a*) a single recording unit with no more than 12 traces or (*b*) up to 4 recording units where shot-to-detector distance exceeded 2 miles (to cover respectively reflection and refraction work). The Mayne CDP patent (see §1.2.7) was one of the last important patents involved before the group disbanded entirely.

In 1929 a new Amerada president decided that GRC would no longer do reflection work for other companies (Karcher, 1974). While Petty and McCollum offered independent alternatives to accomplish geophysical work and some oil companies, such as Humble and Gulf, operated their own crews, the oil companies in general encouraged the formation of more new geophysical enterprises. Thus the early 1930s saw the advent of many geophysical contractors, including those which dominate today. Most of

these were formed by people who left GRC since it dominated the industry until then; some of these are shown in fig. 1.10. In addition a few companies (such as Rogers and General) were formed by people who left other companies such as Petty, and still other companies (such as Heiland) were formed without clear connections to preceding industry.

Among today's major companies, Geophysical Service Inc. (GSI) was formed in 1930, Seismograph Service Corp. (SSC) in 1931, Independent Exploration (now merged into Teledyne) in 1932, Western Geophysical in 1933, United Geophysical in 1935. SGRM (subsequently to become part of Compagnie Générale de Géophysique or CGG) began refraction work in France in 1930 and reflection work in 1934. The field work of Seismos declined to zero in 1931 but Seismos revived refraction work in Germany in 1934 and began reflection work about the same time. Prakla was founded in 1936 and subsequently merged with Seismos to form Prakla-Seismos. The rapid growth of exploration geophysics was almost entirely due to private enterprise with intense rivalry and competition

Fig.1.11. Growth of multi-channel recording. The starting point in 1914 is Mintrop's first portable seismograph. The vertical scale gives number of channels.

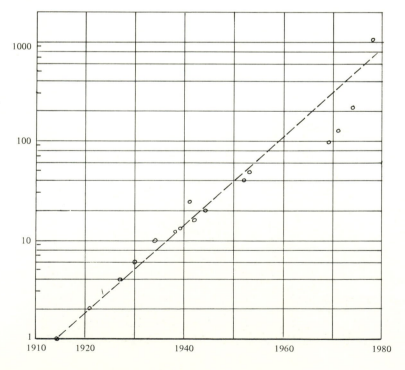

and extreme secrecy between the individual companies involved. Since the early 1930s, no single company has dominated geophysical exploration.

1.2.6 *Evolution of reflection equipment and methods*

The first GRC reflection work in 1926 employed the same 2-galvanometer arrangement used in refraction work, but a third galvanometer was added soon. A 4-channel system was built in 1928 and before long 6-channel instruments were in use; the standard was 6 to 8 in 1937 and by 1940 most crews were 10 to 12 channel. The number of channels has continued to increase (fig. 1.11). For many years after World War II 24 channels were standard, then in the late 1960s 48 channels became common and today (1981) most crews use 48 to 96 channels while some use appreciably more.

The mechanical seismograph was soon replaced with electrical geophones and vacuum-tube amplifiers. The early electrical geophones were mostly of three types: capacitance, variable-reluctance, and moving-coil electrodynamic; oil damping was generally used. Early electrical geophones (fig. 1.12) had to have high sensitivity because of the high noise level of available vacuum tubes. For the variable reluctance and moving-coil types, this meant large magnets because of the low permeability of the magnetic materials then available. As better magnetic materials and lower-noise vacuum tubes became available, the electromagnetic geophone increased in sensitivity and decreased in weight (from some 15 kg to a few hundred grams), electromagnetic damping replaced oil damping, and the electromagnetic type eventually became dominant (for land work). As a result of these improvements, multiple geophones per channel became practical; this usage was introduced in 1933 and was common practice by 1937.

The gain of early instruments was constant and repeated shots were usually required so that reflections at several arrival times could be mapped. Sometimes the gain was manually changed during the recording by the operator turning a switch. About 1932 automatic gain control was developed, first by changing the grid bias with time after the time-break, later by a feedback circuit. Amplifiers increased in gain, sophistication (initial suppression, automatic gain, mixing, etc.) and reliability. A 10-channel recorder from 1931, the first to change frequency response with time, is shown in fig. 1.13. Some timing wheels are shown in fig. 1.14 and reflection records are shown in fig. 1.15. About 1950, recording instruments became sufficiently reliable that the observer could 'do a

Fig.1.12. Early geophones in the Museum of the Geophysical Society of Houston. The geophones weigh (left to right, back row) 6.1, 8.7, 7.9, 6.7, (front row) 8.8 and 0.8 kg. At the lower right is a modern phone (30 g) for comparison.

day's work rather than instrument repair and adjustment'. The Southwestern Industrial Electronics (SIE) Company's P-11 recording system was a major advance in reliability. A chronology of some instrument developments is given in table 1.1.

The need for weathering corrections (§5.6.2) was recognized very early and shallow refraction shots were often made for this purpose. The first mapping was done by correlating reflections on widely separated profiles (fig. 1.16). Barton (1929) wrote:

> . . . a depth determination is made by each shot and to map the dip, folding or faulting of the surface, . . . it is necessary only to scatter 'shots' over the area to be mapped and draw structure-contours or profiles from the results. Practically the application of the method is somewhat uncertain . . . The impossibility of recognizing the reflecting bed is a serious disadvantage . . .

The correlation method did not work well in the Gulf Coast because the area lacks distinctive reflections.

In 1929 T. I. Harkins

. . . noticed that abnormal stepouts [dip moveout – see §3.1.2] were rather characteristic of the (Darrow dome) area and that these abnormal stepouts reversed. He correctly attributed this phenomenon to dipping beds. [Rosaire and Adler, 1934]

Soon dip shooting was carried out along continuous lines of traverse.

Although from the earliest days crews carried out surveys in water-covered areas, the methods were basically those for land crews, improvised for use in water. Petty (1976) describes a survey in Chacahoula Swamp, Louisiana, in 1926 (fig. 1.17) and in 1927 GRC fielded two crews for work in water-covered areas (Rosaire and Lester, 1932). Sidney Kaufman (personal communication), in following up an onshore lead in 1938, took his Shell shallow-water crew seaward 4 miles into 65 feet of water. Surveyors onshore directed locations and this imposed the 4 mile limit. The survey was conducted from three 35 foot fishing boats. The instruments were 8-channel using one land geophone per channel bolted to an 18 inch steel plate to keep it upright on the seafloor.

Fig.1.13. Petty 10-channel recorder from 1931. Vacuum tubes V amplified the current which passed through a 'harp' H (of the type shown in fig.1.8); light from a source L passed by the harp and was focused onto photographic paper in the take-up magazine M. A timing wheel T driven by a clock (see fig.1.14) interrupted the light to give the timing lines. (Photographed at Museum of the Geophysical Society of Houston.)

Extensive marine operations did not appear until 1944 when Superior and Mobil began refraction fan-shooting for saltdomes offshore Louisiana (Jack Lester: personal communication). A survey to map the offshore extension of Los Angeles basin fields was also carried out about this time (C. C. Bates: personal communication). A surveyor onshore gave instructions to keep lines straight while wire paying out through a counter gave the distance; buoys were set to indicate locations. As work progressed farther offshore the chaining continued on compass bearing. Sighting on shot plumes both visually and with radar was also used. Surveying was the principal operational problem and often constituted the major cost. Shoran came into use about 1946, followed by Raydist about 1951. The early refraction and reflection work used geophones planted on bottom. About 1946 reflection work began using a 12-channel bottom drag cable with gimble-mounted geophones. The floating streamer was first used in 1949–50. Both the radionavigation methods and the floating streamer were based on World War II developments.

1.2.7 *Reproducible recording, the common-depth-point method and non-explosive sources*

Frank Rieber (1936) proposed the 'Sonograph' method of recording seismic data (fig. 1.18) so that it could be 'played back'. His oscillograph recorded in variable density on film. On playback, variations in the intensity of a light beam which passed through the film were detected by a photocell. Rieber used the Sonograph to determine the variation of reflection amplitude with apparent dip.

Despite Rieber's pioneering work, reproducible recording did not become practical until the introduction of magnetic-tape recording. Commercial recording and playback equipment became available about 1952. The principal advantage of magnetic-tape recording was thought to be the ability to replay with different filters. About 1955 moveable heads allowed static and dynamic corrections (§4.1) to be applied. The growth of analog magnetic-tape recording is shown in fig. 1.20.

A very important post-war development was the use of record sections for interpretation. Individual seismic

Fig.1.14. Early timing wheels (paddle wheels). The wheels were rotated by a clock motor so that they cut the light beam to produce timing line shadows.

Observers often had individualized wheels which characterized their records. (Photographed at Museum of the Geophysical Society of Houston.)

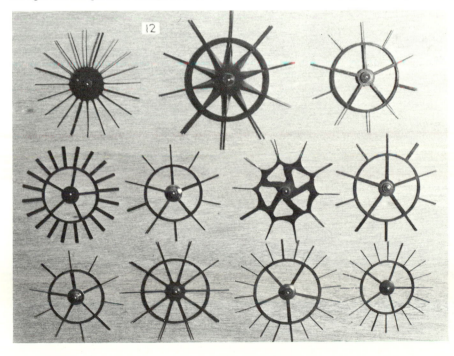

Fig.1.15 (caption on facing page)

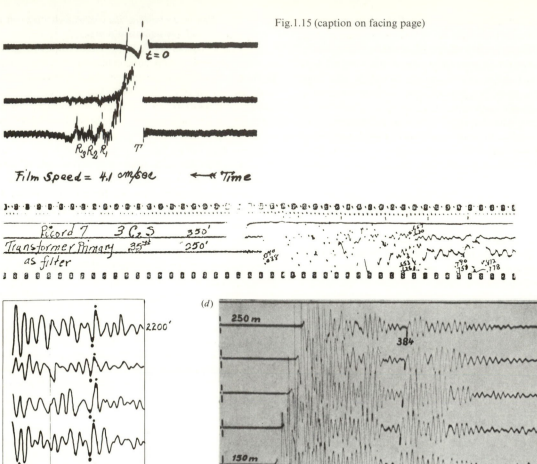

(a)

$t=0$

$R_3 R_2 R_1$ T

Film Speed = 4.1 cm/sec   ← Time

(b)

Record 7   3 C$_2$S   350'
Transformer Primary 35#   250'
as filter

(c)

2200'

1780'

0.7   0.8   0.9   1.0 sec.

(d)

250 m

384

150 m

369

0   0.5
              Sec

(e)

Table 1.1. *Chronology of seismic instrumentation and methods. (Dates are approximate; secrecy and competition often involved development and use of the same feature by several companies without public disclosure.)*

1914	Mintrop's mechanical seismograph
1917	Fessenden patent on seismic method
1921	Seismic reflection work by Geological Engineering Co
1923	Refraction exploration by Seismos in Mexico and Texas
1925	Fan-shooting method
	Electrical refraction seismograph
	Radio used for communications and/or time-break
1926	Reflection correlation method
1929	Reflection dip shooting
1931	Reversed refraction profiling
	Use of uphole phone
	Truck-mounted drill
1932	Automatic gain control
	Interchangeable filters
1933	Use of multiple geophones per group
1936	Rieber sonograph, first reproducible recording
1939	Use of closed loops to check misties
1942	Record sections
	Mixing
1944	Large-scale marine surveying
	Large patterns
1947	Marine shooting with Shoran
1949	Optical mirragraph
1950	Common-depth-point method*
1951	Medium-range radionavigation
1952	Analog magnetic recording*
1953	Vibroseis recording*
	Weight-dropping
1954	Continuous velocity logging
1955	Moveable magnetic heads
1956	Central data processing
1961–2	Analog deconvolution and velocity filtering
1963	Digital data recording*
1965	Airgun seismic source
1967	Depth controllers on marine streamer
1972	Bright spot
1974	Digitization in the field
1975	Seismic stratigraphy
1976	Three-dimensional surveying

*The acceptance of these methods is shown in fig. 1.20.

records had been laid out adjacent to each other in the interpretation process for a long time (fig. 1.19), but the large size of individual records and variations in paper speed and developing quality made it difficult to obtain a synoptic view. Normal moveout (§3.1.1), irregularities in recording or spreads, and the wiggle-trace display mode added to the difficulties. Gulf Oil and Carter (now part of Exxon) and perhaps Shell apparently led in developing variable-density or variable-area displays with uniform horizontal scale and display amplitude. Carter bought Rieber's equipment for this use (among others) about 1946.

Common-depth-point (CDP; see §5.3.1) recording was invented by Harry Mayne (Petty Geophysical) in 1950 as a way of attenuating noise which could not be handled by the use of arrays. Magnetic-tape recording made CDP practical and CDP recording began about 1956, but it did not become used extensively until the early 1960s (fig. 1.20) when its ability to attenuate multiples (§4.2.2) and other kinds of noise led to rapid adoption. Today its use is nearly universal.

Magnetic-tape recording also permitted the addition of traces and thus the use of weaker sources since records from several weak sources could be added together to get the effect of a stronger source. McCollum introduced the use of a dropped weight, the Thumper, as a seismic source about 1953. Weight-dropping expanded seismic work in areas of difficult shothole drilling, such as West Texas, and in desert areas where water for drilling is scarce.

A variety of surface sources for use on land were also developed besides weight-dropping. The most ingenious of these, the Vibroseis™ method (see §5.4.3; a list of trademarks and the companies that hold them, is given in appendix *B*), was developed by John M. Crawford, William Doty and Milford Lee and first used in 1953. Surface sources are now used for about half the land work and Vibroseis is the predominant surface source. Several alternatives to the use of dynamite as a source in the marine environment were developed about 1965. They were generally cheaper and more efficient and in addition environmentally acceptable since they did not injure marine life and consequently they rapidly replaced dynamite as a marine source (fig. 1.20).

Fig.1.15. Early reflection records. (*a*) Photograph of first record made in rock quarry near Washington, DC, April 12, 1919 by Karcher. The upper trace shows the shot instant with time increasing toward the left, the two lower traces the geophone response at different gain; *T* is the arrival of the direct wave, R_1, R_2, R_3 the arrival of reflections. (From Schriever, 1952.)

(*b*) Photographs of two portions of 2-channel record in Oklahoma obtained in 1930. (Courtesy Conoco.) (*c*) Tracing of records such as in (*b*) showing a reflection being mapped. (*d*) Photograph of a 5-channel Seismos record from 1935, without automatic gain control. (*e*) Photograph of a 6-channel SEI string-galvanometer record from Mississippi, about 1938.

While some sophisticated playback processing was done with magnetic-tape recording and some digital processing was done on analog data, the full potential of data processing was not achieved until digital recording was introduced in the 1960s. Digital recording not only resulted in higher fidelity but also in the large-scale application of the digital computer in the processing and interpretation of seismic data. The 'digital revolution' was probably the most far-reaching development in seismic exploration since the pioneering days. For example, obtaining useful data in the North Sea is almost impossible without deconvolution (§8.2.1).

1.3 Geophysical activity

1.3.1 *History of seismic activity*

The number of seismic field crews is shown in fig. 1.21*b*. Seismic activity is tabulated each year by a committee of the Society of Exploration Geophysicists. The reports are regarded as relatively accurate, taking into account that activity in some areas, especially the Soviet Union and China, has not been included. A graph of the number of wildcat wells drilled in the US is also shown in fig. 1.21*b*; it generally parallels the number of seismic crews with a $2\frac{1}{2}$ to 3 year lag so that seismic activity data are a 'leading indicator' of petroleum industry activity.

Fig.1.16. Portion of a dip map which resulted from correlation shooting, January, 1935. Dips were expressed in ft/mile and the arrow lengths indicated the spacing for 50 ft contours. (Courtesy Conoco.)

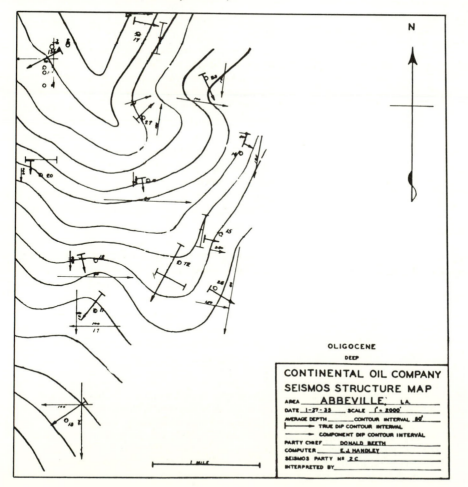

Fig.1.17. Photographs of early refraction work. (From Petty, 1976.) (a) D. E. Petty washes a refraction record in Chacahoula Swamp; the geophone is on the cypress stump in the background. (b) Petty (in boat) with his crew in Chacahoula Swamp.

(a)

(b)

The mean wellhead costs of oil and natural gas priced in constant US dollars are shown in fig. 1.21*a*. Economics and technology have been the governing factors in seismic activity. A surplus of oil about 1937 produced a decline in activity which lasted until the United States became involved in World War II. A doubling of petroleum prices between 1945 and 1948 resulted in seismic activity continuing to rise. However, major finds of oil in the Middle East after World War II resulted in another world surplus of oil. From 1948 to 1973 the price of petroleum remained almost constant and activity generally declined for most of this period, the decline being slowed and occasionally temporarily reversed by new developments. Natural gas reserves peaked in 1970 and thereafter exploration increasingly had gas rather than oil as an objective. In 1973–4 oil prices increased sharply as a result of the cutoff of supplies during the Arab–Israel war and the formation of the OPEC cartel. Uncertainties about dependence on foreign supplies thereafter became an important factor in stimulating seismic exploration.

Outside the United States, activity increased steadily until 1958 and then leveled off. During the 1958–74 period the geography of activity changed several times in response to political and economic factors and discoveries in new areas. Activity in Latin America declined sharply after 1959 because of discouragement in several countries. The discovery of significant hydrocarbon reserves in North Africa, the beginning of North Sea exploration, nationalization threats in Indonesia, the opening of tropical African waters to exploration, and repeated political disruptions in North Africa and the Middle East, were probably the most significant of the factors.

1.3.2 *Data for 1980*

Expenditure for geophysical data acquisition and processing in 1980 were US$3002 million, according to the annual report of the Society of Exploration Geophysicists (Senti, 1981). This amount was up sharply (46%) from the previous year, and up 367% from 1971, when geophysical expenditure began to increase significantly (fig. 1.22). Expenditures have been increasing especially rapidly since 1977.

Seismic work constitutes about 95% of the geophysical expenditure. The report gives various statistics about seismic work, as shown in tables 1.2 through 1.5. An interesting feature of the report for the year 1980 is the

Fig.1.18. Rieber's sonograph, 1936. Field data were recorded on film in variable-density mode. In playback the total light through a slit was summed to give a single output trace. By changing the slit angle, data with various angles of approach (also called apparent dip; see §3.1.2) could be emphasized, each slit angle giving an additional trace. Thus the sonogram record displayed amplitude in the angle-of-approach versus arrival-time domain. (Two views from advertisements in *Geophysics* vols. 1 and 2.)

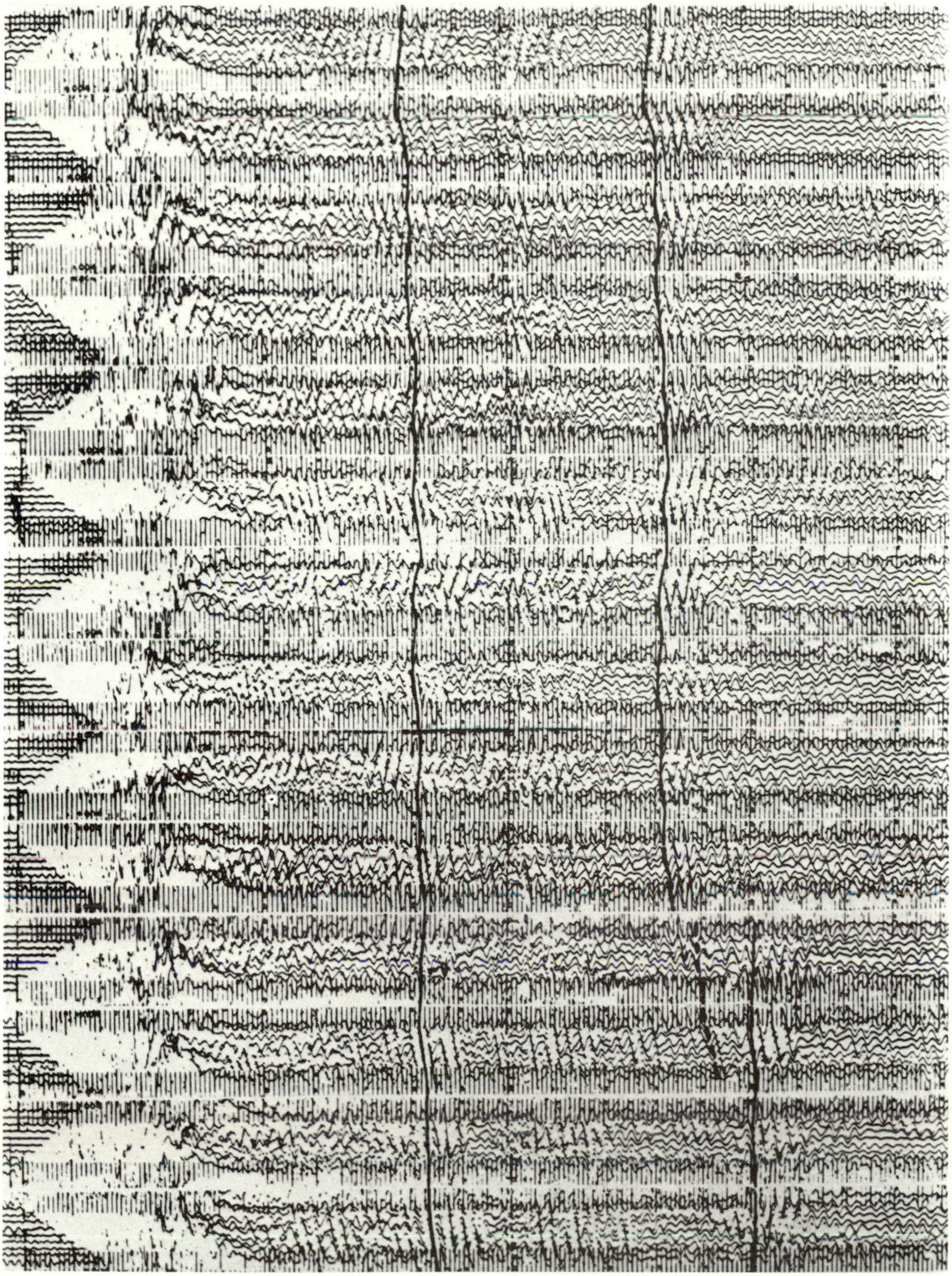

Fig.1.19. Early record section made by splicing individual records together. The records are made with a 10-trace camera and are exceptionally uniform for the period. The lower marked horizon is clearly cut by a large fault. (Courtesy Conoco.)

breakout of S-wave reflection and refraction work, indicating surprisingly large expenditures in these areas.

Non-petroleum seismic expenditures are growing rapidly, although they are small compared to those for petroleum exploration. Senti (1981) feels that the non-petroleum figures are under-reported and should be appreciably larger.

Data on cost per unit of work are available in the annual reports since 1968 (and the report for 1966 included estimates); these are plotted in fig. 1.23.

1.3.3 *Economic justification*

Economic justification of seismic exploration is difficult to document since the figures are not available. Prior to the mid-1950s the American Association of Petroleum Geologists (AAPG) attempted to assess bases to new oil finds but abandoned this because multiple bases were invariably present. Halbouty (1970) indicated that between 1930 and 1960 80% of the giant fields discovered in the United States were at least partially based on seismic

exploration. Today almost all well locations are based on seismic data, so that the consensus clearly indicates justification for seismic work. The seismic activities listed in table 1.5 are growing very rapidly suggesting that the value of seismic data is being recognized more.

The amount of seismic work prior to drilling oil and gas wells is far from uniform. The sixteen major oil companies do 62% of the geology and geophysical work (called 'G & G'), drill 10% of the wildcat wells, and find 44% of the new oil (*AAPG Explorer*, 1980). M. M. Backus (in his presidential address to the SEG, 1980) calculated 7.6×10^6 barrels of oil-equivalent per seismic crew year in the United States and about 10^7 for work outside the US ('barrels of oil-equivalent' includes the fuel value of natural gas found). Backus said this figure should be reduced to about 6×10^6 (for the US) to allow for findings based on the rework of seismic data acquired in prior years. In 1978 each new oil or gas find cost an average of $1.67 million for drilling and $0.47 million for seismic work, so that about four times as much money is spent on

Fig.1.20. Percentage of US seismic activity involving various techniques. (Data from SEG annual Geophysical Activity Reports.)

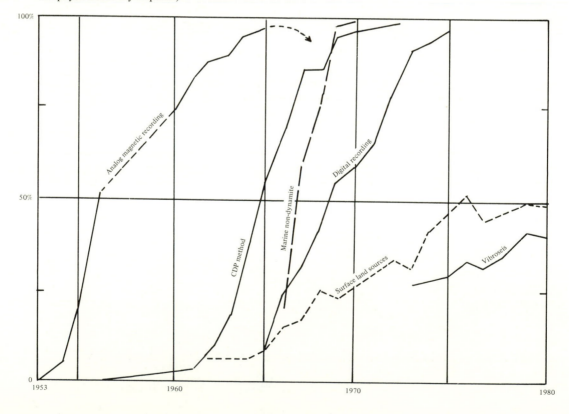

drilling a well as on the studies which determine its location.

Backus calculated that there are on the average 21 geophysical society members for every seismic crew and that the number has been increasing at about 0.4 per year, with the increase since 1976 being greater than this. More G & G work is being done before drilling. The success ratio of exploratory holes improved during the 1972–6 period from 15% (a value which had been declining annually) to about 26%; the figure for 1979 was 29% (Johnston, 1980). The improvement is attributed partially to increased G & G effort and improved seismic technology.

1.4   The literature of exploration seismology

The seismologist who reads English is especially fortunate in that almost all important references are in this language. Most of the important papers and books which have appeared in other languages have either English equivalents or English translations. Furthermore, almost all of the important technical papers are contained in two journals, *Geophysics*, published by the Society of Exploration Geophysicists (SEG), and *Geophysical Prospecting*, published by the European Association of Exploration Geophysicists (EAEG), the latter including occasional articles in French and German.

Fig.1.21. History of seismic exploration activity. (*a*) Mean US wellhead price of crude oil and natural gas in 1978 dollars (data from American Petroleum Institute), (*b*) Mean number of seismic crews (data from SEG Geophysical Activity Reports) and new-field wildcat wells drilled (data from AAPG activity reports); activity in some areas, especially the Soviet Union and China, is not included.

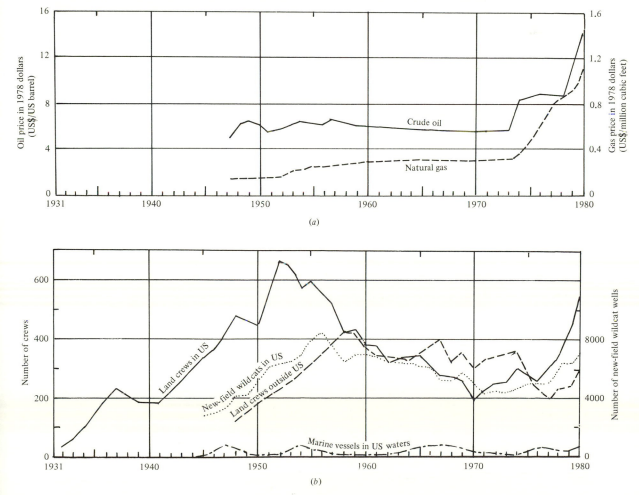

The Society of Economic Geophysicists was founded in Houston in 1930; the name was changed that same year to the Society of Petroleum Geophysicists and in 1937 to the Society of Exploration Geophysicists. It continues to be the largest professional geophysical society today. The Society began publication of *Geophysics* in 1936. Prior to this, papers were published in issues of the *AAPG Bulletin* and *Physics*; many of the most important papers prior to 1936 were republished in *Early Geophysical Papers* in 1947. The European Association of Exploration Geophysicists was founded in 1951 and began publishing *Geophysical Prospecting* in 1953.

A Cumulative Index of *Geophysics* is published every few years (most recently as a supplement to the December 1980 issue of *Geophysics*) which lists the papers in both of these journals, as well as those in other Society publications. The *AAPG Bulletin* contains articles on the geological interpretation of geophysical data, and the *Offshore Technology Conference Preprints* (each year since 1969) provide other important papers.

A. S. Eve and D. A. Keys in September 1928 wrote in their preface to *Applied Geophysics*, '... we know of no book in English which deals with the theoretical and practical sides of all of the many schemes of exploration now available.' This was only four years after the first discovery of hydrocarbons based on seismic refraction.

Eve and Keys noted that 'in 1928 there were thirty or more groups or "troops" at work ..., each consisting of three to five trained men, with an equal number of helpers.' Extreme secrecy was common at this time and their book gives only a brief sketch of methods. As late as the early 1950s some 'black-box' elements remained, that is, details were not disclosed.

Literature on earthquake seismology preceded that dealing with prospecting applications. H. Jeffreys' classic *The Earth* appeared in 1924 (3rd edition in 1952). L. D. Leet's *Practical Seismology and Seismic Prospecting* (1938) combined earthquake and exploration seismology.

The most important books on exploration seismology are tabulated on pages 30–1 following:

Table 1.2. *Percentages of geophysical acquisition expenditures. Data are for the year 1980 (numbers in parentheses for 1979).*

(a) By objective:	Petroleum	96.9%	(94.9%)
	Minerals	1.8	(2.1)
	Engineering	0.5	(1.1)
	Geothermal	0.2	(0.2)
	Ground water	0.2	(0.3)
	Oceanography	0.2	(1.2)
	Research	0.3	(0.3)
(b) By type:	Land	69.6	(68.9%)
	Marine	28.7	(28.6)
	Airborne	1.5	(1.9)
	Borehole	0.2	(0.5)
(c) By locale:	USA	44.3%	(45.9%)
	Canada	11.0	(13.2)
	Mexico	2.4	(2.5)
	South America	7.6	(6.1)
	Europe	9.1	(9.5)
	Africa	10.2	(9.4)
	Middle East	5.1	(4.3)
	Far East	6.9	(7.2)
	Australia, N. Zealand	2.1	(0.4)
	Unspecified	0.9	(1.5)
(d) By method:	Seismic P-wave	92.6%	
	S-wave	1.0	}(95.1%)
	refraction	1.5	
	Gravity	2.6	(0.5)
	Magnetic	0.9	(1.1)
	Resistivity	0.2	(0.3)
	Electromagnetic	0.1	(0.4)
	Induced polarization	0.2	(0.1)
	Magnetotellurics	0.3	(0.2)
	Borehole	0.2	(0.5)
	Other	0.4	(0.5)

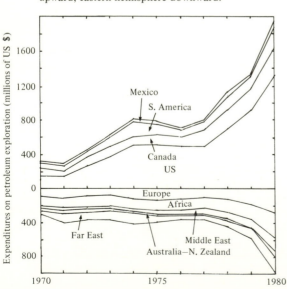

Fig.1.22. Expenditures on petroleum seismic exploration in various parts of the world; western hemisphere expenditures are plotted cumulatively upward, eastern hemisphere downward.

Table 1.3. *Statistics about petroleum seismic activity for 1980. (Figures in parentheses show 1980 figures as percentages of 1979 figures).*

	Land	Marine
Acquisition costs (US$)	1806×10^6 (150%)	420×10^6 (162%)
Line miles	392×10^3 $\Big\}$(117%)	658×10^3 $\Big\}$(116%)
Line kilometers	631×10^3	1059×10^3
Crew months	7517 (110%)	1047 (154%)
Miles/crew month	52 $\Big\}$(30%)	629 $\Big\}$(76%)
Kilometers/crew month	84	1012
Cost/mile (acquisition only)	4612 $\Big\}$(127%)	638 $\Big\}$(139%)
Cost/kilometer (acquisition only)	2866	396
Cost/mile (inc. processing)	5223 $\Big\}$(125%)	1158 $\Big\}$(118%)
Cost/kilometer (inc. processing)	3245	720
Cost/month (US$)	240×10^3	401×10^3

Table 1.4. *Percentage of line-miles by source types (petroleum only). Data are for the year 1980 (figures in parentheses for 1979).*

	Land		Marine	
Dynamite	55.9%	(55.6%)		
Compressed air	2.8	(2.8)	65.3%	(61.7%)
Gas exploder	0.6	(0.9)	18.8	(29.3)
Weight drop	2.1	(1.7)		
Solid chemical	0.4	(0.4)	4.9	(2.9)
Vibrator	37.0	(37.3)		
Electrical			1.9	(0.4)
Implosive			9.3	(5.7)
Other	0.9			

Table 1.5. *Non-petroleum seismic activity. Data are for the year 1980 (figures in parentheses for 1979).*

	Crew-months		Costs (10^3 US$)		% of total costs*	
Mineral exploration	110	(81)	12659	(6134)	54.9%	(13.9%)
Civil engineering[†] (land)	105	(183)	2612[†]	(3140)	57.7[†]	(48.8)
Civil engineering[†] (marine)	192	(147)	10461[†]	(12455)	100.0[†]	(100.0)
Geothermal	15	(4)	1810	(377)	28.6	(7.7)
Ground water	10	(28)	203	(234)	4.5	(4.6)
Oceanography	22	(92)	5187	(22479)	93.6	(95.6)
Research	53		5890		57.6	

*Percentages relate seismic costs to total geophysical costs for the respective activity.

[†] Excluding airborne and borehole costs

1940 L. L. Nettleton, *Geophysical Prospecting for Oil*
C. A. Heiland, *Geophysical Exploration*
J. J. Jakosky, *Exploration Geophysics* (2nd edition, 1950)
1948 SEG *Geophysical Case Histories* (2nd volume in 1956)
1952 C. H. Dix, *Seismic Prospecting for Oil*
M. B. Dobrin, *Introduction to Geophysical Prospecting* (3rd edition, 1976)
1959 M. M. Slotnick, *Lessons in Seismic Computing*

1965 J. E. White, *Seismic Waves: Radiation, Transmission and Attenuation*
F. S. Grant and G. F. West, *Interpretation Theory in Applied Geophysics*
1967 A. W. Musgrave (ed.), *Seismic Refraction Prospecting*
1970 N. A. Anstey, *Seismic Prospecting Instruments – I: Signal Characteristics and Instrument Specifications*
1971 B. S. Evenden and D. R. Stone, *Seismic Prospecting Instruments – II: Instrument Performance and Testing*

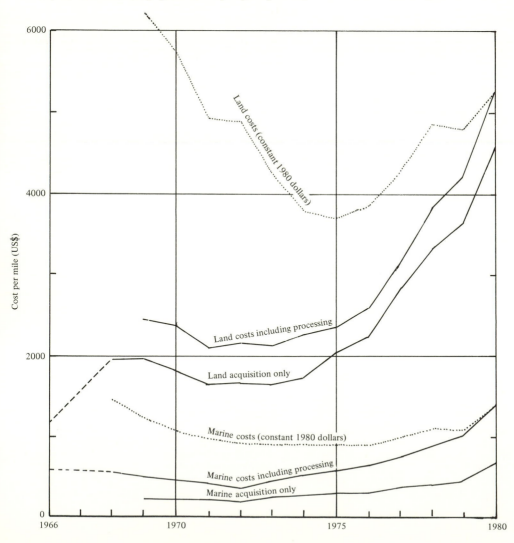

Fig.1.23. Seismic costs per mile (western hemisphere). The dotted curves show costs in constant 1980 dollars. (Data from SEG Geophysical Activity Reports.)

1973 R. E. Sheriff, *Encyclopedic Dictionary of Exploration Geophysics*

1976 W. M. Telford, L. P. Geldart, R. E. Sheriff and D. A. Keys, *Applied Geophysics*

A. A. Fitch, *Seismic Reflection Interpretation*

1977 N. A. Anstey, *Seismic Interpretation*

H. M. Mooney, *Handbook of Engineering Geophysics*

C. E. Payton (ed.), *Seismic Stratigraphy: Applications to Hydrocarbon Exploration*

1978 R. E. Sheriff, *A First Course in Geophysical Exploration and Interpretation*

K. H. Waters, *Reflection Seismology*

J. A. Coffeen, *Seismic Exploration Fundamentals*

1979 R. McQuillin, M. Bacon and W. Barclay, *An Introduction to Seismic Interpretation*

1980 K. Aki and P. G. Richards, *Quantitative Seismology: Theory and Methods* (2 volumes)

R. E. Sheriff, *Seismic Stratigraphy*

In addition to the above, several books have appeared which have important applications in seismic data processing: Blackman and Tukey (1958), Cheng (1959), Lee (1960), Cassand *et al.* (1971), Robinson (1967), Kanasewich (1973), Båth (1974), Claerbout (1976), Kulhánek (1976), Silvia and Robinson (1979), and Robinson and Treitel (1980). Several important educational works have been published privately: *A Pictorial Digital Atlas* (Peterson and Dobrin, 1966), *Seismic Energy Sources* (Kramer *et al.*, 1968), and *Seismic Imaging Atlas* (Peterson and Walter, 1976, 1977, 1978) by United Geophysical Corp., and the *Robinson–Treitel Reader* (Robinson and Treitel, 1973) by Seismograph Service Corp. Patents are also a useful literature source; those relating to seismic devices and methods are discussed by Sittig (1980).

The International Association of Geophysical Contractors (IAGC) publishes a safety guide and other materials to assist in field data acquisition.

2

Theory of seismic waves

Overview

The seismic method utilizes the propagation of waves through the Earth. Since this propagation depends upon the elastic properties of the rocks, we shall discuss first some basic concepts of elasticity (§2.1). (For more thorough treatments, see Jaeger (1958) or Sokolnikoff (1958).)

The size and shape of a solid body can be changed by applying forces to the external surface of the body. These external forces are opposed by internal forces which resist the changes in size and shape. As a result the body tends to return to its original condition when the external forces are removed. Similarly, a fluid resists changes in size (volume) but not changes in shape. This property of resisting changes in size or shape and of returning to the undeformed condition when the external forces are removed is called *elasticity*. A perfectly elastic body is one which recovers completely after being deformed. Many substances including rocks can be considered perfectly elastic without appreciable error provided the deformations are small.

The theory of elasticity relates the forces which are applied to the external surface of a body to the resulting changes in size and shape. The relations between the applied forces and the deformations are most conveniently expressed in terms of the concepts, stress and strain. Strain, a change in shape or dimensions, is generally proportional to the stress (force per unit area) which produces it, as stated in Hooke's law. The constant of proportionality is called an elastic constant or modulus, and moduli for different types of stress and strain are interrelated.

Section 2.2 concerns seismic wave motion. Newton's second law of motion, that an unbalanced force on a mass produces an acceleration, is used to derive two forms of the wave equation. The wave equation is expressed in vector as well as the more conventional scalar notation. Methods of including a source of disturbance and Kirchhoff's formula are also given in this section.

Plane- and spherical-wave solutions to the wave equation are given next. Waves are the disturbances which travel through the medium. The concepts of phase, wavefronts and raypaths are introduced, as is the more general Huygens' principle approach, and the terms amplitude, wavelength, wavenumber, period, frequency and angular frequency are defined.

The two forms of the wave equation which had been derived earlier are related to two types of disturbances which can travel through the body of solids. These involve changes in volume (P-waves) and rotations (S-waves). Discussion of potential functions, from which particle displacements and velocities can be derived, follows. At interfaces both stress and particle displacement must be continuous; these boundary conditions are discussed in §2.2.8.

Surface waves are examined next. Rayleigh waves are important because of the ground-roll noise which they produce on seismic records. Love and Stoneley waves are encountered occasionally.

Section 2.3 examines what happens to seismic body waves as they travel in the Earth. Intensity decreases because of geometrical spreading (divergence), absorption, and partitioning at interfaces. Divergence is the most important factor affecting the change of intensity for the first few kilometers, but eventually absorption becomes dominant. Absorption increases approximately linearly with frequency and hence changes the waveshape with distance. Various expressions for absorption are interrelated. Dispersion and the concepts of group and phase velocity are discussed, although dispersion is not an important factor in seismic exploration.

Reflection and refraction are discussed in §2.3.4. The treatment of diffraction (§2.3.5) involves somewhat complex mathematics; however, the construction of diffraction wavefronts using Huygens' principle is nonmathematical.

The partitioning of energy at interfaces is one of the key sections (§2.4). Boundary conditions permit calculating how wave energy is divided among reflected and transmitted waves. The solution in terms of potentials

yields Knott's equations and that in terms of amplitudes yields Zoeppritz' equations. For the simple but very important case of normal incidence these equations reduce to the familiar normal reflection-coefficient equation which states that the reflection amplitude compared to the incident amplitude varies directly as the change in acoustic impedance (the product of density and velocity) and inversely as the sum of the acoustic impedances. Examples of the magnitudes of reflection coefficients are also given. Further attention is given to incidence other than normal, including the rapidly changing effects when incidence is near the critical angles. This section concludes with a discussion of head waves, waves which have been refracted critically, and normal-mode propagation.

2.1   Theory of elasticity
2.1.1   *Stress*

Stress is defined as force per unit area. Thus, when a force is applied to a body, the stress is the ratio of the force to the area on which the force is applied. If the force varies from point to point, the stress also varies and its value at any point is found by taking an infinitesimally small element of area centered at the point and dividing the total force acting on this area by the magnitude of the area. If the force is perpendicular to the area, the stress is said to be a *normal stress* (or *pressure*). In this book, positive values correspond to tensile stresses (the opposite convention of signs is sometimes used). When the force is tangential to the element of area, the stress is a *shearing stress*. When the force is neither parallel nor perpendicular to the element of area, it can be resolved into components parallel and perpendicular to the element, hence any stress can be resolved into component normal and shearing stresses.

If we consider a small element of volume inside a stressed body, the stresses acting upon each of the six faces of the element can be resolved into components, as shown in fig. 2.1 for the two faces perpendicular to the x-axis. Subscripts denote the x-, y- and z-axes respectively and σ_{yx} denotes a stress parallel to the y-axis acting upon a surface perpendicular to the x-axis. When the two subscripts are the same (as with σ_{xx}), the stress is a normal stress; when the subscripts are different (as with σ_{yx}), the stress is a shearing stress.

When the medium is in static equilibrium, the stresses must be balanced. This means that the three stresses σ_{xx}, σ_{yx} and σ_{zx} acting upon the face *OABC* must be equal and opposite to the corresponding stresses shown on the opposite face *DEFG*, with similar relations for the remaining four faces. In addition, a pair of shearing stresses, such as σ_{yx}, constitute a *couple* tending to rotate the element about the z-axis, the magnitude of the couple being

$$(\text{force} \times \text{lever arm}) = (\sigma_{yx}\,\mathrm{d}y\,\mathrm{d}z)\,\mathrm{d}x.$$

If we consider the stresses on the other four faces, we find that this couple is opposed solely by the couple due to the pair of stresses σ_{xy} with magnitude $(\sigma_{xy}\,\mathrm{d}x\,\mathrm{d}z)\,\mathrm{d}y$. Since the element is in equilibrium, the total moment must be zero; hence $\sigma_{xy} = \sigma_{yx}$. In general, we must have

$$\sigma_{ij} = \sigma_{ji}.$$

2.1.2 *Strain*

When an elastic body is subjected to stresses, changes in shape and dimensions occur. These changes, which are called *strains*, can be resolved into certain fundamental types.

Consider a rectangle $PQRS$ in the xy-plane (see fig. 2.2). When the stresses are applied, let P move to P', PP' having components u and v. If the other vertices Q, R and S have the same displacement as P, the rectangle is merely displaced as a whole by the amounts u and v; in this case there is no change in size or shape and no strain exists. However, if u and v are different for the different vertices, the rectangle will undergo changes in size and shape and strains will exist.

Let us assume that $u = u(x, y)$, $v = v(x, y)$. Then the coordinates of the vertices of $PQRS$ and $P'Q'R'S'$ are as follows:

$$P(x, y): \quad P'(x + u, y + v);$$

$$Q(x + \mathrm{d}x, y): \quad Q'\left(x + \mathrm{d}x + u + \frac{\partial u}{\partial x}\mathrm{d}x,\ y + v + \frac{\partial v}{\partial x}\mathrm{d}x\right);$$

$$S(x, y + \mathrm{d}y): \quad S'\left(x + u + \frac{\partial u}{\partial y}\mathrm{d}y,\ y + \mathrm{d}y + v + \frac{\partial v}{\partial y}\mathrm{d}y\right);$$

$$R(x + \mathrm{d}x, y + \mathrm{d}y): \quad R'\left(x + \mathrm{d}x + u + \frac{\partial u}{\partial x}\mathrm{d}x + \frac{\partial u}{\partial y}\mathrm{d}y,\right.$$
$$\left. y + \mathrm{d}y + v + \frac{\partial v}{\partial x}\mathrm{d}x + \frac{\partial v}{\partial y}\mathrm{d}y\right).$$

In general, the changes in u and v are much smaller than the quantities $\mathrm{d}x$ and $\mathrm{d}y$; accordingly we shall assume that the terms $(\partial u/\partial x)$, $(\partial u/\partial y)$, and so on are small enough that powers and products can be neglected. With this assumption, we see that:

(1) PQ increases in length by the amount $(\partial u/\partial x)\mathrm{d}x$ and PS by the amount $(\partial v/\partial y)\,\mathrm{d}y$; hence $\partial u/\partial x$ and $\partial v/\partial y$ are the fractional increases in length in the direction of the axes;

(2) the infinitesimal angles δ_1 and δ_2 are equal to $\partial v/\partial x$ and $\partial u/\partial y$ respectively;

(3) the right angle at P decreases by the amount $(\delta_1 + \delta_2) = (\partial v/\partial x + \partial u/\partial y)$;

(4) the rectangle as a whole has been rotated counter-clockwise through the angle $(\delta_1 - \delta_2) = (\partial v/\partial x - \partial u/\partial y)$.

Strain is defined as the relative change (that is, the fractional change) in a dimension or shape of a body. The quantities $\partial u/\partial x$ and $\partial v/\partial y$ are the relative increases in length in the directions of the x- and y-axes and are referred to as *normal strains*. The quantity $(\partial v/\partial x + \partial u/\partial y)$

Fig.2.1. Components of stress for faces perpendicular to the x-axis.

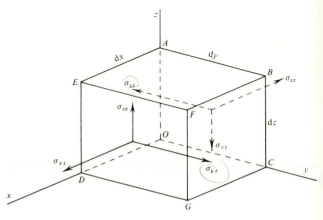

Fig.2.2. Analysis of two-dimensional strain.

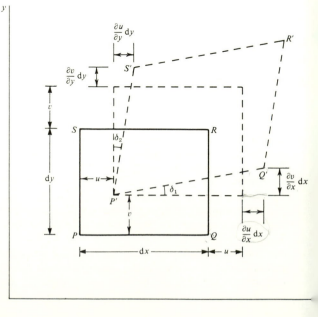

is the amount by which a right angle in the xy-plane is reduced when the stresses are applied, hence is a measure of the change in shape of the medium; it is known as a *shearing strain* and will be denoted by the symbol ε_{xy}. The quantity $(\partial v/\partial x - \partial u/\partial y)$, which represents a rotation of the body about the z-axis, does not involve change in size or shape and hence is not a strain; we shall denote it by the symbol θ_z.

Extending the above analysis to three dimensions, we write (u, v, w) as the components of displacement of a point $P(x, y, z)$. The elementary strains are thus:

$$\text{normal strains}\quad \left.\begin{aligned}\varepsilon_{xx} &= \frac{\partial u}{\partial x}, \\[1mm] \varepsilon_{yy} &= \frac{\partial v}{\partial y}, \\[1mm] \varepsilon_{zz} &= \frac{\partial w}{\partial z};\end{aligned}\right\} \quad (2.1)$$

$$\text{shearing strains}\quad \left.\begin{aligned}\varepsilon_{xy} = \varepsilon_{yx} &= \frac{\partial v}{\partial x} + \frac{\partial u}{\partial y}, \\[1mm] \varepsilon_{yz} = \varepsilon_{zy} &= \frac{\partial w}{\partial y} + \frac{\partial v}{\partial z}, \\[1mm] \varepsilon_{zx} = \varepsilon_{xz} &= \frac{\partial u}{\partial z} + \frac{\partial w}{\partial x}.\end{aligned}\right\} \quad (2.2)$$

In addition to these strains, the body is subjected to simple rotation about the three axes given by

$$\left.\begin{aligned}\theta_x &= \frac{\partial w}{\partial y} - \frac{\partial v}{\partial z}, \\[1mm] \theta_y &= \frac{\partial u}{\partial z} - \frac{\partial w}{\partial x}, \\[1mm] \theta_z &= \frac{\partial v}{\partial x} - \frac{\partial u}{\partial y}.\end{aligned}\right\} \quad (2.3)$$

Equations (2.3) can be written in vectorial form (see §10.1.2)

$$\boldsymbol{\Theta} = \theta_x \mathbf{i} + \theta_y \mathbf{j} + \theta_z \mathbf{k} = \boldsymbol{\nabla} \times \boldsymbol{\zeta}, \quad (2.4)$$

where $\boldsymbol{\zeta} = u\mathbf{i} + v\mathbf{j} + w\mathbf{k} =$ vector displacement of the point $P(x, y)$ and $\mathbf{i}$, $\mathbf{j}$, $\mathbf{k}$ are unit vectors in the x, y, z directions.

The changes in dimensions given by the normal strains result in volume changes when a body is stressed. The change in volume per unit volume is called the *dilatation* and represented by Δ. If we start with a rectangular parallelepiped with edges dx, dy, and dz in the unstrained medium, in the strained medium the dimensions are $dx(1 + \varepsilon_{xx})$, $dy(1 + \varepsilon_{yy})$, $dz(1 + \varepsilon_{zz})$; hence the increase in

volume is approximately $(\varepsilon_{xx} + \varepsilon_{yy} + \varepsilon_{zz})\,dx\,dy\,dz$. Since the original volume was $(dx\,dy\,dz)$, we see that

$$\Delta = \varepsilon_{xx} + \varepsilon_{yy} + \varepsilon_{zz} = \frac{\partial u}{\partial x} + \frac{\partial v}{\partial y} + \frac{\partial w}{\partial z} = \boldsymbol{\nabla} \cdot \boldsymbol{\zeta}. \quad (2.5)$$

2.1.3 *Hooke's law*

In order to calculate the strains when the stresses are known, we must know the relationship between stress and strain. When the strains are small, this relation is given by *Hooke's law* which states that a given strain is directly proportional to the stress producing it. When several stresses exist, each produces strains independently of the others; hence the total strain is the sum of the strains produced by the individual stresses. This means that each strain is a linear function of all of the stresses and vice versa. In general, Hooke's law leads to complicated relations but when the medium is *isotropic*, that is, when properties do not depend upon direction, it can be expressed in the following relatively simple form:

$$\sigma_{ii} = \lambda\Delta + 2\mu\varepsilon_{ii}, \quad i = x, y, z; \quad (2.6)$$

$$\sigma_{ij} = \mu\varepsilon_{ij}, \quad i, j = x, y, z, \quad i \neq j. \quad (2.7)$$

The quantities λ and μ are known as *Lamé's constants*. If we write $\varepsilon_{ij} = (\sigma_{ij}/\mu)$, it is evident that ε_{ij} is smaller the larger μ is. Hence μ is a measure of the resistance to shearing strain and is often referred to as the *modulus of rigidity* or *shear modulus*.

Although Hooke's law has wide application, it does not hold for large stresses. When the stress is increased beyond an *elastic limit* (fig. 2.3a), Hooke's law no longer holds and strains increase more rapidly. Strains resulting from stresses which exceed this limit do not entirely disappear when the stresses are removed. With further stress, a plastic yield point may be reached at which plastic flow begins and the plastic yielding may result in decreasing the strain. Some materials do not pass through a plastic flow phase but rupture first.

Some materials also have a time-dependent behavior to stress (fig. 2.3b). When subjected to a steady stress, such materials creep until eventually they rupture. Creep strain does not disappear if the stress is removed.

2.1.4 *Elastic constants*

Although Lamé's constants are convenient when we are using (2.6) and (2.7) other elastic constants are also used. The most common are *Young's modulus* (E), *Poisson's ratio* (σ) and the *bulk modulus* (k) (the symbol σ is more-or-less standard for Poisson's ratio – the subscripts should prevent any confusion with the stress σ_{ij}). To define the first two we consider a medium in which all stresses are zero except σ_{xx}. Assuming σ_{xx} is positive (that is, a tensile stress), dimensions parallel to σ_{xx} will increase while dimensions normal to σ_{xx} will decrease; this means that ε_{xx} is positive (elongation in the x-direction) while ε_{yy} and ε_{zz} are negative. Also, we can show (see problem 2.2) that $\varepsilon_{yy} = \varepsilon_{zz}$. We now define E and σ by the relations

Fig.2.3. Stress–strain–time relationships. (*a*) Stress versus strain; (*b*) strain versus time.

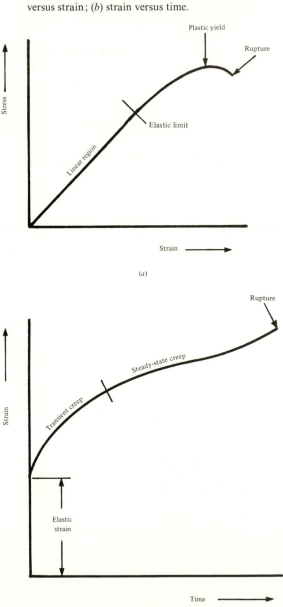

(*a*)

(*b*)

$$E = \sigma_{xx}/\varepsilon_{xx}, \tag{2.8}$$

$$\sigma = -\varepsilon_{yy}/\varepsilon_{xx} = -\varepsilon_{zz}/\varepsilon_{xx}, \tag{2.9}$$

the minus signs being inserted to make σ positive.

To define k, we consider a medium subjected only to a hydrostatic pressure $\mathscr{P}$; this is equivalent to the statements

$$\sigma_{xx} = \sigma_{yy} = \sigma_{zz} = -\mathscr{P}; \quad \sigma_{xy} = \sigma_{yz} = \sigma_{zx} = 0.$$

Then, k is defined as the ratio of the pressure to the dilatation,

$$k = -\mathscr{P}/\Delta, \tag{2.10}$$

the minus sign being inserted to make k positive. Sometimes the *compressibility*, $1/k$, is used as an elastic constant rather than its reciprocal, the bulk modulus.

By substituting the above values in Hooke's law we can obtain the following relations between E, σ and k and Lamé's constants, λ and μ (see problem 2.2):

$$E = \frac{\mu(3\lambda + 2\mu)}{(\lambda + \mu)}, \tag{2.11}$$

$$\sigma = \frac{\lambda}{2(\lambda + \mu)}, \tag{2.12}$$

$$k = \tfrac{1}{3}(3\lambda + 2\mu). \tag{2.13}$$

By eliminating different pairs of constants among the three equations many different relations can be derived expressing one of the five constants in terms of two others (see problem 2.3).

The elastic constants are defined in such a way that they are positive numbers. As a consequence of this, σ must have values between 0 and 0.5 (this follows from (2.12) since both λ and μ are positive and hence $\lambda/(\lambda + \mu)$ is less than unity). Values range from 0.05 for very hard, rigid rocks to about 0.45 for soft, poorly consolidated materials. Liquids have no resistance to shear and hence for them $\mu = 0$ and $\sigma = 0.5$. For most rocks, E, k, and μ lie in the range from 20 to 120 GPa (2×10^{10} to 12×10^{10} N/m^2), E generally being the largest and μ the smallest of the three. Tables of elastic constants of rocks have been given by Birch (1966). (See also problem 2.5.)

Most of the preceding theory assumes an isotropic medium. In fact, rocks are usually in layers with different elastic properties, these properties often varying with direction. Nevertheless in discussing wave propagation we generally ignore such differences and treat sedimentary rocks as isotropic media; when one does so the results are useful and to do otherwise leads to extremely complex and cumbersome mathematical equations, except for the case of *transversely isotropic media*, that is, media in which the properties are the same in one plane but different along the normal to the plane. Some rocks, especially shales, are transversely isotropic, and more importantly, a series of parallel beds, each of which is isotropic but where the properties vary from bed to bed, behaves as though it were transversely isotropic (Postma, 1955; Uhrig and van Melle, 1955).

Taking the z-axis as the axis of symmetry, we write $\lambda_{\parallel}$, $\mu_{\parallel}$ and $\lambda_{\perp}$, $\mu_{\perp}$ for Lamé's constants in the xy-plane (bedding plane) and perpendicular to this plane, respectively. Love (1944) showed that for this case Hooke's law reduces to the following:

$$\left. \begin{aligned} \sigma_{xx} &= (\lambda_{\parallel} + 2\mu_{\parallel})\varepsilon_{xx} + \lambda_{\parallel}\varepsilon_{yy} + \lambda_{\perp}\varepsilon_{zz} \\ \sigma_{yy} &= \lambda_{\parallel}\varepsilon_{xx} + (\lambda_{\parallel} + 2\mu_{\parallel})\varepsilon_{yy} + \lambda_{\perp}\varepsilon_{zz} \\ \sigma_{zz} &= \lambda_{\perp}\varepsilon_{xx} + \lambda_{\perp}\varepsilon_{yy} + (\lambda_{\perp} + 2\mu_{\perp})\varepsilon_{zz}, \end{aligned} \right\} \quad (2.14)$$

$$\left. \begin{aligned} \sigma_{xy} &= \mu_{\parallel}\varepsilon_{xy} \\ \sigma_{yz} &= \mu^{*}\varepsilon_{yz} \\ \sigma_{zx} &= \mu^{*}\varepsilon_{zx}, \end{aligned} \right\} \quad (2.15)$$

where μ^{*} is a new elastic constant independent of the others.

2.1.5   *Strain energy*

When an elastic medium undergoes deformation work is done and an equivalent amount of potential energy is stored in the medium; this energy is intimately related to elastic wave propagation.

If the stress σ_{xx} results in a displacement ε_{xx}, we assume that the stress is increased uniformly from zero to σ_{xx} and hence the average stress is $\frac{1}{2}\sigma_{xx}$. Thus

$$\begin{aligned} E &= \text{work done per unit volume} \\ &= \text{energy per unit volume} \\ &= \tfrac{1}{2}\sigma_{xx}\varepsilon_{xx}. \end{aligned}$$

Summing the effects of all the independent stresses and using (2.6) and (2.7) gives

$$\begin{aligned} E &= \tfrac{1}{2}\sum_{i}\sum_{j}\sigma_{ij}\varepsilon_{ij} \\ &= \tfrac{1}{2}(\sigma_{xx}\varepsilon_{xx} + \sigma_{yy}\varepsilon_{yy} + \sigma_{zz}\varepsilon_{zz}) + \sigma_{xy}\varepsilon_{xy} \\ &\quad + \sigma_{yz}\varepsilon_{yz} + \sigma_{zx}\varepsilon_{zx} \\ &= \tfrac{1}{2}\{\sum(\lambda\Delta + 2\mu\varepsilon_{ii})\varepsilon_{ii}\} + \mu\sum\varepsilon_{ij}^{2} \\ &= \tfrac{1}{2}\lambda\Delta^{2} + \mu(\varepsilon_{xx}^{2} + \varepsilon_{yy}^{2} + \varepsilon_{zz}^{2} + \varepsilon_{xy}^{2} + \varepsilon_{yz}^{2} + \varepsilon_{zx}^{2}). \end{aligned}$$

$$(2.16)$$

2.2   Wave equations and their solutions
2.2.1   *Wave equations*

(*a*) *Scalar wave equation*. Up to this point we have been discussing a medium in static equilibrium. We shall now remove this restriction and consider what happens when the stresses are not in equilibrium. In fig. 2.1 we now assume that the stresses on the rear face of the element of volume are as shown in the diagram but that the stresses on the front face are respectively

$$\sigma_{xx} + \frac{\partial \sigma_{xx}}{\partial x}dx, \quad \sigma_{yx} + \frac{\partial \sigma_{yx}}{\partial x}dx, \quad \sigma_{zx} + \frac{\partial \sigma_{zx}}{\partial x}dx.$$

Since these stresses are opposite to those acting on the rear face, the net (unbalanced) stresses are

$$\frac{\partial \sigma_{xx}}{\partial x}dx, \quad \frac{\partial \sigma_{yx}}{\partial x}dx, \quad \frac{\partial \sigma_{zx}}{\partial x}dx.$$

These stresses act on a face having an area $(dy\,dz)$ and affect the volume $(dx\,dy\,dz)$; hence we get for the net forces per unit volume in the directions of the x-, y- and z-axes the respective values:

$$\frac{\partial \sigma_{xx}}{\partial x}, \quad \frac{\partial \sigma_{yx}}{\partial x}, \quad \frac{\partial \sigma_{zx}}{\partial x}.$$

Similar expressions hold for the other faces; hence we find for the total force in the direction of the x-axis the expression

$$\left(\frac{\partial \sigma_{xx}}{\partial x} + \frac{\partial \sigma_{xy}}{\partial y} + \frac{\partial \sigma_{xz}}{\partial z} \right).$$

Newton's second law of motion states that the unbalanced force equals the mass times the acceleration; thus we obtain the equation of motion along the x-axis,

$$\rho\frac{\partial^{2}u}{\partial t^{2}} = \text{Unbalanced force in the x-direction on a unit volume}$$

$$= \frac{\partial \sigma_{xx}}{\partial x} + \frac{\partial \sigma_{xy}}{\partial y} + \frac{\partial \sigma_{xz}}{\partial z}, \quad (2.17)$$

where ρ is the density (assumed to be constant). Similar equations can be written for the motion along the y- and z-axes.

Equation (2.17) relates the displacements to the stresses. We can obtain an equation involving only displacements by using Hooke's law to replace the stresses with strains and then expressing the strains in terms of the displacements, using (2.1), (2.2), (2.5), (2.6) and (2.7). Thus,

$$\rho\frac{\partial^{2}u}{\partial t^{2}} = \frac{\partial \sigma_{xx}}{\partial x} + \frac{\partial \sigma_{xy}}{\partial y} + \frac{\partial \sigma_{xz}}{\partial z},$$

$$= \lambda\frac{\partial\Delta}{\partial x} + 2\mu\frac{\partial\varepsilon_{xx}}{\partial x} + \mu\frac{\partial\varepsilon_{xy}}{\partial y} + \mu\frac{\partial\varepsilon_{xz}}{\partial z},$$

$$= \lambda\frac{\partial\Delta}{\partial x} + \mu\left\{2\frac{\partial^2 u}{\partial x^2} + \left(\frac{\partial^2 v}{\partial x\,\partial y} + \frac{\partial^2 u}{\partial y^2}\right) \right.$$
$$\left. + \left(\frac{\partial^2 w}{\partial x\,\partial z} + \frac{\partial^2 u}{\partial z^2}\right)\right\},$$

$$= \lambda\frac{\partial\Delta}{\partial x} + \mu\nabla^2 u + \mu\frac{\partial}{\partial x}\left(\frac{\partial u}{\partial x} + \frac{\partial v}{\partial y} + \frac{\partial w}{\partial z}\right),$$

$$= (\lambda + \mu)\frac{\partial\Delta}{\partial x} + \mu\nabla^2 u, \qquad (2.18)$$

where $\nabla^2 u = Laplacian$ of $u = (\partial^2 u/\partial x^2 + \partial^2 u/\partial y^2 + \partial^2 u/\partial z^2)$ (see (10.14)). By analogy we can write the equations for v and w

$$\rho\frac{\partial^2 v}{\partial t^2} = (\lambda + \mu)\frac{\partial\Delta}{\partial y} + \mu\nabla^2 v, \qquad (2.19)$$

$$\rho\frac{\partial^2 w}{\partial t^2} = (\lambda + \mu)\frac{\partial\Delta}{\partial z} + \mu\nabla^2 w. \qquad (2.20)$$

To obtain the wave equation, we differentiate these three equations with respect to x, y and z respectively and add the results together. This gives

$$\rho\frac{\partial^2}{\partial t^2}\left(\frac{\partial u}{\partial x} + \frac{\partial v}{\partial y} + \frac{\partial w}{\partial z}\right) = (\lambda + \mu)\left(\frac{\partial^2\Delta}{\partial x^2} + \frac{\partial^2\Delta}{\partial y^2} + \frac{\partial^2\Delta}{\partial z^2}\right)$$
$$+ \mu\nabla^2\left(\frac{\partial u}{\partial x} + \frac{\partial v}{\partial y} + \frac{\partial w}{\partial z}\right),$$

that is,

$$\rho\frac{\partial^2\Delta}{\partial t^2} = (\lambda + 2\mu)\nabla^2\Delta,$$

or

$$\left.\begin{array}{c}\dfrac{1}{\alpha^2}\dfrac{\partial^2\Delta}{\partial t^2} = \nabla^2\Delta,\\[1.5em]\text{where}\\[1em]\alpha^2 = (\lambda + 2\mu)/\rho.\end{array}\right\} \qquad (2.21)$$

By subtracting the derivative of (2.19) with respect to z from the derivative of (2.20) with respect to y, we get

$$\rho\frac{\partial^2}{\partial t^2}\left(\frac{\partial w}{\partial y} - \frac{\partial v}{\partial z}\right) = \mu\nabla^2\left(\frac{\partial w}{\partial y} - \frac{\partial v}{\partial z}\right),$$

that is,

$$\left.\begin{array}{c}\dfrac{1}{\beta^2}\dfrac{\partial^2\theta_x}{\partial t^2} = \nabla^2\theta_x,\\[1.5em]\text{where}\\[1em]\beta^2 = \mu/\rho.\end{array}\right\} \qquad (2.22)$$

By subtracting appropriate derivatives, we obtain similar results for θ_y and θ_z. These equations are different examples of the *wave equation* which we can write in the general form

$$\frac{1}{V^2}\frac{\partial^2\psi}{\partial t^2} = \nabla^2\psi, \qquad (2.23)$$

where V is a constant.

(b) *Vector wave equation.* The wave equation can also be obtained using vector methods. Equations (2.18), (2.19) and (2.20) are equivalent to the *vector wave equation*,

$$\rho\frac{\partial^2\zeta}{\partial t^2} = (\lambda + \mu)\nabla\Delta + \mu\nabla^2\zeta. \qquad (2.24)$$

If we take the divergence of (2.24) and use (2.5) we get (2.21). Taking the curl of (2.24) and using (2.4) gives the vector wave equation for S-waves,

$$\frac{1}{\beta^2}\frac{\partial^2\Theta}{\partial t^2} = \nabla^2\Theta, \qquad (2.25)$$

which is equivalent to the three scalar equations,

$$\frac{1}{\beta^2}\frac{\partial^2\theta_i}{\partial t^2} = \nabla^2\theta_i, \quad i = x, y, z, \qquad (2.26)$$

((2.22) being one of these).

(c) *Wave equation including source term; Kirchhoff's formula.* The foregoing discussion of the wave equation has made no mention of the sources of the waves, and in fact the equations discussed are only valid in a source-free region. Sources can be taken into account in two ways in general: (a) include in the wave equation terms which represent the forces generating the waves; or (b) surround the point of observation P by a closed surface $\mathcal{S}$ and regard the effect at P as being given by a volume integral throughout the interior of $\mathcal{S}$ to take into account sources inside $\mathcal{S}$ plus a surface integral over $\mathcal{S}$ to give the effect of sources outside $\mathcal{S}$. To apply the first method, we note that (2.18), (2.19) and (2.20) are equivalent to Newton's second law, and these three equations are combined in (2.24). Therefore, a source can be taken into account by adding to the right-hand side of (2.24) the term $\rho\mathbf{F}$ where $\mathbf{F}$ is the external non-elastic force per unit mass (often called *body force*) which gives rise to the wave motion. Thus (2.21), (2.24) and (2.25) become

$$\rho \frac{\partial^2 \zeta}{\partial t^2} = (\lambda + \mu)\nabla\Delta + \mu\nabla^2\zeta + \rho\mathbf{F}, \tag{2.27}$$

$$\frac{\partial^2 \Delta}{\partial t^2} = \alpha^2\nabla^2\Delta + \nabla\cdot\mathbf{F}, \tag{2.28}$$

$$\frac{\partial^2 \mathbf{\Theta}}{\partial t^2} = \beta^2\nabla^2\mathbf{\Theta} + \nabla\times\mathbf{F}. \tag{2.29}$$

These equations are difficult to solve as they stand. The solution is greatly simplified by using the Helmholtz separation method which involves expressing both ζ and $\mathbf{F}$ in terms of new scalar and vector functions. Thus we write

$$\zeta = \nabla\phi + \nabla\times\chi, \quad \nabla\cdot\chi = 0, \tag{2.30}$$

$$\mathbf{F} = \nabla\Upsilon + \nabla\times\Omega, \quad \nabla\cdot\Omega = 0. \tag{2.31}$$

Then

$$\Delta = \nabla\cdot\zeta = \nabla^2\phi,$$

$$\mathbf{\Theta} = \nabla\times\zeta = -\nabla^2\chi,$$

$$\nabla\cdot\mathbf{F} = \nabla^2\Upsilon,$$

$$\nabla\times\mathbf{F} = -\nabla^2\Omega,$$

(see problem 10.5). Substituting in (2.28) and (2.29) we get

$$\nabla^2\left(\alpha^2\nabla^2\phi + \Upsilon - \frac{\partial^2\phi}{\partial t^2}\right) = 0,$$

$$\nabla^2\left(\beta^2\nabla^2\chi + \Omega - \frac{\partial^2\chi}{\partial t^2}\right) = 0.$$

Whenever ϕ, χ, Υ or Ω contain powers of x, y, z higher than the first, the above equations can only be satisfied for all values of x, y, z if the expressions inside the parentheses are identically zero at all points. Since a linear function of x, y, z corresponds to a uniform translation and/or rotation of the medium, we can ignore this possibility and write (Savarensky, 1975, p. 199)

$$\frac{\partial^2\phi}{\partial t^2} = \alpha^2\nabla^2\phi + \Upsilon, \tag{2.32}$$

$$\frac{\partial^2\chi}{\partial t^2} = \beta^2\nabla^2\chi + \Omega. \tag{2.33}$$

In the second method we take Υ in (2.32) as the source density inside $\mathscr{S}$ and specify at each point Q on the surface of $\mathscr{S}$ the function $\phi(x, y, z, t_Q)$ corresponding to sources outside $\mathscr{S}$, t_Q being the instant $(t_0 - r/V)$ where r is the distance between P and Q, and V is the velocity

(assumed constant). Thus we must specify the wave motion at different points on the surface, such that the waves from all points on the surface arrive at P at the same instant t_0. The result, known as *Kirchhoff's formula* (see Ewing *et al.*, 1957, p. 16), is

$$4\pi\phi_P(x, y, z, t_0)$$
$$= \iiint_{\mathscr{V}}\left(\frac{\Upsilon}{r}\right)\mathrm{d}\mathscr{V} + \iint_{\mathscr{S}}\left\{\left(\frac{1}{Vr}\right)\frac{\partial r}{\partial\eta}\left[\frac{\partial\phi}{\partial t}\right]\right.$$
$$\left. - [\phi]\frac{\partial(1/r)}{\partial\eta} + \left(\frac{1}{r}\right)\left[\frac{\partial\phi}{\partial\eta}\right]\right\}\mathrm{d}\mathscr{S}, \tag{2.34}$$

where η is the outward-drawn unit normal and the square brackets denote functions evaluated at the point Q at the time $t_Q = t_0 - r/V$; $[\phi]$ is often referred to as a *retarded potential*. For points P outside $\mathscr{S}$, the volume integral is zero (see Bullen, 1965, p. 67). If we assume that each source emits spherical waves (§2.2.3), (2.34) becomes (Savarensky, 1975, p. 234)

$$4\pi\phi_P(x, y, z, t_0)$$
$$= \left(\frac{1}{V^2}\right)\iiint_{\mathscr{V}}\left(\frac{[\Upsilon]}{r}\right)\mathrm{d}\mathscr{V} + \iint_{\mathscr{S}}\left\{\xi\left[\frac{\partial\phi}{\partial\eta}\right] - [\phi]\frac{\partial\xi}{\partial\eta}\right\}\mathrm{d}\mathscr{S} \tag{2.35}$$

where in the integrand,

$$[\phi] = (1/r)\,\mathrm{e}^{j\omega(t_0 - r/V)} = \xi\,\mathrm{e}^{j\omega t_0}, \quad \xi = (1/r)\,\mathrm{e}^{-j\omega r/V}, \tag{2.36}$$

ω being angular frequency (see §2.2.5).

2.2.2 *Plane-wave solutions*

Let us consider first the case where ψ is a function only of x and t, so that (2.23) reduces to

$$\frac{1}{V^2}\frac{\partial^2\psi}{\partial t^2} = \frac{\partial^2\psi}{\partial x^2}. \tag{2.37}$$

Any function of $(x - Vt)$,

$$\psi = f(x - Vt), \tag{2.38}$$

is a solution of (2.37) (see problem 2.6) provided that ψ and its first two derivatives are finite and continuous. This solution (known as D'Alembert's solution) furnishes an infinite number of particular solutions (for example, $\mathrm{e}^{k(x - Vt)}$, $\sin(x - Vt)$, $(x - Vt)^3$, where we must exclude points at which these functions and their first three derivatives cease to exist or are discontinuous). The answer to a specific problem consists of selecting the appropriate combination of solutions which also satisfies the boundary conditions for the problem.

A *body wave* is defined as a 'disturbance' which travels through the medium. In our notation, the disturbance ψ is a volume change when $\psi = \Delta$ and a rotation when $\psi = \theta_i$. Obviously the disturbance in (2.38) is traveling along the x-axis. We shall now show that it travels with a speed equal to the quantity V.

In fig. 2.4 a certain part of the wave has reached the point P_0 at the time t_0. If the coordinate of P_0 is x_0, then the value of ψ at P_0 is $\psi_0 = f(x_0 - Vt_0)$. If this same portion of the wave reaches P_1 at the time $t_0 + \Delta t$, then we have for the value of ψ at P_1

$$\psi_1 = f\{x_0 + \Delta x - V(t_0 + \Delta t)\}.$$

But, since this is the same portion of the wave which was at P_0 at time t_0, we must have $\psi_0 = \psi_1$, this is,

$$x_0 - Vt_0 = x_0 + \Delta x - V(t_0 + \Delta t).$$

Thus, the quantity V is equal to $\Delta x/\Delta t$ and is therefore the speed with which the disturbance travels.

A function of $(x + Vt)$, for example, $\psi = g(x + Vt)$, is also a solution of (2.37); it denotes a wave traveling in the negative x-direction. The general solution of (2.37)

$$\psi = f(x - Vt) + g(x + Vt), \tag{2.39}$$

represents two waves traveling along the x-axis in opposite directions with the velocity V.

Since the value of ψ is independent of y and z, the disturbance must be the same everywhere in a plane perpendicular to the x-axis. This type of wave is called a *plane wave*.

The quantity $(x - Vt)$ (or $(x + Vt)$) is known as the *phase*. The surfaces on which the wave motion is the same, that is, surfaces on which the phase has the same value, are known as *wavefronts*. In the case which we are considering, the wavefronts are planes perpendicular to the x-axis. Note that the wave is traveling in the direction normal to the wavefront; this holds for all waves in isotropic media. A line denoting the direction of travel of the wave energy is called a *raypath*.

It is convenient at times to have an expression for a plane wave traveling along a straight line inclined at an angle to each of the axes. Assume that the wave is traveling along the x'-axis which has direction cosines (l, m, n) relative to the x-, y- and z-axes (fig. 2.5). Then, at a point P on the x'-axis at a distance x' from the origin, we have

$$x' = lx + my + nz,$$

where the coordinates of P are (x, y, z). Then,

$$\psi = f(lx + my + nz - Vt) + g(lx + my + nz + Vt). \tag{2.40}$$

2.2.3 *Spherical-wave solutions*

In addition to plane waves, we shall have occasion to use another important type of wave, the *spherical wave* where the wavefronts are a series of concentric spherical surfaces. We express (2.23) in spherical coordinates (r, θ, ϕ), where θ is the colatitude and ϕ the longitude (see problem 2.7).

$$\frac{1}{V^2}\frac{\partial^2 \psi}{\partial t^2} = \frac{1}{r^2}\left\{\frac{\partial}{\partial r}\left(r^2\frac{\partial \psi}{\partial r}\right) + \frac{1}{\sin\theta}\frac{\partial}{\partial\theta}\left(\sin\theta\frac{\partial\psi}{\partial\theta}\right)\right.$$
$$\left. + \frac{1}{\sin^2\theta}\frac{\partial^2\psi}{\partial\phi^2}\right\}. \tag{2.41}$$

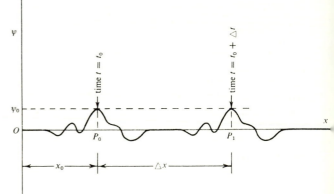

Fig.2.4. Illustrating the velocity of a wave.

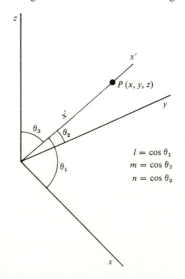

Fig.2.5. Wave direction not along an axis.

$$l = \cos\theta_1$$
$$m = \cos\theta_2$$
$$n = \cos\theta_3$$

We consider only the special case when the wave motion is independent of θ and ϕ, hence is a function only of r and t. Then we get the simplified equation,

$$\frac{1}{V^2}\frac{\partial^2 \psi}{\partial t^2} = \frac{1}{r^2}\frac{\partial}{\partial r}\left(r^2 \frac{\partial \psi}{\partial r}\right). \qquad (2.42)$$

A solution of the above equation is

$$\psi = (1/r)f(r - Vt) \qquad (2.43)$$

(see (2.36)). Obviously

$$\psi = (1/r)g(r + Vt)$$

is also a solution and the general solution of (2.42) is (see problem 2.6c)

$$\psi = (1/r)f(r - Vt) + (1/r)g(r + Vt) \qquad (2.44)$$

in which the first term represents a wave expanding outward from a central point and the second term a wave collapsing toward the central point.

When r and t are fixed, $(r - Vt)$ is constant and hence ψ is constant. Thus, at the instant t the wave has the same value at all points on the spherical surface of radius r. The spherical surfaces are therefore wavefronts and the radii are rays. Obviously the rays are normal to the wavefront as in the case of plane waves.

As the wave progresses outward from the center, the radius increases by the amount V during each unit of time. Eventually the radius becomes very large and the portion of the wavefront near any particular point will be approximately plane. If we consider fig. 2.6 we see that the error which we introduce when we replace the spherical wavefront PQR by the plane wavefront $P'QR'$ is due to the divergence between the true direction of propagation given by the direction of the radius and the assumed direction normal to the plane. By taking OQ very large or PR very small (or both), we can make the error as small as desired. Since plane waves are easy to visualize and also the simplest to handle mathematically, we generally assume that conditions are such that the plane-wave assumption is valid.

2.2.4 *Huygens' principle*

The solutions of the wave equation given by (2.39), (2.40) and (2.44) are restricted to plane and spherical waves. On the other hand, the Kirchhoff formula is valid for any type of body wave. As expressed in (2.34) (assuming no sources inside $\mathscr{S}$), it states that the effect at a point P is the sum of effects which took place earlier at all points on a surface $\mathscr{S}$ enclosing P, allowance being made for the time for these effects to travel from $\mathscr{S}$ to P. Thus, each point on $\mathscr{S}$ behaves at though it were a new wave source.

To obtain Huygens' principle, we take $\mathscr{S}$ coincident with that portion of the wavefront which we wish to take into account in finding the effect at P, and then complete the closed surface by passing it through space where the effect has not yet arrived so that ϕ is zero over this part.

Huygens' principle is important in understanding wave travel and is often useful in drawing successive positions of wavefronts. *Huygens' principle* states that every point on a wavefront can be regarded as a new source of waves. The physical rationale behind this is that each particle located on a wavefront has moved from its equilibrium position in approximately the same manner, that the elastic forces on neighboring particles are thereby changed, and that the resultant of the changes in force because of the motion of all the points on the wavefront thus begins to produce the motion which forms the next wavefront. In this way Huygens' principle helps explain how information about seismic disturbances are communicated in the Earth. Specifically, given the location of a wavefront at a certain instant, future positions of the wavefront can be found by considering each point on the first wavefront as a new wave source. In fig. 2.7, AB is the wavefront at the time t_0 and we wish to find the wavefront at a later time $(t_0 + \Delta t)$. During the interval Δt, the wave will advance a distance $V\Delta t$, V being the velocity (which may vary from point to point). We select points on the wavefront, P_1, P_2, P_3 and so on, from which we draw arcs of radii $V\Delta t$. Provided we select enough points, the envelope of the arcs ($A'B'$) will define as

Fig.2.6. Relation between spherical and plane waves.

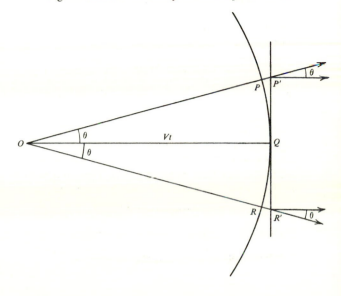

accurately as we wish the position of the wavefront at the time $(t_0 + \Delta t)$. Except on the envelope, the elemental waves interfere destructively with each other so that their effects cancel. When AB is plane, and V constant, we need draw only two arcs and the straight line tangent to the two arcs defines the new wavefront.

2.2.5 Harmonic waves

So far we have discussed only the geometrical aspect of waves, that is, the way in which the wave depends upon the space coordinates. However, ψ is a function of t also, and so we must consider the time dependence of waves as well.

The simplest form of time variation is that of the *harmonic wave*, that is, waves involving sine or cosine expressions such as

$$\psi = A \cos \kappa (x - Vt), \tag{2.45}$$

$$\psi = A \sin \kappa (lx + my + nz - Vt), \tag{2.46}$$

$$\psi = (B/r) \cos \kappa (r + Vt). \tag{2.47}$$

At a fixed point, ψ varies as the sine or cosine of the time; hence the motion is simple harmonic. The values of ψ range from $+A$ to $-A$ for the plane wave of (2.45) and (2.46) and from $+B/r$ to $-B/r$ for the spherical wave of (2.47). The limiting value, A or B/r, is known as the *amplitude* of the wave ψ.

For a fixed value of t, whenever x in (2.45) increases by the amount $2\pi/\kappa$, the argument of the cosine increases by 2π and hence the value of ψ repeats. This distance, $2\pi/\kappa$, is called the *wavelength*, usually represented by the symbol, λ. The number of waves per unit distance, $\kappa/2\pi = 1/\lambda$, is called the *wavenumber* (though some writers call κ the wavenumber). In (2.46) and (2.47) $(lx + my + nz)$ and r represent the distance the wave has travelled from the origin, hence are equivalent to x in (2.45); therefore, κ has the same significance here as in (2.45).

If the space coordinates in (2.45), (2.46) and (2.47) are kept fixed and t is allowed to increase, the value of ψ

repeats each time that t increases by the amount T where $\kappa VT = 2\pi = 2\pi VT/\lambda$; consequently

$$\left.\begin{aligned} T &= \lambda/V, \\ v &= 1/T = V/\lambda, \\ V &= v\lambda = \omega/\kappa, \end{aligned}\right\} \tag{2.48}$$

where T is called the *period*, v the *frequency* of the wave, and ω the *angular frequency* defined as $\omega = 2\pi v = \kappa V$ (see fig. 2.8). Using the above symbols, we can write (2.45) in the following equivalent forms:

$$\left.\begin{aligned} \psi &= A \cos \kappa (x - Vt) = A \cos (2\pi/\lambda)(x - Vt), \\ &= A \cos (\kappa x - \omega t) = A \cos \omega \{(x/V) - t\}, \\ &= A \cos (\kappa x - 2\pi v t) = A \cos 2\pi \{(x/\lambda) - vt\}. \end{aligned}\right\} \tag{2.49}$$

It is often convenient to use an exponential form such as

$$\psi = A \, e^{j\omega[\{(lx + my + nz)/V\} - t]} = A \, e^{j(\kappa r - \omega t)}, \tag{2.50}$$

where r is the distance from the source in the direction of the ray; we can obtain the cosine or sine expressions by taking the real or imaginary parts.

The quantities (l, m, n) in (2.40), (2.46) and (2.50) represent the direction cosines of the ray. In analytic geometry it is shown that $l^2 + m^2 + n^2 = 1$. Although ordinarily each of the cosines has a maximum value of unity, satisfying the wave equation requires only that the sum of the squares be unity. If we admit pure imaginary numbers, some of the 'direction cosines' can be greater than unity. Let us take in fig. 2.5, $\theta_1 = j\theta$, $\theta_2 = \frac{1}{2}\pi$, $\theta_3 = \frac{1}{2}\pi - j\theta$, θ being real and positive; then

$$l = \cos j\theta = \cosh \theta, \quad m = 0,$$
$$n = \cos(\tfrac{1}{2}\pi - j\theta) = \sin j\theta = j \sinh \theta,$$
$$l^2 + m^2 + n^2 = \cosh^2 \theta - \sinh^2 \theta = 1,$$
$$\psi = A \, e^{-(\omega z/V)\sinh\theta} \, e^{j\omega\{(x/V)\cosh\theta - t\}}. \tag{2.51}$$

This represents a plane wave traveling parallel to the x-axis with velocity $V/\cosh\theta < V$ and amplitude $A \, e^{-(\omega z/V)\sinh\theta}$. If we had taken $\theta_1 = -j\theta$, l would be unchanged while n would change sign; this would give a wave traveling in the negative z-direction. Since the amplitude decreases exponentially with z, these waves are called *evanescent waves*. We shall refer again to these waves in §2.3.4.

In exploration seismology the range of frequencies recorded with appreciable energy is generally from about 2 to 120 Hz, while the dominant frequencies lie in a narrower range from 15 to 50 Hz for reflection work and from 5 to 20 Hz for refraction work. Since velocities generally range from 1.6 to 6.5 km/s, dominant wave-

Fig.2.7. Using Huygens' principle to locate new wavefronts.

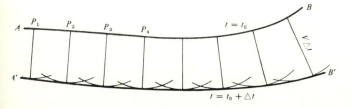

Fig.2.8. Nomogram of (2.48); a straight line relates the velocity, frequency and wavelength (the outside scales are metric, the inner ones English). For example, a velocity of 2 km/s and frequency of 50 Hz give a wavelength of 40 m.

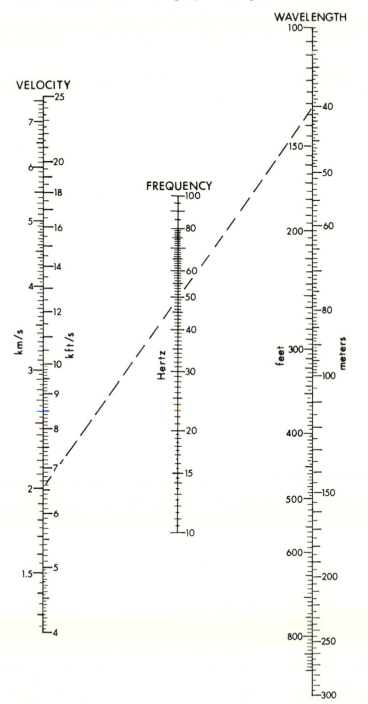

lengths range from about 30 to 400 m for reflection work and from 80 to 1300 m for refraction (see fig. 2.8).

2.2.6 *P-waves and S-waves*

Up to this point our discussion of wave motion has been based upon (2.23). The quantity ψ has not been defined; we have merely inferred that it is some disturbance which is propagated from one point to another with the speed V. However, in a homogeneous isotropic medium, (2.21) and (2.25) must be satisfied. We can identify the functions Δ and θ with ψ and conclude that two types of waves can be propagated in a homogeneous isotropic medium, one corresponding to changes in the dilatation, Δ, the other to changes in one or more components of the rotation given in (2.3).

The first type is variously known as a *dilatational, longitudinal, irrotational, compressional* or P-*wave*, the latter name being due to the fact that this type is usually the first (primary) event on an earthquake recording. The second type is referred to as the *shear, transverse, rotational* or S-*wave* (since it is usually the second event observed on earthquake records). The P-wave has the velocity α in (2.21) and the S-wave the velocity β in (2.25) where

$$\alpha = \{(\lambda + 2\mu)/\rho\}^{\frac{1}{2}}, \tag{2.52}$$

$$\beta = (\mu/\rho)^{\frac{1}{2}}. \tag{2.53}$$

Since the elastic constants are positive, α is always greater than β, and

$$\beta^2/\alpha^2 = \mu/(\lambda + 2\mu) = (\tfrac{1}{2} - \sigma)/(1 - \sigma), \tag{2.54}$$

using (2.12). As σ decreases from 0.5 to zero, β/α increases from zero to its maximum value, $1/\sqrt{2}$; thus, the velocity of the S-wave ranges from zero up to 70% of the velocity of the P-wave.

For fluids μ is zero and hence β is also zero; therefore S-waves do not propagate through fluids.

The seismic velocity in actual rocks depends on many factors, including porosity, lithology, cementation, depth, age, pressure regime, interstitial fluids, etc., which are discussed in chapter 7. The velocity of water-saturated sedimentary rocks is generally in the 1.5 to 6.5 km/s range, increasing with loss of porosity, cementation, depth, and age. (Velocity versus depth relations for three situations are shown in fig. 6.23.) The velocity of P-waves in water is approximately 1.5 km/s. P-wave velocity is lowered, often markedly, when a gas replaces water as the interstitial fluid. This is especially important in the near-surface, generally above the water table, where a low-velocity layer (LVL, also called the weathering layer) typically has a velocity in the 0.4 to 0.8 km/s range, occasionally as low as 150 m/s, sometimes as high as 1.2 km/s.

Let us investigate the nature of the motion of the medium corresponding to the two types of wave motion. Consider a spherical P-wave of the type given by (2.43). Fig. 2.9 shows wavefronts drawn at quarter-wavelength intervals, t being chosen so that $\kappa V t$ is an integer. The arrows represent the direction of motion of the medium at the wavefront. The medium is undergoing maximum compression at B (that is, the dilatation Δ is a minimum) and minimum compression (maximum Δ) at the wavefront D; particle velocity is zero at each of these points.

We can visualize the plane-wave situation by imagining that the radius in fig. 2.9 has become very large so that the wavefronts are practically plane surfaces. The displacements will everywhere be perpendicular to these planes so that there will no longer be convergence or divergence of the particles of the medium as they move back and forth parallel to the direction of propagation of the wave. Such a displacement is longitudinal which explains why P-waves are sometimes called longitudinal waves. P-waves are the dominant waves involved in seismic exploration. A plane P-wave is illustrated in fig. 2.10a.

To determine the motion of a medium during the passage of an S-wave, we return to (2.26) and consider the case where a rotation θ_z which is a function of x and t only is being propagated along the x-axis. We have

$$\frac{1}{\beta^2} \frac{\partial^2 \theta_z}{\partial t^2} = \frac{\partial^2 \theta_z}{\partial x^2}.$$

Since

$$\theta_z = \frac{\partial v}{\partial x} - \frac{\partial u}{\partial y} = \frac{\partial v}{\partial x}$$

from (2.3), we see that the wave motion consists solely of a displacement v of the medium in the y-direction, v being a function of both x and t. Since v is independent of y and z, the motion is everywhere the same in a plane perpendicular to the x-axis; thus the case we are discussing is that of a plane S-wave traveling along the x-axis (fig. 2.10b).

We can visualize the above relations using fig. 2.11. When the wave arrives at P, it causes the medium in the vicinity of P to rotate about the axis $Z'Z''$ (parallel to the z-axis) through an angle ε. Since we are dealing with infinitesimal strains, ε must be infinitesimal and we can ignore the curvature of the displacements and consider that points such as P' and P'' are displaced parallel to the y-axis to the points Q' and Q''. Thus, as the wave travels along the x-axis, the medium is displaced transversely to the direction of propagation, hence the name transverse wave. Moreover, since the rotation varies from point to point at any given instant, the medium is subjected to varying shearing stresses as the wave moves along; this accounts for the name shear wave.

Since we might have chosen to illustrate θ_y in fig. 2.11 instead of θ_z, it is clear that shear waves have two degrees of freedom, unlike P-waves which have only one – along the radial direction. In practice S-wave motion is usually resolved into components parallel and perpendicular to the surface of the ground, these are known respectively as SH and SV *waves*. (When the wave is traveling neither horizontally nor vertically, the motion is resolved into a horizontal (SH) component and a component in the vertical plane through the direction of propagation.) Henceforth S-wave will mean SV-wave unless otherwise noted.

Because the two degrees of freedom of S-waves are independent, we can have an S-wave which involves motion in only one plane, for example, SH or SV motion; such a wave is said to be *plane polarized*. We can also have a wave in which the SH and SV motion have the same frequency and a fixed phase difference; such a wave is *elliptically polarized*. Polarization of S-waves is a factor in their exploration use (see §5.7).

In the case of a medium which is not homogeneous and isotropic, it may not be possible to resolve wave motion into separate P-waves and S-waves. However, inhomogeneities and anisotropy in the Earth are small

Fig.2.9. Displacements for a spherical P-wave.

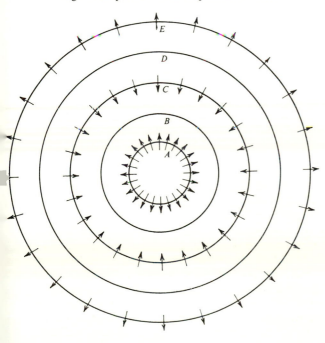

enough that assumption of separate P- and S-waves is valid for practical purposes.

2.2.7 *Displacement and velocity potentials*

Solutions of the wave equations such as those in (2.40) and (2.44) furnish expressions for Δ and θ_i. However, often we need to know the displacements u, v, w, or the velocities $\dot{u}$, $\dot{v}$, $\dot{w}$, and reference to (2.1)–(2.5) will show that these are not easily found given only values of Δ and θ_i. This difficulty is often resolved by using potential functions $\phi(x, y, z, t)$ and $\chi(x, y, z, t)$, which are solutions of the P- and S-wave equations respectively and which are so chosen that u, v, w (or $\dot{u}$, $\dot{v}$, $\dot{w}$) can be found by differentiation.

A simple example of such functions is the following:

$$\chi = 0, \quad \nabla\phi = \zeta = (u\mathbf{i} + v\mathbf{j} + w\mathbf{k}),$$

so that

$$u = \frac{\partial\phi}{\partial x}, \quad v = \frac{\partial\phi}{\partial y}, \quad w = \frac{\partial\phi}{\partial z}. \right\} \tag{2.55}$$

This procedure is valid only if it corresponds with Δ being a solution of the P-wave equation. Since ζ is a solution and $\Delta = \nabla\cdot\zeta = \nabla^2\phi$, Δ is also a solution (because derivatives of a solution are also solutions). Setting $\chi = 0$ is equivalent to saying that S-waves do not exist and this choice of potential functions is suitable for discussing wave motion in fluids.

For wave motion in three-dimensional solids, ϕ and χ can be defined so that

$$\zeta = \nabla\left(\phi + \frac{\partial\chi}{\partial z}\right) - \nabla^2\chi\mathbf{k}. \tag{2.56}$$

This ensures that Δ and Θ are solutions of the P- and S-wave equations respectively (see problem 2.9).

For two-dimensional wave motion in the xz-plane, ϕ and χ can be defined by

$$\zeta = \nabla\phi + \nabla\times\chi, \quad \chi = -\chi\mathbf{j},$$

$$u = \frac{\partial\phi}{\partial x} + \frac{\partial\chi}{\partial z}, \quad w = \frac{\partial\phi}{\partial z} - \frac{\partial\chi}{\partial x}. \right\} \tag{2.57}$$

It is easy to show that (2.5) and (2.4) can be expressed as

$$\Delta = \nabla\cdot\zeta = \nabla^2\phi, \quad \Theta = \nabla\times\zeta = \nabla^2\chi\mathbf{j}, \tag{2.58}$$

so that Δ and Θ are again solutions of the P- and S-wave equations.

Since the wave equations are still valid if both sides are differentiated with respect to time t, it follows that velocity potentials will be obtained in each of the above cases if u, v, w, and ζ are replaced with $\dot{u}$, $\dot{v}$, $\dot{w}$, and $\dot{\zeta}$.

2.2.8 *Boundary conditions*

When a wave arrives at a surface separating two media having different elastic properties, it gives rise to reflected and refracted waves as described in §2.3.4. The relationships between the various waves can be found from the relations between the stresses and displacements on the two sides of the interface. At the boundary between two media, the stresses and displacements must be continuous.

Two neighboring points R and S, which lie on opposite sides of the boundary as shown in fig. 2.12, will in general have different values of normal stress. This difference results in a net force which accelerates the layer between them. However, if we choose points closer and closer together, the stress values must approach each other and in the limit when the two points coincide on the boundary, the two stresses must be equal. If this were not so, the infinitesimally thin layer at the boundary would be acted upon by a finite force and hence have an acceleration which would approach infinity as the two points approach each other. Since the same reasoning applies to a tangential stress, we see that the normal and tangential components of stress must be continuous (cannot change abruptly) at the boundary.

The normal and tangential components of displacement must also be continuous. If the normal displacement were not continuous, one medium would either separate from the other leaving a vacuum in between or else would penetrate into the other so that the two media would occupy the same space. If the tangential displacement were not continuous, the two media would move differently on opposite sides of the boundary and one would slide over the other. Such relative motion is assumed to be impossible and so displacement must be continuous.

When one or both of the solid media are replaced by a fluid or a vacuum, the boundary conditions are reduced in number (see problem 2.10).

Fig.2.10. Motion during passage of a plane wave. (*a*) P-wave; (*b*) S-wave.

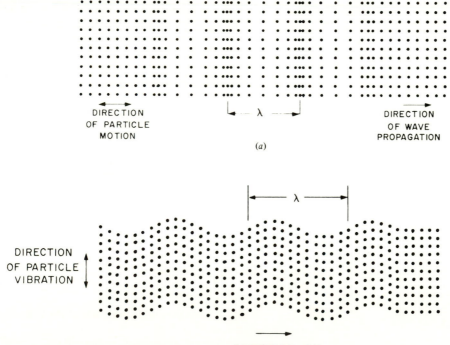

DIRECTION OF PARTICLE MOTION

λ

DIRECTION OF WAVE PROPAGATION

(*a*)

λ

DIRECTION OF PARTICLE VIBRATION

DIRECTION OF WAVE PROPAGATION

(*b*)

2.2.9 *Waves generated by a spherically symmetrical source*

The potential function $\phi = (1/r)f(t - r/V)$ is a solution to the wave equation when there is spherical symmetry (see (2.42)); hence the radial displacement $u(r, t)$ is

$$u(r, t) = \frac{\partial \phi}{\partial r} = -\left(\frac{1}{r^2}\right)f\left(t - \frac{r}{V}\right) + \left(\frac{1}{r}\right)\frac{\partial}{\partial r}\left\{f\left(t - \frac{r}{V}\right)\right\}$$

$$(2.59)$$

(using (2.55) with the x-axis in the radial direction). For harmonic waves the two terms have equal importance at a distance $r = \lambda/2\pi$ (see problem 2.13), but the first term decays rapidly in importance at greater distances. The second term is the *far-field effect* whereas the *near-field*

Fig.2.11. Rotation of medium during passage of an S-wave.

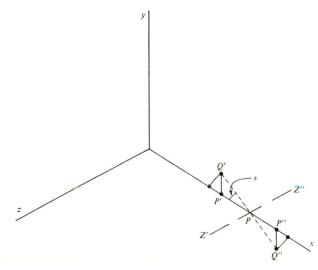

Fig.2.12. Continuity of normal stress.

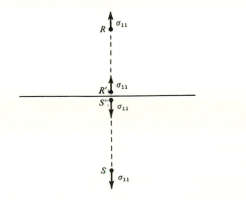

effect depends on both terms. This distinction is important when calculating a far-field waveshape from near-field recordings.

Equation (2.59) can be used to derive the wave motion created by symmetrical displacement of the medium outward from a point source. When the wave is created by very high pressures, as in an explosion of dynamite, the wave equation is not valid near the source because the medium does not obey Hooke's law there; this difficulty is usually resolved by surrounding the source by a spherical surface of radius r_0 such that the wave equation is valid for $r \geq r_0$, then specifying the displacement or pressure on this surface due to the source.

Let us consider the case where the displacement $u(r, t)$ is to be found, given the displacement $u_0(t)$ of the surface $r = r_0$. We let $\zeta = t - (r - r_0)/V$ and write

$$\phi(r, t) = (1/r)f(\zeta), \quad \zeta \geq 0, r \geq r_0 \\ = 0 \qquad\qquad \zeta < 0;$$

$$(2.60)$$

then

$$u(r, t) = \frac{\partial \phi}{\partial r} = -\left(\frac{1}{r^2}f(\zeta) + \frac{1}{rV}\frac{df(\zeta)}{d\zeta}\right).$$

$$(2.61)$$

At $r = r_0$, $\zeta = t$ and $u(r, t) = u_0(t)$ where $u_0(t)$ depends on the specific source,

$$u_0(t) = -\left(\frac{1}{r_0^2}f(t) + \frac{1}{r_0 V}\frac{df(t)}{dt}\right).$$

$$(2.62)$$

Using these values and multiplying both sides of (2.62) by the integrating factor e^{Vt/r_0}, we get

$$\frac{d}{dt}\{e^{Vt/r_0}f(t)\} = e^{Vt/r_0}\left(\frac{df(t)}{dt} + \frac{V}{r_0}f(t)\right) = -r_0 Vu_0(t)e^{Vt/r_0},$$

$$f(t) = -r_0 Ve^{-Vt/r_0}\int_0^t u_0(t)e^{Vt/r_0}\,dt.$$

$$(2.63)$$

Note that the lower limit of the integral means that $t = 0$ is the instant at which the wave first reaches the surface r_0, $u_0(t)$ being zero before this.

To carry the calculation further we must know $u_0(t)$. Let us approximate an explosion by the expression

$$u_0(t) = ke^{-at}, \quad t \geq 0, a > 0, \\ = 0 \qquad\quad t < 0.$$

$$(2.64)$$

Then

$$f(t) = -r_0 Ve^{-Vt/r_0}\int_0^t k e^{(V/r_0 - a)t}\,dt$$

$$= \frac{r_0 Vk}{(V/r_0 - a)}(e^{-Vt/r_0} - e^{-at}).$$

We replace t in this expression by $\zeta = t - (r - r_0)/V$ and (2.61) becomes

$$u(r, t) = \frac{\partial \phi}{\partial r} = \frac{r_0 k}{r(V/r_0 - a)} \left\{ \frac{V}{r_0} e^{-V\zeta/r_0} - a e^{-a\zeta} \right.$$

$$\left. - \frac{V}{r} e^{-V\zeta/r_0} + \left(\frac{V}{r} \right) e^{-a\zeta} \right\} \tag{2.65}$$

$$\approx \frac{r_0 k}{r(V/r_0 - a)} \left\{ \frac{V}{r_0} e^{-V\zeta/r_0} - a e^{-a\zeta} \right\}, \quad r \gg r_0, \tag{2.66}$$

the latter equation giving the far-field solution.

The fact that (2.65) and (2.66) are valid only for $\zeta > 0$ merely means that $u(r, t)$ is zero until $t = (r - r_0)/V$, that is, until the disturbance reaches the point. At this instant $\zeta = 0$ and $u(r, t) = k(r_0/r)$; hence the initial displacement is the same as that of the surface r_0 except that it is reduced by the factor (r_0/r), that is, $u(r, t)$ falls off inversely as the distance (see §2.3.1 and (2.92)). Moreover, $u = 0$ at $t = \infty$ and also when (see (2.65))

$$V\{(1/r_0) - (1/r)\} e^{-V\zeta/r_0} + \{(V/r) - a\} e^{-a\zeta} = 0,$$

that is, when

$$t = \frac{r - r_0}{V} + \frac{1}{V/r_0 - a} \ln \left\{ \frac{V(r - r_0)}{r_0 r(a - V/r)} \right\}.$$

Provided $V/r_0 > a > V/r$, this equation has a real positive root and $u(r, t)$ will vanish, that is, the displacement must change sign. Since V/r_0 is large in practice and V/r rapidly becomes small, in general the unidirectional pulse in (2.64) gives rise to an oscillatory wave.

By using different expressions for $u_0(t)$ in (2.63) or by specifying $\mathscr{P}_0(t)$, the pressure at the cavity, we can investigate the wave motion for various spherically symmetrical sources (see Blake, 1952; Savarensky, 1975, p. 243–55). By finding the limit as a in (2.64) goes to zero (see problem 2.12), we get the result for a unit step, step(t); then the results for other inputs can be found using convolution techniques (see §10.3.6 and §10.5.1).

2.2.10 *Surface waves*

(a) *Rayleigh waves.* In an infinite, homogeneous, isotropic medium, only P- and S-waves exist. However, whenever there is a surface separating media of different elastic properties, surface waves may exist. Their amplitudes decrease with increasing distance from the surface.

The most important surface wave in exploration seismology is the *Rayleigh wave*, which is propagated along a free surface of a solid. While a 'free' surface means contact with a vacuum, the elastic constants and density of air are so low in comparison with values for rocks that the surface of the Earth is approximately a free surface. *Ground roll* is the term commonly used for Rayleigh waves.

We take the x-axis in the surface and the z-axis positive downward. Appropriate potentials for a plane Rayleigh wave traveling along the x-axis (see (2.57)) are

$$\phi = A e^{-m\kappa z} e^{j\kappa(x - V_R t)}, \quad \chi = B e^{-n\kappa z} e^{j\kappa(x - V_R t)}, \tag{2.67}$$

where m and n must be real positive constants so that the wave decreases in amplitude away from the surface; V_R is, of course, the velocity of the wave. Substituting ϕ and χ in the wave equations gives

$$m^2 = (1 - V_R^2/\alpha^2), \quad n^2 = (1 - V_R^2/\beta^2). \tag{2.68}$$

Since m and n are real, $V_R < \beta < \alpha$, so that the velocity of the Rayleigh wave is less than that of an S-wave.

We next apply the boundary conditions that σ_{zz} and σ_{xz} vanish at the boundary $z = 0$. Using the results of problem 2.11, we get for $z = 0$

$$\sigma_{zz} = \lambda \Delta + 2\mu \varepsilon_{zz} = \lambda \nabla^2 \phi + 2\mu \frac{\partial w}{\partial z}$$

$$= \lambda \nabla^2 \phi + 2\mu \left(\frac{\partial^2 \phi}{\partial z^2} - \frac{\partial^2 \chi}{\partial x \, \partial z} \right) = 0$$

$$\sigma_{xz} = \mu \varepsilon_{xz} = \mu \left(\frac{\partial u}{\partial z} + \frac{\partial w}{\partial x} \right)$$

$$= \mu \left(2 \frac{\partial^2 \phi}{\partial x \, \partial z} + \frac{\partial^2 \chi}{\partial z^2} - \frac{\partial^2 \chi}{\partial x^2} \right) = 0. \tag{2.69}$$

Substituting (2.67) into the above and setting $z = 0$ gives

$$\{(\lambda + 2\mu)m^2 - \lambda\}A + 2jn\mu B = 0$$

and

$$-2jmA + (n^2 + 1)B = 0.$$

We can use (2.52), (2.53) and (2.68) to write the first result in the form

$$(2\beta^2 - V_R^2)A + 2jn\beta^2 B = 0.$$

Eliminating the ratio B/A from the two equations gives

$$(2 - V_R^2/\beta^2)(n^2 + 1) = 4mn;$$

hence

$$V_R^6 - 8\beta^2 V_R^4 + (24 - 16\beta^2/\alpha^2)\beta^4 V_R^2 + 16(\beta^2/\alpha^2 - 1)\beta^6 = 0. \tag{2.70}$$

Since the left side of (2.70) is negative for $V_R = 0$ and positive for $V_R = +\beta$, a real root must exist between these two values, this root giving the Rayleigh wave velocity

V_R. However, we cannot find this root without knowing β/α.

For many rocks $\sigma \approx \frac{1}{4}$, that is, $(\beta/\alpha)^2 \approx \frac{1}{3}$ from (2.54). If we use this value, the three roots of (2.70) are $V_R^2 = 4\beta^2$, $2(1 \pm 1/\sqrt{3})\beta^2$. Since V_R/β must be less than unity, the only permissible solution is

$$V_R^2 = 2(1 - 1/\sqrt{3})\beta^2, \quad \text{or} \quad V_R = 0.919\beta.$$

We now find that $V_R/\alpha = 0.531$, $m = 0.848$, $n = 0.393$, $B/A = +1.468j$; hence

$$\phi = A\,e^{-0.848\kappa z}\,e^{j\kappa(x - V_R t)},$$

$$\chi = 1.468jA\,e^{-0.393\kappa z}\,e^{j\kappa(z - V_R t)}.$$

Using (2.57) we get for the displacements at the surface

$$u = 0.423j\kappa A\,e^{j\kappa(x - V_R t)}, \quad w = 0.620\kappa A\,e^{j\kappa(x - V_R t)}.$$

Taking the real part of the solution (which corresponds to a displacement at the source of $\cos \omega t$), we obtain finally

$$\left. \begin{aligned} u &= -0.423\kappa A \sin \kappa(x - V_R t), \\ w &= 0.620\kappa A \cos \kappa(x - V_R t). \end{aligned} \right\} \tag{2.71}$$

At a given point on the surface a particle describes an ellipse in the vertical xz-plane as shown in fig. 2.13a, the horizontal axis being about two-thirds the vertical axis. The angle θ is given by

$$\tan \theta = -w/u = 1.465 \cot \kappa(x - V_R t). \tag{2.72}$$

As t increases, $\cot \kappa(x - V_R t)$ and θ increase, that is, P moves around the ellipse in a counter-clockwise (retrograde) direction for a wave moving from left to right.

Rayleigh-wave velocity as a function of Poisson's ratio is shown in fig. 2.14. Since V_R as given by (2.70) is independent of frequency, Rayleigh waves on the surface of a homogeneous medium do not exhibit dispersion (see §2.3.3). Field observations (fig. 2.13b) agree roughly with the type of motion shown in fig. 2.13a, differences being attributed to the Earth being layered and anisotropic rather than an ideal, homogeneous, isotropic medium. Measurements also show that Rayleigh waves are dispersive (Dobrin, 1951). Rayleigh waves are low-velocity, low-frequency waves with a spectrum which is not sharply peaked, and hence involve a broad range of wavelengths. Since $m\kappa$ and $n\kappa$ determine the penetration (penetration showing the exponential fall-off predicted by (2.67) is illustrated in fig. 2.13c), there is a large variation for different frequency components, most of the energy being confined to a zone one or two wavelengths thick. Since the elastic constants vary considerably near

the surface, especially at the base of the LVL (see §5.6.2), the velocity varies with wavelength, the waves are dispersive, and the shape of the wavetrain changes with distance (see §2.3.3).

(b) *Love waves.* In earthquake seismology *Love waves* are sometimes observed; these involve transverse motion (SH) parallel to the surface of the ground. Energy sources normally used in seismic work do not generate Love waves to significant degree and hence they are usually unimportant in seismic exploration. Also, geophones designed to respond only to vertical motion would not detect Love waves. However, they are sometimes of interest (they contribute noise in SH-wave exploration – see §5.7).

We consider a semi-infinite medium bounded by the plane $z = 0$ and overlain by a slab (layer) of thickness h whose upper surface is a free surface. The density, elastic constants, velocities and displacements in the upper layer will be distinguished by primes. We consider propagation along the x-axis of a wave which has only SH displacement, that is,

$$v = A\,e^{m\kappa z}\,e^{j\kappa(x - V_L t)}, \quad z < 0 \ (\text{in the lower medium})$$

$$v' = (B\,e^{n\kappa z} + C\,e^{-n\kappa z})e^{j\kappa(x - V_L t)}, \quad z > 0 \ (\text{in the slab}).$$

Since v, v' must satisfy the S-wave equation, we substitute in (2.26) and find

$$m^2 = 1 - V_L^2/\beta^2, \quad n^2 = 1 - V_L^2/\beta'^2. \tag{2.73}$$

We must have m real so that $e^{m\kappa z}$ approaches zero as z approaches $-\infty$; hence $V_L \leq \beta$. However, n need not be real since z is always finite in the slab.

The boundary conditions are that the stress $\sigma_{yz} = 0$ at the free surface $z = h$ (see problem 2.10), and that σ_{yz} and v are continuous at the interface $z = 0$. Using (2.2) and (2.7), we get

$$\sigma_{yz} = \mu'\varepsilon_{yz} = \mu'\frac{\partial v'}{\partial z} = 0 \text{ at } z = h \left(\text{since } \frac{\partial w}{\partial y} = 0 \right);$$

$$v' = v, \quad \mu'\left(\frac{\partial v'}{\partial z} \right) = \mu\left(\frac{\partial v}{\partial z} \right) \text{ at } z = 0.$$

Substituting the values of v, v' and letting $a = e^{-2n\kappa h}$, $b = \mu m/\mu' n$, we obtain

$$\left. \begin{aligned} B - aC &= 0, \\ A - B - C &= 0, \\ bA - B + C &= 0. \end{aligned} \right\} \tag{2.74}$$

For the equations to be solvable, the determinant of the coefficients must vanish:

Fig.2.13. Rayleigh waves. (*a*) Predicted motion of a particle on the surface of a semi-infinite solid; (*b*) actual motion of a particle on surface of the Earth (from Howell, 1959); (*c*) cross-section showing motion of particles on the surface and at depth for a semi-infinite solid.

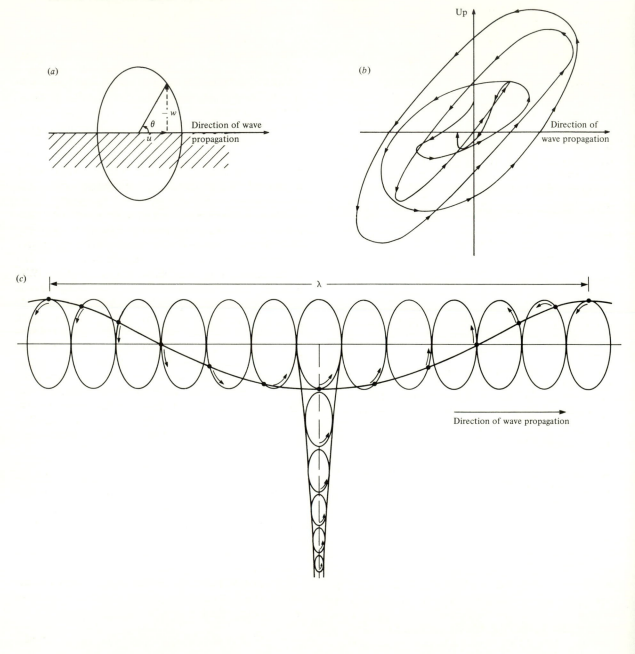

$$\begin{vmatrix} 0 & +1 & -a \\ 1 & -1 & -1 \\ b & -1 & +1 \end{vmatrix} = 0.$$

This gives

$$-b = \left(\frac{1-a}{1+a}\right) = \frac{1-e^{-2n\kappa h}}{1+e^{-2n\kappa h}} = \frac{-\mu m}{\mu' n} = \tanh(n\kappa h),$$

using the identity,

$$\tanh x = \frac{\sinh x}{\cosh x} = \frac{e^x - e^{-x}}{e^x + e^{-x}} = \frac{1 - e^{-2x}}{1 + e^{-2x}}.$$

Since $\tanh x$ is positive for all real values of x, the above equation can only be satisfied by taking n as pure imaginary, that is, $n = j\xi$ where ξ is a real number. Since $\tanh jx = j\tan x$, we get

$$\mu m - \mu'\xi\tan\kappa\xi h = 0. \tag{2.75}$$

From (2.73)

$$V_L^2/\beta'^2 = 1 - n^2 = 1 + \xi^2, \quad \text{so } V_L \geq \beta'.$$

Thus

$$\beta \geq V_L \geq \beta',$$

and the velocity of transverse waves must be higher in the deeper layer than in the surface layer for Love waves to exist.

Since $\kappa = 2\pi/\lambda = \omega/V_L$, as the frequency increases from zero, $\tan\kappa\xi h$ increases and approaches infinity; thus for (2.75) to hold, as the frequency increases, ξ must approach zero and V_L must approach β'. Conversely, as κ approaches zero, ξ approaches its maximum value and V_L approaches β. Hence at high frequencies the Love-wave velocity approaches the velocity of S-waves in the surface layer, and as the frequency approaches zero the Love-wave velocity approaches the S-wave velocity in the lower layer (Dobrin, 1951).

Fig.2.14. Rayleigh-wave velocity, V_R, as a function of Poisson's ratio, σ.

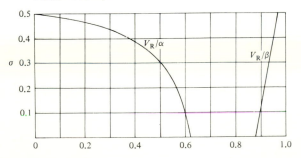

The expression for v in the surface layer can be written

$$\begin{aligned} v &= (Be^{n\kappa z} + Ce^{-n\kappa z})e^{j\kappa(x - V_L t)} \\ &= B(e^{j\kappa\xi z} + e^{2j\kappa\xi h}e^{-j\kappa\xi z})e^{j\kappa(x - V_L t)} \\ &= B(e^{j\kappa\xi(z-h)} + e^{-j\kappa\xi(z-h)})e^{j\kappa(x + \xi h - V_L t)} \\ &= 2B\cos\kappa\xi(z - h)e^{j\kappa(x + \xi h - V_L t)}, \end{aligned} \tag{2.76}$$

where C has been eliminated using the first equation in (2.74). Thus, v vanishes at horizontal nodal planes at depths z_r where

$$\kappa\xi(z_r - h) = \pi(r + \tfrac{1}{2}), \quad r \text{ being an integer.} \tag{2.77}$$

The existence of nodal planes is characteristic of normal-mode propagation (§2.4.8) and in fact Love waves can be explained in terms of normal-mode propagation.

(c) *Stoneley waves.* Rayleigh and Love waves are propagated along a free surface. Generalized (or modified) Rayleigh waves, usually called *Stoneley waves*, are surface waves propagated along the interface between two media under certain stringent conditions (see Stoneley, 1924; Scholte, 1947; Ewing et al., 1957). A Stoneley wave is always possible at a solid – liquid interface, its velocity being less than the Rayleigh-wave velocity at a free surface of the solid medium. When a solid surface layer of thickness h overlies a solid semi-infinite medium, Stoneley waves are possible only when $\beta_1 \approx \beta_2$ and $\lambda \ll h$. In this case the Stoneley-wave velocity is given by a fourth-order equation involving the elastic constants of both media, the equation having a solution only for certain ranges of values of μ_1/μ_2 and ρ_1/ρ_2; when a solution exists, the Stoneley-wave phase velocity is intermediate between the Rayleigh-wave velocity and the higher of β_1 or β_2.

(d) *Tube waves.* Waves traveling in the direction of the axis of a fluid-filled borehole (*tube waves*) are of considerable interest in connection with velocity surveys in wells and also because potentially they can furnish information about the elastic properties of the surrounding medium.

We assume a homogeneous fluid in a borehole in a homogeneous isotropic medium (fig. 2.15). Using $\mathscr{P}$ for the pressure and w for the displacement, Newton's second law, net force = mass × acceleration, applied to a volume element of the fluid, $\mathscr{V} = \pi r^2 \Delta z$, is

$$\left(\frac{\partial\mathscr{P}}{\partial z}\Delta z\right)\pi r^2 = -(\rho\pi r^2\Delta z)\frac{\partial^2 w}{\partial t^2},$$

or

$$\frac{\partial\mathscr{P}}{\partial z} = -\rho\frac{\partial^2 w}{\partial t^2}. \tag{2.78}$$

From (2.10),

$$\mathcal{P} = -k\Delta = -k\Delta\mathcal{V}/\mathcal{V}.$$

The change in volume $\Delta\mathcal{V}$ is due to expansion both along the axis and radially, that is,

$$\Delta\mathcal{V} = \pi r^2 \frac{\partial w}{\partial z}\Delta z + (2\pi r u_r)\Delta z,$$

where u_r is the change in the radius of the hole. Thus we get

$$\mathcal{P} = -k\left(\frac{\partial w}{\partial z} + \frac{2u_r}{r}\right). \tag{2.79}$$

Lamb (1960, §157) derived the following relation between u_r and $\mathcal{P}$ for an annulus of inner and outer radii r and R, where E, σ and μ are respectively Young's modulus, Poisson's ratio and the shear modulus for the annulus material:

$$\frac{u_r}{r} = \frac{\mathcal{P}}{E}\left\{\frac{(1+\sigma)(R^2+r^2)-2\sigma r^2)}{R^2-r^2}\right\}.$$

If we let $R \to \infty$, we obtain for a cylindrical hole in an infinite medium

$$u_r/r = \mathcal{P}(1+\sigma)/E = \mathcal{P}/2\mu$$

(using problem 2.3). Substitution in (2.79) gives

$$\mathcal{P}\left(\frac{1}{k}+\frac{1}{\mu}\right) = -\frac{\partial w}{\partial z}$$

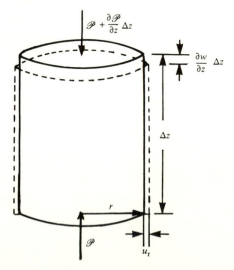

Fig.2.15. Changes involved in passage of a tube wave.

and substitution of this result in (2.78) gives the wave equation,

$$\frac{\partial^2 w}{\partial z^2} = (1/V_T^2)\frac{\partial^2 w}{\partial t^2}, \quad V_T^2 = \frac{1}{\rho}\left(\frac{1}{\kappa}+\frac{1}{\mu}\right)^{-1}. \tag{2.80}$$

Thus measurement of the tube-wave velocity V_T and the density ρ and bulk modulus k of the fluid permit us to calculate μ for the surrounding medium. White (1965, pp. 153–6) discusses tube waves in boreholes penetrating two solid media and in cased boreholes.

2.2.11 *Wave equation for transversely isotropic media*

Up to this point our discussion of wave motion applies only to a homogeneous isotropic medium. When the medium is not isotropic, the mathematics become more complex the more anisotropic the medium. The case of a transversely isotropic medium (§2.1.4) can be treated without great difficulty. We consider wave motion in the xz-plane so that the displacement v and derivatives with respect to y are zero. We substitute (2.14) and (2.15) into (2.17) and, using (2.1) and (2.2), we get the wave equations for transversely isotropic media:

$$\rho\frac{\partial^2 u}{\partial t^2} = \frac{\partial \sigma_{xx}}{\partial x} + \frac{\partial \sigma_{xz}}{\partial z}$$

$$= \frac{\partial}{\partial x}\left\{(\lambda_{\parallel}+2\mu_{\parallel})\frac{\partial u}{\partial x}+\lambda_{\perp}\frac{\partial w}{\partial z}\right\}+\frac{\partial}{\partial z}\left\{\mu^*\left(\frac{\partial u}{\partial z}+\frac{\partial w}{\partial x}\right)\right\}$$

$$= (\lambda_{\parallel}+2\mu_{\parallel})\frac{\partial^2 u}{\partial x^2}+\mu^*\frac{\partial^2 u}{\partial z^2}+(\lambda_{\perp}+\mu^*)\frac{\partial^2 w}{\partial z \partial x}, \tag{2.81}$$

$$\rho\frac{\partial^2 w}{\partial t^2} = (\lambda_{\perp}+\mu^*)\frac{\partial^2 u}{\partial x \partial z}+\mu^*\frac{\partial^2 w}{\partial x^2}+(\lambda_{\perp}+2\mu_{\perp})\frac{\partial^2 w}{\partial z^2}. \tag{2.82}$$

We simplify the problem by assuming a plane wave traveling in the xz-plane in the direction of increasing x and decreasing z, the angle between the raypath and the x-axis being θ. We now use the potential functions of (2.57) in the form

$$\phi = A\,e^{j\omega\zeta}, \quad \chi = B\,e^{j\omega\zeta}$$

where

$$\zeta = (lx-nz)/V - t, \quad l = \cos\theta, \quad n = \sin\theta.$$

Then,

$$u = \frac{\partial\phi}{\partial x}+\frac{\partial\chi}{\partial z} = (j\omega/V)(lA-nB)\,e^{j\omega\zeta},$$

$$w = \frac{\partial \phi}{\partial z} - \frac{\partial \chi}{\partial x} = -(j\omega/V)(nA + lB)\,e^{j\omega\zeta}.$$

When we substitute these into (2.81) and (2.82), the following factors appear in every term and hence can be ignored: $j\omega/V$, $(j\omega)^2$ and $e^{j\omega\zeta}$. Equations (2.81) and (2.82) become

$$\{\rho V^2 l - (\lambda_\parallel + 2\mu_\parallel)l^3 - (\lambda_\perp + 2\mu^*)ln^2\}A$$
$$- \{\rho V^2 n - (\lambda_\parallel + 2\mu_\parallel - \lambda_\perp - \mu^*)l^2 n - \mu^* n^3\}B = 0,$$
$$\{\rho V^2 n - (\lambda_\perp + 2\mu_\perp)n^3 - (\lambda_\perp + 2\mu^*)l^2 n\}A$$
$$+ \{\rho V^2 l + (\mu^* - 2\mu_\perp)ln^2 - \mu^* l^3\}B = 0.$$

Writing

$$\alpha_\parallel^2 = (\lambda_\parallel + 2\mu_\parallel)/\rho, \quad \alpha_\perp^2 = (\lambda_\perp + 2\mu_\perp)/\rho,$$
$$\alpha^{*2} = (\lambda_\perp + 2\mu^*)/\rho, \quad \beta^{*2} = \mu^*/\rho,$$

these become

$$(V^2 - \alpha_\parallel^2 l^2 - \alpha^{*2} n^2)lA$$
$$- \{V^2 - (\alpha_\parallel^2 - \alpha^{*2} + \beta^{*2})l^2 - \beta^{*2} n^2\}nB = 0,$$
$$(V^2 - \alpha_\perp^2 n^2 - \alpha^{*2} l^2)nA$$
$$+ \{V^2 - (\alpha_\perp^2 - \alpha^{*2} + \beta^{*2})n^2 - \beta^{*2} l^2\}lB = 0.$$

Eliminating A and B gives the following quadratic equation in V^2:

$$\frac{\{V^2 - (\alpha_\parallel^2 - \alpha^{*2} + \beta^{*2})l^2 - \beta^{*2} n^2\}n}{(V^2 - \alpha_\parallel^2 l^2 - \alpha^{*2} n^2)l}$$
$$= \frac{-\{V^2 - (\alpha_\perp^2 - \alpha^{*2} + \beta^{*2})n^2 - \beta^{*2} l^2\}l}{(V^2 - \alpha_\perp^2 n^2 - \alpha^{*2} l^2)n},$$

or

$$\{V^2 - (\alpha_\parallel^2 - \alpha^{*2} + \beta^{*2})l^2 - \beta^{*2} n^2\}$$
$$(V^2 - \alpha_\perp^2 n^2 - \alpha^{*2} l^2)n^2$$
$$+ \{V^2 - (\alpha_\perp^2 - \alpha^{*2} + \beta^{*2})n^2 - \beta^{*2} l^2\}$$
$$(V^2 - \alpha_\parallel^2 l^2 - \alpha^{*2} n^2)l^2 = 0. \tag{2.83}$$

The solution has been given by Stoneley (1949), Grant and West (1965, p. 42) and White (1965, p. 46). The roots are always real and positive and approach α and β of (2.52) and (2.53) as the anisotropy approaches zero. When the wave is traveling vertically, $l = 0$, $n = 1$ and $V = \alpha_\perp$ or β^* for vertically traveling P- or SH-waves. When $l = 1$, $n = 0$, $V = \alpha_\parallel$ or β^*, corresponding to horizontally traveling P- or SH-waves. However, when the wave is traveling at an angle to the vertical, the roots are complicated functions of the elastic constants and the motion is not separated into distinct P- and S-waves.

Anisotropy measurements are usually expressed in terms of the ratio of the velocities parallel and perpendicular to the bedding. Uhrig and van Melle (1955) give a table showing anisotropy values of 1.2 to 1.4 for rocks at the surface and values of 1.1 to 1.2 for sediments at depths of 2.1–2.4 km in West and Central Texas. Stoep (1966) found average values between 1.00 and 1.03 for Texas Gulf Coast sediments. Segonzac and Laherrere (1959) obtained values for sediments in the North Sahara area which ranged from 1.00 for sandstones to 1.08–1.12 for limestones and 1.15–1.20 for anhydrites.

2.2.12 *Wave equation in fluid media*

In fluids only P-waves are propagated and we are generally interested in pressure variations rather than displacements or velocities, as in solid media. Equation (2.55) can be expressed in terms of pressure $\mathscr{P}$. We redefine ϕ in the form

$$\mathbf{V}\phi = \dot{u}\mathbf{i} + \dot{v}\mathbf{j} + \dot{w}\mathbf{k}, \quad \dot{u} = \frac{\partial u}{\partial t}, \text{ etc.} \tag{2.84}$$

In (2.17) we set

$$\sigma_{xy} = \sigma_{yz} = \sigma_{zx} = 0, \quad \sigma_{xx} = \sigma_{yy} = \sigma_{zz} = -\mathscr{P};$$

hence

$$\rho \frac{\partial^2 u}{\partial t^2} = -\frac{\partial \mathscr{P}}{\partial x} = \text{acceleration along the } x\text{-axis,} \tag{2.85}$$

and similarly for the y- and z-axes. Adding the three components of acceleration gives

$$\rho \mathbf{V} \frac{\partial \phi}{\partial t} = -\mathbf{V}\mathscr{P}.$$

Ignoring an additive constant (hydrostatic pressure),

$$\mathscr{P} = -\rho \frac{\partial \phi}{\partial t} = j\omega\rho\phi, \tag{2.86}$$

if we consider only harmonic waves (as in (2.50)) where ϕ contains the factor $e^{-j\omega t}$. Thus both ϕ and $\mathscr{P}$ satisfy the P-wave equation as in (2.21), the velocity reducing to

$$\alpha = (k/\rho)^{\frac{1}{2}} \tag{2.87}$$

in fluids.

In the case of a gas, k depends upon the way the gas is compressed, isothermally or adiabatically (i.e., with no transfer of heat during the wave passage). For sound waves in air, the compression is essentially adiabatic so that the pressure and volume obey the law,

$$\mathscr{P}\mathscr{V}^\gamma = \text{constant}, \quad \gamma = c_\mathscr{P}/c_\mathscr{V} \approx 1.4 \text{ for air,}$$

where $c_{\mathscr{P}}$ and $c_{\mathscr{V}}$ are the specific heats at constant pressure and volume, respectively (Shortley and Williams, 1950, p. 542). Equation (2.10) can be written

$$k = -\frac{\Delta\mathscr{P}}{\Delta\mathscr{V}/\mathscr{V}} = -\frac{\mathscr{V}\,\mathrm{d}\mathscr{P}}{\mathrm{d}\mathscr{V}},$$

where $\Delta\mathscr{P}$ is the pressure change created by the wave. Using the adiabatic law, logarithmic differentiation gives $k = \gamma\mathscr{P}$ and hence

$$\alpha = (\gamma\mathscr{P}/\rho)^{\frac{1}{2}} \tag{2.88}$$

2.3 Effects of the medium on wave propagation

2.3.1 *Energy density; intensity*

Probably the single most important feature of any wave is the energy associated with the motion of the medium as the wave passes through it. Usually we are not concerned with the total energy of a wave but rather with the energy in the vicinity of the point where we observe it; the *energy density* is the energy per unit volume in the neighborhood of a point.

Consider a spherical harmonic P-wave for which the radial displacement for a fixed value of r is given by

$$u = \mathscr{A}\cos(\omega t + \gamma),$$

where γ is a *phase angle*. The displacement u ranges from $-\mathscr{A}$ to $+\mathscr{A}$. Since the displacement varies with time, each element of the medium has a velocity, $\dot{u} = \partial u/\partial t$, and an associated kinetic energy. The kinetic energy δE_k contained within each element of volume $\delta\mathscr{V}$ is

$$\delta E_k = \tfrac{1}{2}(\rho\delta\mathscr{V})\dot{u}^2.$$

The kinetic energy per unit volume is

$$\frac{\delta E_k}{\delta\mathscr{V}} = \tfrac{1}{2}\rho\dot{u}^2 = \tfrac{1}{2}\rho\omega^2\mathscr{A}^2\sin^2(\omega t + \gamma).$$

This expression varies from zero to a maximum of $\tfrac{1}{2}\rho\omega^2\mathscr{A}^2$.

The wave also involves potential energy resulting from the elastic strains created during the passage of the wave. As the medium oscillates back and forth, the energy is converted back and forth from kinetic to potential form, the total energy remaining fixed. When a particle is at zero displacement, the potential energy is zero and the kinetic energy is a maximum, and when the particle is at its extreme displacement the energy is all potential. Since the total energy equals the maximum value of the kinetic energy, the energy density E for a harmonic wave is

$$E = \tfrac{1}{2}\rho\omega^2\mathscr{A}^2 = 2\pi^2\rho v^2\mathscr{A}^2. \tag{2.89}$$

Thus we see that the energy density is proportional to the first power of the density of the medium and to the second power of the frequency and amplitude of the wave. (See Braddick, 1965, for a different derivation of (2.89).)

We are interested also in the rate of flow of energy and we define the *intensity* as the quantity of energy which flows through a unit area normal to the direction of wave propagation in unit time. Take a cylinder of infinitesimal cross-section, $\delta\mathscr{S}$, whose axis is parallel to the direction of propagation and whose length is equal to the distance traveled in the time, δt. The total energy inside the cylinder at any instant t is $EV\delta t\,\delta\mathscr{S}$; at the time $t + \delta t$ all of this energy has left the cylinder through one of the ends. Dividing by the area of the end of the cylinder, $\delta\mathscr{S}$, and by the time interval, δt, we get I, the amount of energy passing through unit area in unit time:

$$I = EV. \tag{2.90}$$

For a harmonic wave, this becomes

$$I = \tfrac{1}{2}\rho V\omega^2\mathscr{A}^2 = 2\pi^2\rho Vv^2\mathscr{A}^2. \tag{2.91}$$

In fig. 2.16 we show a spherical wavefront diverging from a center O. By drawing sufficient radii we can define two portions of wavefronts, $\mathscr{S}_1$ and $\mathscr{S}_2$, of radii r_1 and r_2, such that the energy which flows outward through the spherical cap $\mathscr{S}_1$ in one second must be equal to that passing outward through the spherical cap $\mathscr{S}_2$ in one second (since the energy is moving only in the radial direction). The flow of energy per second is the product of the intensity and the area; hence

$$I_1\mathscr{S}_1 = I_2\mathscr{S}_2.$$

Since the areas $\mathscr{S}_1$ and $\mathscr{S}_2$ are proportional to the square of their radii, we get

$$I_2/I_1 = \mathscr{S}_1/\mathscr{S}_2 = (r_1/r_2)^2.$$

Fig.2.16. Dependence of intensity upon distance.

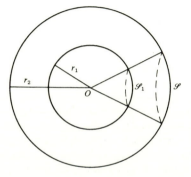

Moreover, it follows from (2.90) that E is proportional to I and hence

$$I_2/I_1 = E_2/E_1 = (r_1/r_2)^2. \tag{2.92}$$

Thus, geometrical spreading causes the intensity and the energy density of spherical waves to decrease inversely as the square of the distance from the source (Newman, 1973). This is called *spherical divergence*.

For a plane wave the rays do not diverge and hence the intensity of a plane wave is constant. Figure 2.16 could represent a cross-section of a cylindrical wave, that is, a wave generated by a very long linear source, the arcs $\mathscr{S}_1$ and $\mathscr{S}_2$ being cylindrical wavefronts. Since the arcs are proportional to the radii, *cylindrical divergence* causes the intensity to vary inversely as the radius. Thus we can write

$$I_2/I_1 = E_2/E_1 = (r_1/r_2)^m, \tag{2.93}$$

where $m = 0$, 1 or 2 according as the wave is plane, cylindrical or spherical.

Ratios of intensity, energy or power are usually expressed in decibels, the value in dB being $10\log_{10}$ of the intensity, energy or power ratio. Because these vary as the square of the amplitude, dB is also given as $20\log_{10}$ of the amplitude ratio. The natural log of the amplitude ratio (in nepers) is also used (see problem 2.17).

2.3.2 *Absorption*

(*a*) *General.* We shall also consider two other mechanisms, absorption and partitioning at interfaces, which cause the energy density of a wave to decrease. In the preceding section we considered variations of the energy distribution as a function of geometry. Implicit in the discussion was the assumption that none of the wave energy was transformed into other forms. In reality, as the wave motion passes through the medium, the elastic energy associated with the wave motion is gradually absorbed by the medium, reappearing ultimately in the form of heat. This process is called *absorption* and is responsible for the eventual complete disappearance of the wave motion.

The measurement of absorption is very difficult. Absorption varies with frequency and laboratory measurements, which are invariably made at high frequencies, may not apply to actual field conditions. Field measurements must be corrected for reflection and refraction effects and the path usually involves several media. Measurement difficulties have resulted in wide divergence in absorption measurements. Toksöz and Johnston (1981) summarize much of the literature regarding absorption.

(*b*) *Expressions for absorption.* The decrease of amplitude due to absorption appears to be exponential with distance for elastic waves in rocks. Thus, we can write for the decrease in amplitude because of absorption only,

$$A = A_0 e^{-\eta x}, \tag{2.94}$$

where A and A_0 are values of the amplitude of a wavefront at two points a distance x apart and η is the *absorption coefficient*.

Other measures of absorption are based on the decrease in amplitude with time; to relate these to η we assume a cyclic waveform,

$$A = A_0 e^{-ht} \cos \omega t, \tag{2.95}$$

and make measurements at a fixed location; h is called the *damping factor*. The *logarithmic decrement*, δ, is defined by

$$\delta = \ln\left(\frac{\text{amplitude}}{\text{amplitude one cycle later}}\right). \tag{2.96}$$

It can be expressed in terms of the damping factor as

$$\delta = hT = h/v = 2\pi h/\omega, \tag{2.97}$$

where $T = $ period; δ is measured in nepers. Quality factor, Q, can be defined as

$$Q = 2\pi/(\text{Fraction of energy lost per cycle})$$
$$= 2\pi(E/\Delta E). \tag{2.98}$$

Since energy is proportional to amplitude squared, $E = E_0 e^{-2ht}$ and $\Delta E/E_0 = -2h\Delta t$. Setting $\Delta t = T$, we get $\Delta E/E = 2hT = 2\delta$ and $Q = \pi/\delta$. Q can also be expressed as

$$Q = \pi n, \tag{2.99}$$

where $n = $ number of oscillations for the amplitude to decrease by the factor e; for $e^{hnT} = e$, $n = 1/hT$ and

$$Q = \pi/hT = \pi v/h = \pi/\delta. \tag{2.100}$$

Still another manner of expressing Q is $Q = \cot \phi$ where $\phi = $ *loss angle*.

During one period a wave travels one wavelength so that if the loss of energy is due to absorption only, $hT = \eta\lambda$ and we can interrelate η, δ and Q:

$$Q = \pi/\eta\lambda = \pi/\delta. \tag{2.101}$$

Absorption in the form given by (2.94) appears naturally in solutions of the type given in (2.50) if we permit the elastic constants to be complex numbers. Real elastic-constant values correspond to media without absorption while complex values imply exponential absorption. Complex values of λ and μ result in *complex velocity* values (see (2.52) and (2.53)). If the velocity is written as $a + jb$ where $1/(a + jb) = 1/V + j\eta/\omega$, substitution in (2.50) for a wave along the x-axis gives

$$\psi = A\,e^{j\omega\{x(1/V + j\eta/\omega) - t\}} = A\,e^{-\eta x}e^{j\omega(x/V - t)},$$

which agrees with (2.94).

(*c*) *Measurements of absorption.* Experimental evidence suggests that the absorption coefficient η is approximately proportional to frequency, that is, $\eta\lambda$ is roughly constant for a particular rock. The increase of absorption with frequency provides one mechanism for the observed loss of high frequencies with distance. Measurements of absorption have been summarized by Bradley and Fort (1966) and Attewell and Ramana (1966); the latter found a best fit value of $\eta = 0.2$ dB/km for the average of values from 26 authors. Waters (1978, p. 24) gives a fairly extensive table of Q values. Values fall generally in the ranges shown in table 2.1. Q values for S-waves appear to be one-half to one-third those for P-waves. Tullos and Reid (1969) report measurements in the first 3 m of Gulf Coast sediments of $\eta = 13$ dB/λ ($Q = 0.24$) but 0.15 to 0.36 dB/λ for the next 300 m ($Q = 20$ to 9). Often quoted measurements in the Pierre shale by McDonal *et al.* (1958) were $\eta = 0.39$ dB/km for P-waves ($\delta = 0.9$ dB, $Q = 3.5$) and $\eta = 3.3$ dB/km for S-waves; the Pierre shale is a massive formation in Colorado about 1200 m thick with a P-wave velocity of 2330 m/s.

(*d*) *Absorption mechanisms.* The mechanisms by which elastic wave energy is transformed into heat are not clearly understood. Various loss mechanisms have been proposed (White, 1965, 1966) but none appear adequate. Internal friction in the forms of sliding friction (or sticking, then sliding) and viscous losses in the interstitial fluids are probably the most important mechanisms, the latter being more important in high permeability rocks. Other effects, probably of minor significance in general, are the loss by conduction of part of the heat generated during the compressive part of the wave, piezoelectric and thermoelectric effects, and the energy used to create new surfaces (of importance only near the source). Many of the postulated mechanisms predict that Q is dependent upon frequency (it is proportional to frequency in liquids). The loss mechanism in rocks must be regarded as an unsolved problem.

(*e*) *Relative importance of absorption and spreading.* To compare the loss by absorption with the loss of intensity by geometrical spreading (see (2.92)), we have calculated the losses in going various distances from a point 200 m from the source assuming $\eta = 0.15$ dB/λ. The results

shown in table 2.2 were calculated using the following relations:

Absorption: Intensity loss in dB $= 10\log_{10}(I_0/I)$
$= 20\log_{10}(A_0/A) = 0.3(x/\lambda) = 0.3(x_s - 200)/\lambda$
$= 0.3v(x_s - 200)/2000$,
Spreading: Intensity loss in dB $= 10\log_{10}(I_0/I)$
$= 20\log_{10}(x_s/200)$,

where $x_s =$ distance to the shotpoint. The table shows that losses by spreading are more important than losses by absorption for low frequencies and short distances. As the frequency and distance increase, absorption losses increase and eventually become dominant.

The increased absorption at higher frequencies results in change of waveshape with distance. Peg-leg multiples (§4.2.2*b*) and possibly other mechanisms also produce waveshape changes. Fig. 2.17 shows the energy decreasing with distance and with frequency; the frequency-dependent attenuation is greater than expected from absorption alone.

2.3.3 *Dispersion; group velocity*

The velocity, V, α or β, which appears in (2.21)–(2.37) is known as the *phase velocity* because it is the distance traveled per unit time by a point of constant phase, such as a peak or trough. This is not necessarily the same as the velocity with which a pulse travels, which

Table 2.1. *Absorption constants for rocks.*

	Q	δ(dB) $= \eta\lambda$	η/v (dB/km Hz)
Igneous rocks	75–150	0.04–0.02	0.008–0.003
Sedimentary rocks	20–150	0.16–0.02	0.10–0.004
Rocks with gas in pore space	5–50	0.63–0.06	1.3–0.03

Table 2.2. *Energy losses by absorption and spreading* ($\eta = 0.15$ dB/wavelength *and* $V = 2.0$ km/s).

		Distance from shotpoint (x_s)			
	Frequency (v)	1200 m	2200 m	4200 m	8200 m
Absorption	1 Hz	0.075 dB	0.15 dB	0.3 dB	0.6 dB
	3	0.22	0.45	0.9	1.8
	10	0.75	1.5	3	6
	30	2.2	4.5	9	18
	100	7.5	15	30	60
Spreading	All	16	21	26	32

is the *group velocity* and will be denoted by *U*. Consider, for example, the wavetrain shown in fig. 2.18; we could determine the group velocity *U* by drawing the envelope of the pulse (the double curve *ABC*, *AB'C*) and measuring the distance which the envelope travels in unit time. The relation between *U* and *V* is shown in fig. 2.18 where *V* is given by the rate of advance of a certain phase (such as a trough) while *U* is measured by the speed of the maximum amplitude of the envelope.

If we decompose a pulse into its component frequencies by Fourier analysis, we find a spectrum of frequencies. If the velocity is the same for all frequencies the pulse shape will remain the same and the group velocity will be the same as the phase velocity. However, if the velocity varies with frequency, the pulse changes shape as it travels and the group velocity is different from the phase velocity, that is, the medium is *dispersive*. It can be shown (see problem 2.19) that the group velocity *U* is

$$U = V - \lambda \frac{\mathrm{d}V}{\mathrm{d}\lambda} = V + \omega \frac{\mathrm{d}V}{\mathrm{d}\omega}, \tag{2.102}$$

where *V*, λ, ω, $\mathrm{d}V/\mathrm{d}\lambda$ and $\mathrm{d}V/\mathrm{d}\omega$ are average values for

Fig. 2.17. Loss of amplitude as a function of one-way traveltime, based on measurements with a geophone clamped in borehole with source at surface. The curves labeled 'divergence' and 'divergence plus transmission loss' are calculated from sonic-log data allowing for loss of energy in transmission through reflecting interfaces. The 20, 40, 60 Hz curves show attenuation at those frequencies. (Courtesy SSC.)

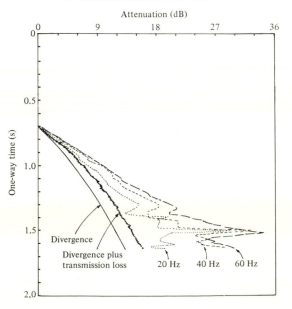

the range of frequencies which make up the principal part of the pulse.

When *V* increases with frequency, *V* is smaller than *U* as illustrated in fig. 2.18*a* where the envelope travels faster than the individual cycles. When *V* decreases with frequency, the opposite is true (as in fig. 2.18*b*).

Dispersion is important for several reasons, perhaps the most important being that the energy of a pulse travels with the velocity *U* (except where there is appreciable absorption; see Brillouin, 1960, pp. 98–100). Also, dispersion of body waves is a consequence of most of the theories proposed to account for absorption, but such dispersion has not been observed. Most rocks simply exhibit little variation of velocity with frequency in the seismic frequency range. Ward and Hewitt (1977) found the same velocity at 35 Hz as at 55 Hz in a monofrequency well survey to about 800 m. Futterman (1962) shows that the dispersion expected for seismic body waves is small for usual situations. Dispersion is, however, important in connection with surface waves (see §2.2.10) and certain other phenomena.

2.3.4   *Reflection and refraction; Snell's law*

Whenever a wave encounters an abrupt change in the elastic properties, as when it arrives at a surface separating two beds, part of the energy is *reflected* and remains in the same medium as the original energy; the balance of the energy is *refracted* into the other medium with an abrupt change in the direction of propagation occurring at the interface. Reflection and refraction are fundamental in exploration seismology and we shall discuss these in some detail.

We can derive the familiar laws of reflection and refraction using Huygens' principle. Consider a plane wavefront *AB* incident on a plane interface as in fig. 2.19 (if the wavefront is curved, we merely take *A* and *B* sufficiently close together that *AB* is a plane to the required degree of accuracy; however, see also §2.3.5); *AB* occupies the position *A'B'* when *A* arrives at the surface; at this instant the energy at *B'* still must travel the distance *B'R* before arriving at the interface. If $B'R = V_1 \Delta t$, then Δt is the time interval between the arrival of the energy at *A'* and at *R*. By Huygens' principle, during the time Δt the energy which reached *A'* will have traveled either upward a distance $V_1 \Delta t$ or downward a distance $V_2 \Delta t$. By drawing arcs with center *A'* and lengths equal to $V_1 \Delta t$ and $V_2 \Delta t$, and then drawing the tangents from *R* to these arcs, we locate the new wavefronts, *RS* and *RT* in the upper and lower media. The angle at *S* is a right angle and $A'S = V_1 \Delta t = B'R$; therefore the triangles *A'B'R* and *A'SR* are equal with the result that the *angle of incidence* θ_1 is equal to the *angle of reflection* θ_1' ; this is the *law of reflection*. For the refracted wave, the angle at T is a right angle and we have

$$V_2 \Delta t = A'R \sin \theta_2,$$

and

$$V_1 \Delta t = A'R \sin \theta_1 ;$$

hence

$$\frac{\sin \theta_1}{V_1} = \frac{\sin \theta_2}{V_2} = p. \qquad (2.103)$$

The angle θ_2 is called the *angle of refraction* and (2.103) is the *law of refraction*, also known as *Snell's law*. The angles are usually measured between the raypaths and a normal to the interface but these angles are the same as those between the interface and the wavefronts in isotropic media. The laws of reflection and refraction can be combined in the single statement: at an interface the quantity $p = (\sin \theta_i)/V_i$ has the same value for the incident, reflected and refracted waves. This generalized form of Snell's law will be understood in future references to Snell's law. The quantity p is called the *raypath parameter*. It will be shown in §2.4.2 that Snell's law also holds for wave conversion

Fig.2.18. Comparison of group and phase velocities. (*a*) Definition of group velocity, U, and phase velocity, V; (*b*) arrival of a dispersive wave at successive geophones.

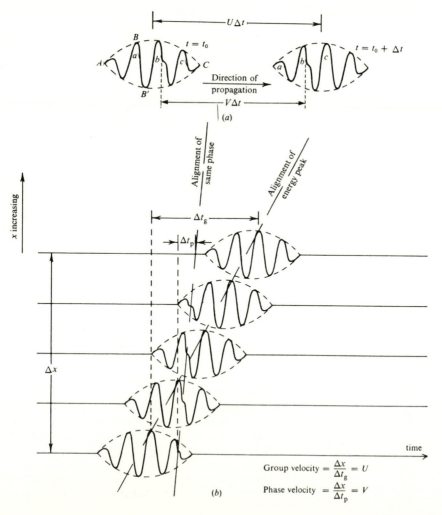

from P- to S-waves (and vice versa) upon reflection or refraction.

   When the medium consists of a number of parallel beds, Snell's law requires that the quantity p have the same value everywhere for all reflected and refracted rays resulting from a given initial ray.

   The foregoing derivation assumed a planar surface and therefore specular reflection. If the surface includes bumps of height d, reflected waves from them will be ahead of those from the rest of the surface by $2d$. These can be neglected where $2d/\lambda < \frac{1}{4}$ (the 'Rayleigh' criterion), i.e., when $d < \frac{1}{8}\lambda$. Most interfaces satisfy this criterion for ordinary seismic waves. For oblique reflection the criteria are less stringent and reflection can be regarded as specular from relatively rough surfaces.

   When V_2 is less than V_1, θ_2 is less than θ_1. However, when V_2 is greater than V_1, θ_2 reaches 90° when $\theta_1 = \sin^{-1}(V_1/V_2)$. For this value of θ_1, the refracted ray is traveling along the interface. The angle of incidence for which

Fig.2.19. Reflection and refraction of a plane wave.

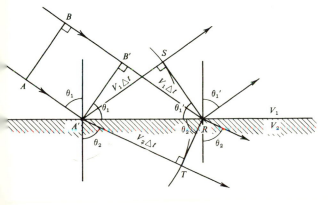

Fig.2.20. Imaginary angles of reflection and refraction.

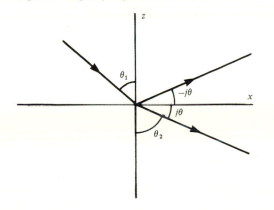

$\theta_2 = 90°$ is the *critical angle*, Θ; obviously, $\sin\Theta = V_1/V_2$. For angles of incidence greater than Θ, it is impossible to satisfy Snell's law (using real angles) since $\sin\theta_2$ cannot exceed unity and *total reflection* occurs. This does not mean that 100% of the energy is reflected, however, since converted S-waves (see §2.4.1) and evanescent waves (see §2.2.5) are generated.

   Noting the method used to derive (2.51) in §2.2.5, we write Snell's law for the case $\theta_1 > \Theta$ (see fig. 2.20) in the form

$$\sin\theta_2 = (V_2/V_1)\sin\theta_1 = \sin(\tfrac{1}{2}\pi - j\theta)$$
$$= \cos j\theta - \cosh\theta = l,$$
$$n = \cos\theta_2 = \sin j\theta = j\sinh\theta;$$

hence (2.50) becomes

$$\psi = A\,e^{-(\omega z/V)\sinh\theta}\,e^{j\omega\{(x/V)\cosh\theta - t\}}.$$

If we take θ negative in fig. 2.20, the only change is in the sign of the first exponential on the right-hand side. Thus, just as in the case of (2.51), evanescent waves can exist on both sides of the interface and their amplitudes decrease as we go away from the interface. The rate of attenuation is proportional to $\sinh\theta$, which has its maximum value at the grazing angle, $\theta_1 = \frac{1}{2}\pi$. The introduction of imaginary angles to satisfy Snell's law for angles exceeding the critical angle means that the reflection coefficient will be complex and phase shifts will occur (see problem 2.23) which will be complicated functions of the angle of incidence.

   Snell's law is very useful in determining raypaths and arrival times and in deriving reflector position from observed arrival times, but it does not give information about the amplitudes of the reflected and transmitted waves. This subject is taken up in §2.4.

2.3.5 *Diffraction*

   (*a*) *General.* In discussing reflection and refraction, we stated that when an interface is curved we merely have to select a portion sufficiently small that it can be considered plane. However, such a simplification is not always possible, for example, when the radius of curvature of an interface is less than a few wavelengths or the reflector is terminated by a fault, pinchout, unconformity, etc. In such cases the simple laws of reflection and refraction are no longer adequate because the energy is *diffracted* rather than reflected or refracted. Since seismic wavelengths are often 100 m or more, many geologic features give rise to diffractions.

   (*b*) *Basic formulae.* The mathematical treatment of diffraction is complex and we shall give only a brief summary of a simplified treatment due to Trorey (1970). We shall assume coincident source and receiver (see Trorey, 1977, for the non-coincident case), constant velocity and neglect S-waves; thus χ in (2.55) is zero. We start with the wave equation (2.23) (writing ϕ in place of ψ) and take the Laplace transform, obtaining

$$\nabla^2\phi = (1/V^2)\partial^2\phi/\partial t^2 \leftrightarrow \nabla^2\Phi = (s/V)^2\Phi$$

where $\Phi(x, y, z, s) =$ the Laplace transform of $\phi(x, y, z, t)$ (see §10.4) and the double-headed arrow indicates equivalence in different domains. Note that we are assuming that ϕ and $\partial\phi/\partial t$ are zero at $t = 0$ for all x, y, z.

The solution of this equation for a point source at the origin is

$$\Phi = (c/r)\,e^{-sr/V} \tag{2.104}$$

where r is the distance from the source to the point of observation and V is the wave velocity. (This can be verified by direct substitution, noting that $r^2 = x^2 + y^2 + z^2$, $\partial r/\partial x = x/r$, etc.) In general, c should include the Laplace transform of the input waveform at the source, but in effect we have taken the transform to be unity, that is, a unit impulse input, $\delta(t)$. The results for other types of sources can be found by time-domain convolution (see (10.179) and (10.180)).

In a source-free region the P-wave potential function ϕ is given by (2.34) with $\Upsilon = 0$, hence we can get another expression for ϕ by taking the Laplace transform of (2.34), the result being

$$4\pi\Phi = \iint_{\mathscr{S}} e^{-sr/V}\left(\frac{s}{rV}\Phi\frac{\partial r}{\partial\eta} - \Phi\frac{\partial(1/r)}{\partial\eta} + \frac{1}{r}\frac{\partial\Phi}{\partial\eta}\right)\,d\mathscr{S}. \tag{2.105}$$

The factor $e^{-sr/V}$ arises because ϕ in the integrand of (2.34) is evaluated at the time $t = t_0 - r/V$ whereas Φ is the transform of $\phi(x, y, z, t)$ (see (10.166)).

(c) Diffraction effect for part of a plane reflector. We shall now calculate the diffraction effect of an area $\mathscr{S}$ which is part of a plane reflector $z = h$ (see fig. 2.21a), both source and detector being at the origin. We enclose the origin with a hemisphere of infinite radius with center $(0, 0, h)$, the base of which is the plane $z = h$. In order to apply (2.105), we replace the source with its image at $(0, 0, 2h)$, thus making the hemisphere a source-free region. We ignore absorption and assume a constant reflection coefficient over $\mathscr{S}$, so that c in (2.104) is constant. Clearly $1/r = 0 = \Phi$ over the hemisphere, hence the contribution to the integral over the hemisphere is zero. We can also set $\Phi = 0$ over the portions of the plane $z = h$ except for the portion $\mathscr{S}$ whose effect we wish to evaluate.

We now substitute (2.104) in (2.105), noting that r in (2.104) is now r_0 in fig. 2.21a since the source is now at the image point O'; hence

$$\frac{\partial r}{\partial\eta} = \frac{\partial r}{\partial z} = \frac{z}{r} = \frac{h}{r};\ \frac{\partial r_0}{\partial\eta} = \frac{\partial r_0}{\partial z} = \frac{-h}{r_0};\ \frac{\partial(1/r)}{\partial\eta} = -\frac{h}{r^3};$$

$$\frac{\partial\Phi}{\partial\eta} = \frac{d\Phi}{dr_0}\frac{\partial r_0}{\partial\eta} = -\frac{c}{r_0}e^{-sr_0/V}\left(\frac{1}{r_0} + \frac{s}{V}\right)\left(\frac{-h}{r_0}\right)$$

$$= \frac{ch}{r^2}e^{-sr/V}\left(\frac{1}{r} + \frac{s}{V}\right),$$

where we have set $r_0 = r$ after differentiating. The result is

$$2\pi\Phi = ch\iint_{\mathscr{S}} e^{-2sr/V}\left(\frac{1}{r^4} + \frac{s}{Vr^3}\right)\,d\mathscr{S}. \tag{2.106}$$

This surface integral can be transformed into a contour integral as follows. In fig. 2.21b, the element of area is $\rho\,d\rho\,d\theta$ in polar coordinates; since $r^2 = \rho^2 + h^2$, $\rho\,d\rho = r\,dr$ and therefore

$$2\pi\Phi = ch\int_\theta\int_r e^{-2sr/V}\left(\frac{1}{r^3} + \frac{s}{Vr^2}\right)\,dr\,d\theta. \tag{2.107}$$

If we integrate the first term by parts with respect to r, we obtain

$$\int_{r_1}^{r_2}\frac{e^{-2sr/V}}{r^3}\,dr = \frac{e^{-2sr/V}}{-2r^2}\bigg|_{r_1}^{r_2} - \int_{r_1}^{r_2}\frac{s\,e^{-2sr/V}}{Vr^2}\,dr.$$

Substituting in (2.107), we get

$$4\pi\Phi = ch\oint_\theta\{(1/r_1^2)e^{-2sr_1/V} - (1/r_2^2)e^{-2sr_2/V}\}\,d\theta. \tag{2.108}$$

If the z-axis does not cut $\mathscr{S}$, we can set $r = \xi$ for points on the boundary, giving

$$\Phi = -(ch/4\pi)\oint(1/\xi^2)e^{-2s\xi/V}\,d\theta, \tag{2.109}$$

where A traverses the boundary of $\mathscr{S}$ in the counterclockwise direction.

If the z-axis cuts $\mathscr{S}$ (see fig. 2.21c), $r_1 = h =$ constant in (2.108), so that

$$\Phi = (c/2h)e^{-2sh/V} - (ch/4\pi)\oint(1/\xi^2)e^{-2s\xi/V}\,d\theta. \tag{2.110}$$

where A again traverses the boundary in the counterclockwise direction. If $\mathscr{S}$ includes the entire xy-plane, i.e., we have a continuous plane reflector, $\xi = \infty$ and the integral vanishes. The first term in this equation thus repre-

Fig.2.21. Diffraction effect of a plane area $\mathscr{S}$. (After Trorey, 1970.) (a) Calculation using a surface integral; (b) calculation using a line integral when the origin is not over the area; (c) calculation when the origin is over the area; (d) evaluating the line integral when ξ is multi-valued.

(a)

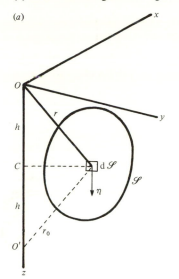

(b)

(c)

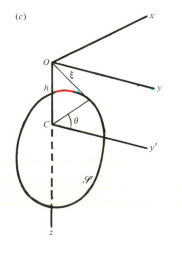

(d)

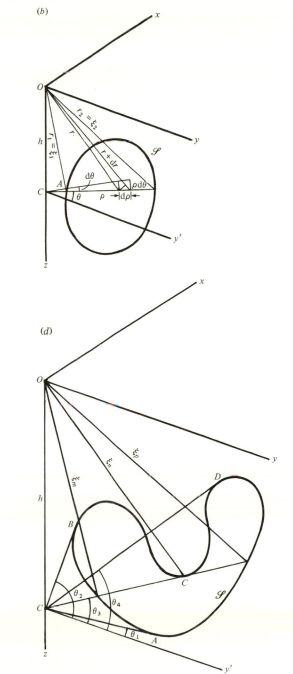

sents the simple reflection from this plane while the integral represents the diffracted wave. Comparison of (2.109) and (2.110) shows that the diffracted wave is given by the same expression in both cases.

An important point to note is that both the reflection and diffraction terms in (2.109) and (2.110) are derived from the integral in (2.106) where the integration is carried over the entire surface. When we use rays and think of reflection and diffraction as occurring at a point or along a line, we are greatly simplifying the actual phenomena. In fact, both reflections and diffractions are the resultants of energy which returns from all parts of the surface. From this point of view a reflection is merely a special type of diffraction, a point of view which has interesting practical applications (see §8.3.2).

(*d*) *Time-domain solution for diffraction.* We shall complete our discussion of diffractions by obtaining the time-domain solution of (2.109) and (2.110). The inverse transform of the reflection term on the right-hand side of (2.110) gives the impulse $(c/2h)\,\delta(t - 2h/V)$, that is, a repetition of the input at the source after a delay of $2h/V =$ two-way traveltime from the source to the plane surface, with the amplitude falling off inversely as the distance. Thus the reflection has the same waveshape as the source. The diffraction terms can be found as follows: we write $t = 2\xi/V =$ two-way traveltime from the shotpoint to the variable point on the boundary. Equation (2.109) now becomes

$$\Phi = \frac{ch}{\pi V^2} \oint \frac{e^{-st}}{t^2}\, d\theta = \frac{ch}{\pi V^2} \oint \frac{e^{-st}}{t^2} \frac{d\theta}{dt}\, dt. \qquad (2.111)$$

We must pay careful attention to the limits of integration since ξ, and hence t, is in general a multi-valued function of θ; for example, when $\theta = \theta_3$ in fig. 2.21*d*, ξ can have any of the values ξ_m, ξ_n or ξ_p. To avoid difficulties the integral is calculated as the point of integration goes from A to B (θ from θ_1 to θ_2), then from B to C, C to D, and finally D to A, the proper values of ξ (and t) being used along each segment of the path. Along a given portion of the path, say between $t = t_1$ and $t = t_2$ ($t_2 > t_1$), we have

$$\Phi = \frac{ch}{\pi V^2} \int_{t_1}^{t_2} \frac{e^{-st}}{t^2} \left(\frac{d\theta}{dt}\right) dt = \int_0^{+\infty} \phi(t) e^{-st}\, dt.$$

where $\phi(t) \leftrightarrow \Phi(s)$

$$\left.\begin{aligned}
\phi(t) &= 0, & t &< t_1, \\
&= (ch/\pi V^2 t^2)(d\theta/dt), & t_1 &< t < t_2, \\
&= 0, & t &> t_2.
\end{aligned}\right\} \qquad (2.112)$$

The derivative, $d\theta/dt$, is finite except when ξ is constant, such as where $\mathscr{S}$ is bounded by an arc of a circle with center at the origin, for which case $dt = 0$. In this special case, (2.111) gives

$$\Phi = (ch/\pi V^2 t_0^2)\, e^{-st_0}(\theta_i - \theta_j),$$

where θ_i and θ_j fix the end points of the circular arc and t_0 is the two-way traveltime to the arc. The inverse transform is

$$\phi = (ch/\pi V^2 t_0^2)(\theta_i - \theta_j)\delta(t - t_0).$$

When $(d\theta/dt)$ is finite, we can get the time-domain solution for the diffracted wave by dividing the boundary of the area $\mathscr{S}$ so that the two-way traveltime t between P and the boundary is a single-valued function of θ in each part, calculating ϕ for each part from (2.112), then summing the various ϕs to get ϕ for the diffracted wave.

(*e*) *Diffraction effect of a half-plane.* A simple illustration of the method is the calculation of the diffraction effect in the important case of a horizontal half-plane at a depth h, the edge being parallel to the x-axis and a distance y_0 from it. Referring to fig. 2.22, the edge of the half-plane is BD where in fact B and D are at infinity, so that θ increases from $-\frac{1}{2}\pi$ to $+\frac{1}{2}\pi$ as the point of integration A traverses the boundary in a clockwise direction. The result (see problem 2.20) is

Fig.2.22. Calculation of the diffraction from a half-plane.

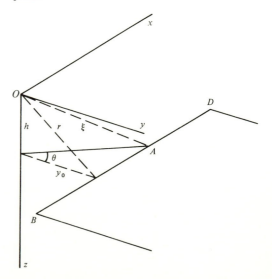

$$\phi = \frac{2(ch/\pi V^2)(1/t^2)(t_y t)}{(t^2 + t_y^2 - t_r^2)(t^2 - t_r^2)^{\frac{1}{2}}}$$

$$= \frac{(4chy_0/\pi V^3 t)}{(t^2 + t_y^2 - t_r^2)(t^2 - t_y^2)^{\frac{1}{2}}}, \quad t > t_r \Bigg\} \qquad (2.113)$$

$$= 0, \quad t < t_r,$$

where $t = 2\xi/V$, $t_r = 2r/V$, $t_y = 2y_0/V$.

The above value of $\phi(t)$ gives the diffracted wave recorded at the point $P(0, 0, 0)$ as the result of a unit impulse, $\delta(t)$, applied at the same point. If the input is $g(t)$ instead of $\delta(t)$, Φ will have the factor $G(s)$ and the response becomes $\phi(t) * g(t)$ (see (10.174)).

Equation (2.113) gives the diffraction effect whether P is off the plane, as in fig. 2.22, or over the plane; since y_0 changes sign as P passes over the edge, the diffracted wave undergoes a 180° phase shift as P passes over the edge. Moreover, if we write D for the value of ϕ for the

diffracted wave observed when P is infinitesimally close to the edge and to the left of it, the total effect observed when P is the same distance to the right of the edge will be $R - D$, R being the value of the reflection term in (2.110). Since $\phi(t)$ is continuous,

$$R - D = D, \text{ or } D = \tfrac{1}{2}R. \qquad (2.114)$$

Thus the maximum amplitude of the diffraction from a half-plane is half the amplitude of the reflected wave (as observed far from the edge). Fig. 2.23 shows what is expected from a half-plane based on (2.113). As the edge of the reflector is approached, the diffraction gains in amplitude while $(R - D)$ decreases in amplitude until at the edge $D = \tfrac{1}{2}R$ and the sum is $\tfrac{1}{2}R$. The phase reversal of the diffraction before the edge is reached (the *backward branch* of the diffraction) from that beyond the edge (the *forward branch*) is evident in fig. 2.23.

(*f*) *Use of Huygen's principle to construct diffracted wavefronts.* The surface integral in (2.106) shows that the diffraction effect at a point is the sum of effects arising from the entire diffracting surface. This suggests the use of Huygens' principle to construct diffracted wavefronts, and this is the case for points more than a few wavelengths away from the diffracting source. Fig. 2.24 illustrates this construction for a faulted reflector. We assume a plane wavefront AB incident normally on the faulted bed CO, the position of the wavefront when it reaches the surface of the bed at $t = t_0$ being COD. At $t = t_0 + \Delta t$, the portion to the right of O has advanced to the position GH while the portion to the left of O has been reflected and has

Fig.2.23. Seismic response of a half-plane. (After Trorey, 1970.) (*a*) Model; (*b*) computed seismic record for coincident sources and geophones. The arrowhead indicates the location of the edge of the half-plane.

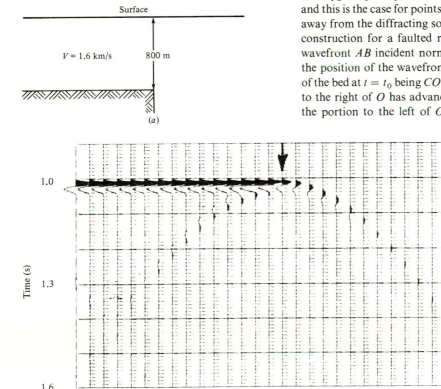

(a)

(b)

reached the position *EF*. We might have constructed the wavefronts *EF* and *GH* by selecting a large number of centers in *CO* and *OD* and drawing arcs of length $V \Delta t$; *EF* and *GH* would then be determined by the envelopes of these arcs. However, for the portion *EF* there would be no centers to the right of *O* to define the envelope while for the portion *GH* there would be no centers to the left of *O* to define the envelope. Thus, *O* marks the transition point between centers which give rise to the upward-traveling wavefront *EF* and centers which give rise to the downward-traveling wavefront *GH*; the arc *FPG* with center *O* is the diffracted wavefront originating at *O* and connecting the two wavefronts, *EF* and *GH*. The diffracted wavefront also extends into the geometrical shadow area *GN* and into the region *FM*.

The characteristics of diffractions in various situations is discussed further in §4.2.1.

2.4   Partitioning of energy at an interface

2.4.1   *General*

The boundary conditions described in §2.2.8 lead to rather complex relations for reflection and refraction at an interface. The nature of the two media fixes the densities and elastic constants and thus the velocities. The angles of reflection and refraction are fixed in terms of the velocities, as will be shown below. The only variables remaining to satisfy the boundary conditions are the amplitudes of the waves generated. When both media are solids, there are four equations resulting from the boundary conditions, so that we must have four variables. A P-wave (or an S-wave) incident on an interface sepa-

rating two solids must in general generate reflected and refracted S-waves as well as reflected and refracted P-waves. Thus for an incident P-wave, as shown in fig. 2.25, we have reflected and refracted P-waves at the angles θ_1 and θ_2 and reflected and refracted S-waves at the angles δ_1 and δ_2. The waves whose mode changes at an interface (the reflected and refracted S-waves in the foregoing example) are called *converted waves*.

Note that S-waves have two degrees of freedom, and motion perpendicular to the plane containing the incident wave and the normal to the interface is not involved in conversion from P- to S-waves nor vice versa. Where the interface is horizontal, this is equivalent to saying that incident P-waves can generate reflected and

Fig.2.25. Waves generated at an interface by an incident P-wave.

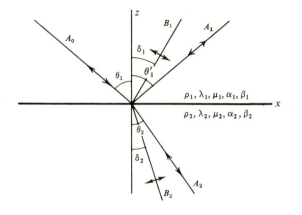

Fig.2.24. Diffracted wavefronts for a faulted bed.

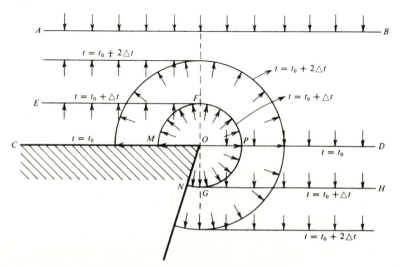

refracted P- and SV-waves but not SH-waves, that incident SV-waves can generate P- and SV-waves, but that incident SH-waves generate only reflected and refracted SH-waves.

2.4.2   *Knott's equations*

We obtain Knott's equations by starting with the following displacement potentials:

$$\left.\begin{aligned}
\text{in medium (1):} \quad \phi_1 &= A_0\,e^{j\omega\zeta_0} + A_1\,e^{j\omega\zeta_1}, \\
\chi_1 &= B_1\,e^{j\omega\zeta_1'}, \\
\text{in medium (2):} \quad \phi_2 &= A_2\,e^{j\omega\zeta_2}, \\
\chi_2 &= B_2\,e^{j\omega\zeta_2'},
\end{aligned}\right\} \quad (2.115)$$

where

$$\left.\begin{aligned}
\zeta_0 &= (x\sin\theta_1 - z\cos\theta_1)/\alpha_1, \\
\zeta_1 &= (x\sin\theta_1' + z\cos\theta_1')/\alpha_1, \\
\zeta_1' &= (x\sin\delta_1 + z\cos\delta_1)/\beta_1, \\
\zeta_2 &= (x\sin\theta_2 - z\cos\theta_2)/\alpha_2, \\
\zeta_2' &= (x\sin\delta_2 - z\cos\delta_2)/\beta_2,
\end{aligned}\right\} \quad (2.116)$$

(note that in (2.50), $l_\phi = \sin\theta_i$, $l_\chi = \sin\delta_i$, $m = 0$, $n_\phi = \pm\cos\theta_i$, $n_\chi = \pm\cos\delta_i$, $i = 1, 2$). It should be noted that the expressions for ϕ_i, χ_i lack the time factor, $e^{-j\omega t}$; the omission is deliberate since the factor always cancels in the boundary conditions; we only need to remember that differentiation with respect to time is equivalent to multiplying by $-j\omega$.

Substitution of the potential functions in the boundary conditions involves differentiation once or twice with respect to x and/or z, and then setting $z = 0$ (see problem 2.11). Each term on both sides of the equations will then have one of the factors $e^{(j\omega x \sin\theta_i)/\alpha_i}$ or $e^{(j\omega x \sin\delta_i)/\beta_i}$, multiplied by various constants. Since the equations must hold for all values of x, we must have

$$\frac{\sin\theta_1}{\alpha_1} = \frac{\sin\theta_1'}{\alpha_1} = \frac{\sin\theta_2}{\alpha_2} = \frac{\sin\delta_1}{\beta_1} = \frac{\sin\delta_2}{\beta_2} = p,$$

a constant.   (2.117)

This generalized form of Snell's law enables us to simplify the potential functions by writing

$$\zeta_0 = p(x - z\cot\theta_1), \quad \zeta_1 = p(x + z\cot\theta_1),$$
$$\zeta_2 = p(x - z\cot\theta_2), \quad \zeta_1' = p(x + z\cot\delta_1),$$
$$\zeta_2' = p(x - z\cot\delta_2).$$

The factor $j\omega p$ will cancel in the final result, hence we can simplify the calculations by taking differentiation with respect to x as equivalent to multiplication by unity and with respect to z by $\pm\cot\theta_i$ or $\pm\cot\delta_i$.

The first boundary condition requires the continuity of normal displacements at the interface, that is, $(w)_1 = (w)_2$ at the interface $z = 0$:

$$\left(\frac{\partial\phi}{\partial z} - \frac{\partial\chi}{\partial x}\right)_1 = \left(\frac{\partial\phi}{\partial z} - \frac{\partial\chi}{\partial x}\right)_2 \quad \text{at } z = 0,$$

hence

$$(-A_0 + A_1)\cot\theta_1 - B_1 = -A_2\cot\theta_2 - B_2.$$

The next condition is that the shear displacements be equal, $u|_1 = u|_2$, at $z = 0$:

$$\left(\frac{\partial\phi}{\partial x} + \frac{\partial\chi}{\partial z}\right)_1 = \left(\frac{\partial\phi}{\partial x} + \frac{\partial\chi}{\partial z}\right)_2 \quad \text{at } z = 0,$$

or

$$(A_0 + A_1) + B_1\cot\delta_1 = A_2 - B_2\cot\delta_2.$$

The continuity of normal stress requires that $\sigma_{zz}|_1 = \sigma_{zz}|_2$ at $z = 0$, and hence

$$\lambda\nabla^2\phi + 2\mu\left(\frac{\partial^2\phi}{\partial z^2} - \frac{\partial^2\chi}{\partial x\,\partial z}\right)$$

is continuous. Thus,

$$\begin{aligned}
&\lambda_1(A_0 + A_1)(1 + \cot^2\theta_1) \\
&\quad + 2\mu_1\{(A_0 + A_1)\cot^2\theta_1 - B_1\cot\delta_1\} \\
&= \lambda_2 A_2(1 + \cot^2\theta_2) + 2\mu_2(A_2\cot^2\theta_2 + B_2\cot\delta_2).
\end{aligned}$$

Using (2.52), (2.53) and (2.117), this becomes

$$\begin{aligned}
&\mu_1(\cot^2\delta_1 - 1)(A_0 + A_1) - 2\mu_1 B_1\cot\delta_1 \\
&= \mu_2(\cot^2\delta_2 - 1)A_2 + 2\mu_2 B_2\cot\delta_2.
\end{aligned}$$

Continuity of the tangential stress means that $\sigma_{xz}|_1 = \sigma_{xz}|_2$ at $z = 0$, that is,

$$\mu\left(2\frac{\partial^2\phi}{\partial x\,\partial z} + \frac{\partial^2\chi}{\partial z^2} - \frac{\partial^2\chi}{\partial x^2}\right)$$

is continuous; thus

$$\begin{aligned}
&\mu_1\{2(-A_0 + A_1)\cot\theta_1 + B_1(\cot^2\delta_1 - 1)\} \\
&= \mu_2\{-2A_2\cot\theta_2 + B_2(\cot^2\delta_2 - 1)\}.
\end{aligned}$$

If we substitute $a_i = \cot\theta_i$, $b_i = \cot\delta_i$, $c_i = b_i^2 - 1$, the above equations become

$$-a_1 A_0 + a_1 A_1 - B_1 = -a_2 A_2 - B_2 \qquad (2.118)$$

$$A_0 + A_1 + b_1 B_1 = A_2 - b_2 B_2 \qquad (2.119)$$

$$\mu_1 c_1 A_0 + \mu_1 c_1 A_1 - 2\mu_1 b_1 B_1 = \mu_2 c_2 A_2 + 2\mu_2 b_2 B_2 \qquad (2.120)$$

$$-2\mu_1 a_1 A_0 + 2\mu_1 a_1 A_1 + \mu_1 c_1 B_1 = -2\mu_2 a_2 A_2 + \mu_2 c_2 B_2. \qquad (2.121)$$

These equations, due to Knott (1899), fix the amplitudes of the reflected and refracted waves generated by a P-wave incident on a plane interface separating two solid media. Similar equations can be derived for an incident S-wave and for other types of media, for example, solid–liquid (Ergin, 1952).

2.4.3 *Distribution of energy*

Knott's equations have a very interesting property. If we multiply together corresponding sides of the first and third equations and also corresponding sides of the second and fourth equations, and then add these products, the result is

$$(\rho_1 \cot \theta_1)A_1^2 + (\rho_1 \cot \delta_1)B_1^2 + (\rho_2 \cot \theta_2)A_2^2$$
$$+ (\rho_2 \cot \delta_2)B_2^2 = (\rho_1 \cot \theta_1)A_0^2. \qquad (2.122)$$

Since the first and third equations relate to the normal displacement and stress on the two sides of the interface while the second and fourth relate to the tangential displacement and stress, the products have the dimensions of energy per unit area. From this we make the correct surmise that (2.122) gives the distribution of energy among the various reflected and refracted waves. To demonstrate this, we repeat the derivation of (2.89) in terms of the potential function ϕ (χ does not enter here because it concerns S-waves). For the incident P-wave we have for the kinetic energy per unit volume E_K

$$\frac{\partial E_K}{\partial \mathscr{V}} = \tfrac{1}{2}\rho_1 \left\{ \left(\frac{\partial u}{\partial t}\right)^2 + \left(\frac{\partial w}{\partial t}\right)^2 \right\}$$
$$= \tfrac{1}{2}\rho_1 \left\{ \left(\frac{\partial^2 \phi}{\partial x\,\partial t}\right)^2 + \left(\frac{\partial^2 \phi}{\partial z\,\partial t}\right)^2 \right\}.$$

Using (2.115) and (2.116) and noting that we must take $\partial/\partial x = j\omega p$, $\partial/\partial z = -(j\omega \cos \theta_1)/\alpha_1$, $\partial/\partial t = -j\omega$, we have for $z = 0$,

$$\frac{\partial E_K}{\partial \mathscr{V}} = \tfrac{1}{2}\rho_1 \left\{ (+\omega^2 p A_0)^2 \right.$$
$$\left. + \left(\frac{+\omega^2 (\cos \theta_1)A_0}{\alpha_1}\right)^2 \right\} e^{j\omega\{(x \sin \theta_1)/\alpha_1 - t\}}.$$

Taking the maximum of the real part, we find the expression for the energy density E, that is,

$$E = \tfrac{1}{2}\rho_1 \omega^4 (A_0/\alpha_1)^2. \qquad (2.123)$$

Comparison with (2.89) shows that

$$A_0 = (\alpha_1/\omega)\mathscr{A}_0 \qquad (2.124)$$

where $\mathscr{A}_0$ is the amplitude of the displacement in the direction of propagation.

The energy brought up to a unit area of the interface per unit time by the incident P-wave will be the energy in a cylinder of length α_1, of unit cross-section (measured parallel to the interface), and inclined at an angle θ_1 to the normal to the surface, that is,

$$(\text{volume of cylinder}) \times E = (\alpha_1 \cos \theta_1)(\rho_1 \omega^4 A_1^2/2\alpha_1^2)$$
$$= \tfrac{1}{2}p\rho_1 \omega^4 A_0^2(\cot \theta_1).$$

Since similar expressions must hold for the energy carried away by the other waves, we see that we have only to multiply each term in (2.122) by $\tfrac{1}{2}p\omega^4$ to obtain the distribution of energy among the various waves.

2.4.4 *Zoeppritz' equations*

If we use (2.124) in a more general form, that is, $A_0 = (\alpha_1/\omega)\mathscr{A}_0$, $A_i = (\alpha_i/\omega)\mathscr{A}_i$, $B_i = (\beta_i/\omega)\mathscr{B}_i$, $i = 1, 2$, we can replace the displacement potential amplitudes in Knott's equations with the displacement amplitudes $\mathscr{A}_0$, $\mathscr{A}_i$, $\mathscr{B}_i$. (Just as in the case of A_0, A_1, etc., the displacement amplitudes do not give directly the amplitudes of u and w; instead, $\mathscr{A}$ is the amplitude of the displacement in the direction of wave propagation while $\mathscr{B}$ is the amplitude of displacement normal to the direction of propagation.) Replacing A_0, A_1, etc. in Knott's equations with $\mathscr{A}_0, \mathscr{A}_1$, etc., we get the following equations, known as Zoeppritz' equations (Zoeppritz, 1919) (see also problem 2.22):

$$\mathscr{A}_1 \cos \theta_1 - \mathscr{B}_1 \sin \delta_1 + \mathscr{A}_2 \cos \theta_2 + \mathscr{B}_2 \sin \delta_2$$
$$= \mathscr{A}_0 \cos \theta_1, \qquad (2.125)$$

$$\mathscr{A}_1 \sin \theta_1 + \mathscr{B}_1 \cos \delta_1 - \mathscr{A}_2 \sin \theta_2 + \mathscr{B}_2 \cos \delta_2$$
$$= -\mathscr{A}_0 \sin \theta_1, \qquad (2.126)$$

$$\mathscr{A}_1 Z_1 \cos 2\delta_1 - \mathscr{B}_1 W_1 \sin 2\delta_1 - \mathscr{A}_2 Z_2 \cos 2\delta_2$$
$$- \mathscr{B}_2 W_2 \sin 2\delta_2 = -\mathscr{A}_0 Z_1 \cos 2\delta_1, \qquad (2.127)$$

$$\mathscr{A}_1 (\beta_1/\alpha_1) W_1 \sin 2\theta_1 + \mathscr{B}_1 W_1 \cos 2\delta_1$$
$$+ \mathscr{A}_2 (\beta_2/\alpha_2) W_2 \sin 2\theta_2 - \mathscr{B}_2 W_2 \cos 2\delta_2$$
$$= \mathscr{A}_0 (\beta_1/\alpha_1) W_1 \sin 2\theta_1, \qquad (2.128)$$

where

$$Z_i = \rho_i \alpha_i, \quad W_i = \rho_i \beta_i, \quad i = 1, 2.$$

The products of density and velocity (Z_i and W_i) are known as *acoustic impedances*. To apply these equations at an interface, we must know the density and velocities in each of the media, hence Z_1, Z_2, W_1 and W_2 are known. For a given $\mathscr{A}_0$ and θ_1, we can calculate θ_2, δ_1 and δ_2 from (2.117) and the four amplitudes, $\mathscr{A}_1$, $\mathscr{A}_2$, $\mathscr{B}_1$ and $\mathscr{B}_2$, from (2.125)–(2.128).

2.4.5 *Partitioning at normal incidence*

Zoeppritz' equations reduce to a very simple form for normal incidence. Since the curves change slowly for small angles of incidence (say up to 20°), the results for normal incidence have wide application. For a P-wave at normal incidence, there are no tangential stresses and displacements; hence $\mathscr{B}_1 = \mathscr{B}_2 = 0$ and (2.125)–(2.128) reduce to

$$\mathscr{A}_1 + \mathscr{A}_2 = \mathscr{A}_0,$$

$$Z_1 \mathscr{A}_1 - Z_2 \mathscr{A}_2 = -Z_1 \mathscr{A}_0.$$

The solution of these equations is

$$R = \frac{\mathscr{A}_1}{\mathscr{A}_0} = \frac{\alpha_2 \rho_2 - \alpha_1 \rho_1}{\alpha_2 \rho_2 + \alpha_1 \rho_1} = \frac{Z_2 - Z_1}{Z_2 + Z_1} \approx \frac{\Delta Z}{2Z}$$

$$= \tfrac{1}{2} \Delta (\ln Z), \tag{2.129}$$

$$T = \frac{\mathscr{A}_2}{\mathscr{A}_0} = \frac{2\alpha_1 \rho_1}{\alpha_2 \rho_2 + \alpha_1 \rho_1} = \frac{2Z_1}{Z_2 + Z_1}. \tag{2.130}$$

Equations (2.129) and (2.130) define *reflection coefficient* R (also called reflectivity) and *transmission coefficient* T. Equation (2.129) shows that the amplitude of a sequence of isolated reflections is a record of changes in the log of acoustic impedances, the viewpoint taken in seismic log manufacture (§9.4.5). The fractions of energy reflected and transmitted are given by E_R and E_T (which are also sometimes called reflection and transmission coefficients):

$$E_R = \frac{\alpha_1 \rho_1 \omega^2 \mathscr{A}_1^2}{\alpha_1 \rho_1 \omega^2 \mathscr{A}_0^2} = \left(\frac{Z_2 - Z_1}{Z_2 + Z_1} \right)^2 = R^2, \tag{2.131}$$

$$E_T = \frac{\alpha_2 \rho_2 \omega^2 \mathscr{A}_2^2}{\alpha_1 \rho_1 \omega^2 \mathscr{A}_0} = \frac{4Z_1 Z_2}{(Z_2 + Z_1)^2} = \frac{Z_2}{Z_1} T^2. \tag{2.132}$$

Obviously $E_R + E_T = 1$. Note that (2.131) and (2.132) are unchanged if Z_1 and Z_2 are interchanged; hence the energy partition does not depend upon which medium contains the incident wave. When $Z_2/Z_1 = 1$, $R = E_R = 0$ and all the energy is transmitted; note that this does not require that $\rho_1 = \rho_2$ and $\alpha_1 = \alpha_2$. As the impedance contrast approaches zero or infinity, T approaches zero and R approaches unity; thus, the farther the impedance contrast is from unity, the stronger the reflected energy.

Table 2.3 shows how the reflected energy varies for impedance contrasts such as may be expected within the Earth. Since both density and velocity contrasts are small for most of the interfaces encountered, only a small portion of the energy is reflected at any one interface; this is illustrated by the first four lines in table 2.3. The 'sandstone-on-limestone' interface is about as large a contrast as is apt to be encountered, whereas the 'shallow interface' and 'deep interface' figures are much more typical of most interfaces in the Earth; hence usually appreciably less than 1% of the energy is reflected at any interface. The major exceptions involve the bottom and surface of the ocean and the base of the weathering (see §2.2.6). A much larger proportion of the energy can be reflected from these and hence they are especially important in the generation of multiple reflections (§4.2.2) and other phenomena with which we shall deal later.

Note that, while the energy fractions E_R and E_T do not depend on which side of an interface the wave is

Table 2.3. *Energy reflected at interface between two media.*

Interface	First medium		Second medium		Z_1/Z_2	R	E_R
	Velocity	Density	Velocity	Density			
Sandstone on limestone	2.0	2.4	3.0	2.4	0.67	0.2	0.040
Limestone on sandstone	3.0	2.4	2.0	2.4	1.5	−0.2	0.040
Shallow interface	2.1	2.4	2.3	2.4	0.93	0.045	0.0021
Deeper interface	4.3	2.4	4.5	2.4	0.97	0.022	0.0005
'Soft' ocean bottom	1.5	1.0	1.5	2.0	0.50	0.33	0.11
'Hard' ocean bottom	1.5	1.0	3.0	2.5	0.20	0.67	0.44
Surface of ocean	1.5	1.0	0.36	0.0012	3800	−0.9994	0.9988
Base of weathering	0.5	1.5	2.0	2.0	0.19	0.68	0.47
Shale over water sand	2.4	2.3	2.5	2.3	0.96	0.02	0.0004
Shale over gas sand	2.4	2.3	2.2	1.8	1.39	−0.16	0.027
Gas sand over water sand	2.2	1.8	2.5	2.3	0.69	0.18	0.034

All velocities in km/s, densities in g/cm³; the minus signs indicate 180° phase reversal.

incident, this is not true of the reflected amplitude $\mathscr{A}_1$ since interchanging Z_1 and Z_2 in (2.129) changes the sign of the ratio $\mathscr{A}_1/\mathscr{A}_0$. A negative value of $\mathscr{A}_1$ means that the reflected wave is 180° out-of-phase with the incident wave; thus, for an incident wave $\mathscr{A}_0 \cos \omega t$ the reflected wave is $\mathscr{A}_1 \cos(\omega t + \pi)$. In table 2.3 phase reversal occurs for the situations where Z_1 exceeds Z_2.

2.4.6 *Partitioning at non-normal incidence*

Turning now to the general case where the angle of incidence is not necessarily 0°, we shall illustrate solutions of Zoeppritz' equations by graphs showing the energy partition as functions of the angle of incidence for certain values of parameters. Many curves would be required to show the variations of energy partitioning as a function of incident angle because of the many parameters which can be varied: incident P-, SH-, or SV-wave, P-wave velocity ratio, density ratio, and S-wave velocities in each medium (or the equivalent of defining Poisson's ratio for each medium). Fig. 2.26 shows several cases representative of the variety of results possible.

Fig. 2.26a shows the partitioning of energy as a function of the angle of incidence when a P-wave is incident in the high-velocity medium for a P-wave velocity ratio $\alpha_2/\alpha_1 = 0.5$, a density ratio $\rho_2/\rho_1 = 0.8$, $\sigma_1 = 0.30$, $\sigma_2 = 0.25$. For small incident angles all of the energy is in the reflected or transmitted P-waves, E_{RP} and E_{TP} respectively, and hence there are essentially no S-waves. As the incident angle increases, some of the energy goes into reflected and transmitted S-waves, E_{RS} and E_{TS} respectively, mostly at the expense of the reflected P-wave. Note that at intermediate angles of incidence the reflected S-wave carries more energy than the reflected P-wave. Such *converted waves* (waves resulting from the conversion of P-waves to S-waves or vice versa at an interface) are sometimes recorded at long offsets where they are evidenced by alignments which disappear as one tries to follow them to shorter offsets (see fig. 4.8b). As grazing incidence is approached the energy of the reflected P-wave increases until at grazing incidence all of the energy is in the reflected P-wave.

The opposite situation is shown in fig. 2.26b where $\alpha_2/\alpha_1 = 2.0$, $\rho_2/\rho_1 = 0.5$, $\sigma_1 = 0.30$, $\sigma_2 = 0.25$. Since $Z_1 = Z_2$, the P-wave reflection coefficient is essentially zero for small incident angles. As the incident angle increases, S-wave energy increases. As the critical angle for P-waves is approached, the transmitted P-wave energy falls rapidly to zero and no transmitted P-wave exists for larger incident angles. Also as the critical angle for P-waves is approached, both reflected P-wave and reflected S-wave become very strong; such a build-up in reflection strength

near the critical angle is called *wide-angle reflection*. Sometimes it is possible to make use of this phenomenon to map reflectors using long offsets where they cannot be followed at short offsets (Meissner, 1967). As the critical angle for S-waves is approached, the transmitted S-wave falls to zero.

If we had not had a density contrast but otherwise the values had been as indicated in fig. 2.26b there would have been a reflected P-wave at small incident angles (as shown by the dashed curve) whose fractional energy would have decreased slightly as the incident angle increased.

Fig. 2.26c shows the P-wave reflection coefficient for various P-wave velocity ratios when $\rho_1 = \rho_2$ and $\sigma_1 = \sigma_2 = 0.25$. The reflected energy is zero for a velocity ratio of one (no impedance contrast) and increases both as the ratio becomes larger than one and as it becomes smaller than one. The two peaks for $\alpha_2/\alpha_1 > 1$ occur at the critical angles for P- and S-waves respectively. Fig. 2.26d shows the energy of the reflected P-wave for various density contrasts when $\alpha_2/\alpha_1 = 1.5$ and $\sigma_1 = \sigma_2 = 0.25$.

Koefoed (1962) gives 100 tables of the longitudinal and transverse reflection and transmission coefficients and the phase shifts for angles greater than the critical angle (see §2.3.4) for incident longitudinal waves.

2.4.7 *Head waves*

In refraction seismology we make use of waves which have been refracted at the critical angle; these waves are often called *head waves*. In fig. 2.27a we see a P-wave incident on the refracting horizon at the critical angle Θ. After refraction it travels along the interface in the lower medium. This produces an oscillatory motion parallel to and immediately below the interface (as shown by the double-headed arrow just below the interface). Since relative motion between the two media is not possible, the upper medium is forced to move in phase with the lower medium. The disturbance in the upper medium travels along the interface with the same velocity V_2 as the refracted wave just below the interface. Let us assume that these disturbances represented by the arrows reach the point P in fig. 2.27b at the time t. According to Huygens' principle, P then becomes a center from which a wave spreads out into the upper medium. After a further time interval Δt, this wave has a radius of $V_1 \Delta t$ while the wave moving along the refractor has reached Q, PQ being equal to $V_2 \Delta t$. Drawing the tangent from Q to the arc of radius $V_1 \Delta t$, we obtain the wavefront RQ. Hence the passage of the refracted wave along the interface in the lower medium generates a plane wave traveling upward in the upper medium at the angle θ, where

$$\sin \theta = V_1 \Delta t / V_2 \Delta t = V_1/V_2.$$

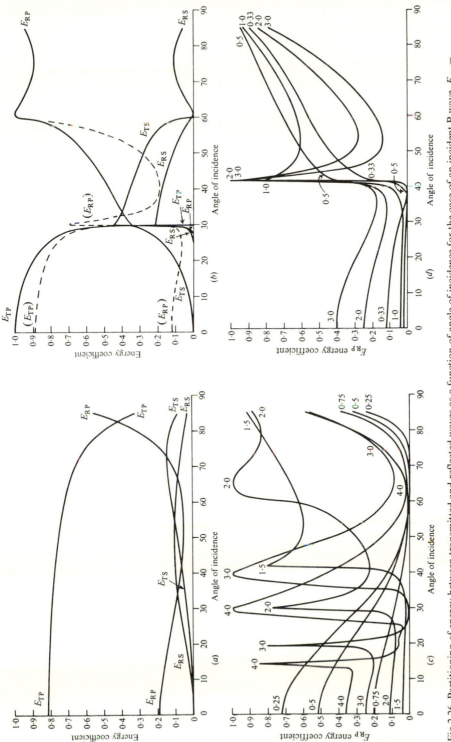

Fig. 2.26. Partitioning of energy between transmitted and reflected waves as a function of angle of incidence for the case of an incident P-wave. E_{TP} = fraction of energy in transmitted P-wave, E_{RP} = fraction in reflected P-wave, E_{RS} = fraction in reflected S-wave, E_{TS} = fraction in transmitted S-wave. (From Tooley et al., 1965.) (a) Case where velocity in incident medium is larger; $\alpha_2/\alpha_1 = 0.5$, $\rho_2/\rho_1 = 0.8$, $\sigma_1 = 0.3$, $\sigma_2 = 0.25$; (b) case where velocity in incident medium is smaller; $\alpha_2/\alpha_1 = 2.0$, $\rho_2/\rho_1 = 0.5$, $\sigma_1 = 0.3$, $\sigma_2 = 0.25$; (c) fraction of energy reflected as P-wave for various P-wave velocity ratios and $\rho_2/\rho_1 = 1.0$, $\sigma_1 = \sigma_2 = 0.25$; (d) fraction of energy reflected as P-wave for various density ratios and $\sigma_2/\alpha_1 = 1.5$, $\sigma_1 = \sigma_2 = 0.25$.

Thus we see that $\theta = \Theta$, so that the two inclined portions of the path are symmetrically disposed with respect to the normal to the refractor.

Head waves are also called *conical waves* because in a constant-velocity medium the wavefront has the shape of a cone whose axis is the perpendicular from the source to the interface, that is, the surface generated by SQ when fig. 2.27b is rotated about OL.

The Knott and Zoeppritz approaches to energy partitioning at an interface, both of which assume a plane incident wave, fail to predict head waves, that is, the transmission coefficient T_p vanishes at the critical angle. Hence one might expect that head waves carry no energy. However, head waves do exist and frequently are very strong, so that we are faced with the paradox of a theory which predicts the correct geometry but states that the intensity is zero. One can argue that T_p vanishes at the critical angle, not because the amplitude of the refracted wave vanishes, but because the width of the refracted beam approaches zero as the angle of refraction approaches $90°$ (see fig. 2.27c). The transmitted wave just prior to

the critical angle carries considerable energy and even a modest velocity gradient in the lower medium will bend rays back to the interface and thus make their energy available to support the head wave (fig. 2.27c). Even without the velocity gradient, diffraction would extend the wavefront upward the short distance separating the geometrical termination of the refraction front and the interface. Similar arguments apply to critically refracted S-waves.

The solution to this paradox was given by Jeffreys (1926) and by Cagniard (1962), who solved the problem of a spherical wave incident on a plane interface; however the mathematics is difficult (Ewing *et al.*, 1957, Grant and West, 1965, and Dix, 1954, summarize the method; Bortfeld, 1962a, b, gives simplified solutions). Cagniard's solution predicts four reflected head waves at a solid–solid interface where $\alpha_1 < \beta_2$ and one transmitted head wave; these are illustrated in fig. 2.28.

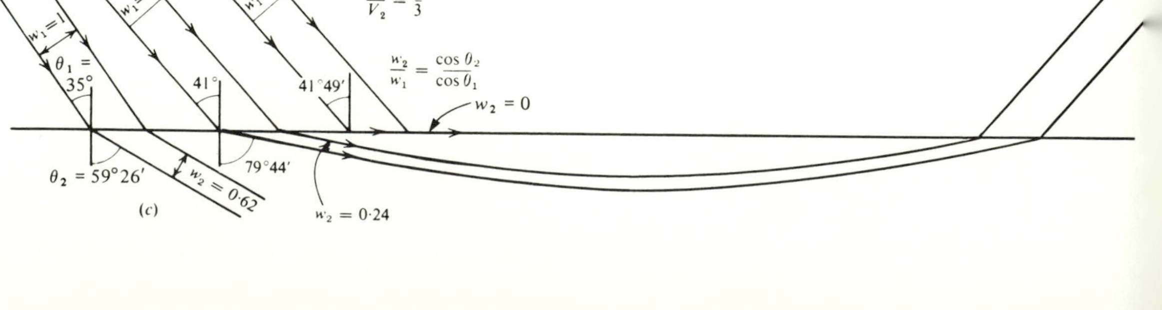

2.4.8 *Normal-mode propagation*

Under certain circumstances wave energy may be trapped within a layer which is then known as a *wave guide*. Two kinds of boundary conditions can produce this situation: (a) the impedance contrast is so great that the reflection coefficient is effectively unity; (b) waves within the wave guide are incident on a boundary at an angle greater than the critical angle so that total reflection occurs and no energy leaks through the boundary (except for evanescent waves which can be neglected). Common examples are a water layer at the surface and a horizontal coal seam at depth (§5.3.9d). Propagation in a wave guide is referred to as *normal-mode propagation* and is analogous to organ pipe reverberation, a water layer corresponding to an open organ pipe and a coal seam to a closed organ pipe. The normal modes are equivalent to the fundamental and its harmonics in an organ pipe. Love waves and SV-waves in the surface layer can be explained as normal-mode propagation (see Grant and West, 1965, pp. 81–5).

Fig. 2.29a shows various waves bouncing back and forth at different angles of reflection within a wave guide.

Fig.2.27. Head waves. (a) Motion at the interface; (b) wavefront emerging from refractor at critical angle; (c) changes in beam width upon refraction and effect of velocity gradient in the lower medium (α_2 increasing downwards).

(a)

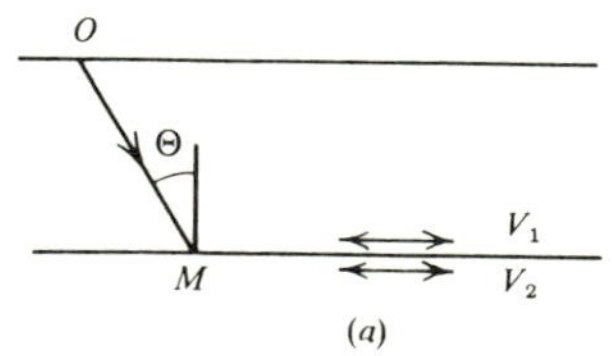

(b)

(c)

For most of the angles there is destructive interference between the different waves but for certain angles there is constructive interference and consequently a strong build-up of energy reflected at these angles. In fig. 2.29b the wavefront AC has been reflected upward at the lower boundary at the angle θ where $\theta > \Theta$, the critical angle. A parallel wavefront which occupied the same position AC earlier, then was later reflected at the upper and lower boundaries, following raypaths such as $EFGH$ and $BDAI$, now coincides with the later wavefront at AC. Since $(EF + FG + GH) = (BD + DA)$ we see that the phase difference between the two waves is $\kappa(BD + DA) + m\pi + \varepsilon$ where $m = 0$ or 1, $m\pi$ = sum of phase reversals on reflection at the two boundaries and ε is a phase shift which occurs when $\theta > \Theta$ (Officer, 1958, pp. 200–1). For a water layer $m = 1$ whereas $m = 0$ for a coal seam.

For constructive interference we must have

$$\kappa_n(BD + DA) + m\pi + \varepsilon = 2n\pi.$$

Since

$$(DA + BD) = (h/\cos \theta) + (h/\cos \theta) \cos 2\theta$$
$$= 2h\cos\theta,$$

we have

$$2\kappa_n h \cos \theta = (4\pi h v_n / V_1) \cos \theta = (2n - m)\pi - \varepsilon,$$

or

$$v_n = \{(2n - m) - (\varepsilon/\pi)\} V_1/(4h\cos\theta). \tag{2.133}$$

Fig.2.28. Five head waves (PP_2P_1, PP_2S_1, PS_2P_1, PS_2S_1, PP_2S_2) for incident P-wave for case where $\alpha_1 < \beta_2$. (After Caignard, 1962.)

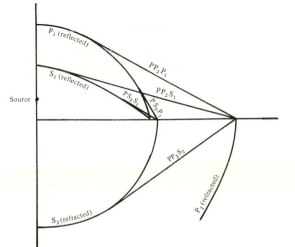

Neglecting ε for the moment, we get constructive interference when

$$v_n = (2n - m)V_1/(4h\cos\theta). \tag{2.134}$$

For a water layer, $m = 1$, hence

$$\left.\begin{array}{l} v_1 = V_1/(4h\cos\theta), \\ v_2 = 3V_1/(4h\cos\theta) = 3v_1, \\ \cdots \\ v_n = (2n - 1)v_1, \end{array}\right\} \tag{2.135}$$

which corresponds to an open organ pipe (except for the factor $\cos\theta$). For a coal seam, $m = 0$, and

$$\left.\begin{array}{l} v_1 = V_1/(2h\cos\theta), \\ v_2 = 2V_1/(2h\cos\theta) = 2v_1, \\ \cdots \\ v_n = nv_1, \end{array}\right\} \tag{2.136}$$

which is analogous to a closed organ pipe. Thus, provided the original wave generated by the source contains the appropriate frequencies, normal-mode propagation consists of a series of waves of frequencies v_1 and its odd or even harmonics propagating along the wave guide by reflection at angles θ which satisfy (2.135) or (2.136).

In addition to the upward propagating set of wavefronts parallel to AC, there is a symmetrical downward-propagating set parallel to PQ in fig. 2.29c, and the interference between the two sets creates a standing wave pattern along the perpendicular to the wave guide. As a result, the wave motion is propagated parallel to the boundaries of the wave guide. The velocity V_1 is the phase velocity normal to the wavefronts but there is a different phase velocity V in the direction of the effective wave propagation. Referring to fig. 2.29c, the wavefronts AC and PQ intersect at R, and there will be a local build-up of energy here. This energy density maximum propagates in the direction RR'; if AC and $A'C'$, also PQ and $P'Q'$, are the wavefront positions one time unit apart, then the phase at R moves to R' in one time unit so that $V = RR'$, that is,

$$V = V_1/\sin\theta. \tag{2.137}$$

Since θ is a function of frequency because of (2.134)–(2.136), V is also frequency dependent, so that the wave motion is dispersive.

The minimum value of θ is Θ, hence there is a minimum cutoff frequency v_0 where (for a water layer)

$$v_0 = V_1/(4h\cos\Theta). \tag{2.138}$$

Fig.2.29. Wave-guide phenomonon. (*a*) Many waves bouncing back and forth in a layer of velocity V_1 because of nearly perfect reflectivity at the boundaries; (*b*) construction to show reinforcement conditions; (*c*) phase and group velocity relationship.

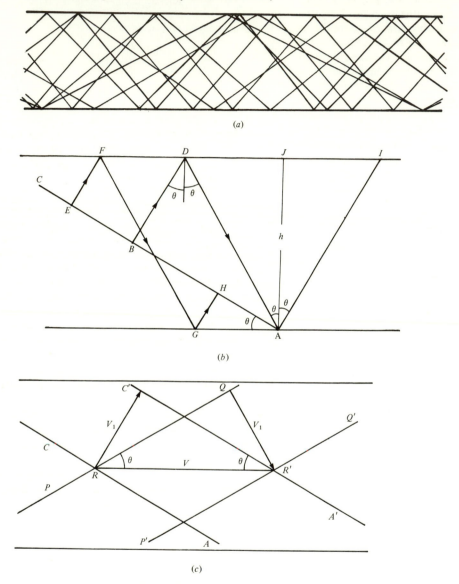

(*a*)

(*b*)

(*c*)

the corresponding phase velocity V being $V = V_1/\sin\Theta = V_2$. As θ increases, v increases but V decreases. In the limit, $\theta \to \frac{1}{2}\pi$ (the grazing angle), $v \to \infty$ and $V \to V_1$.

If we do not neglect ε, the formulas are more complicated but the results are basically the same. Officer (1958) shows that for $\theta > \Theta$,

$$\left.\begin{aligned}\tan\tfrac{1}{2}\varepsilon &= (\rho_1/\rho_2)\{\tan^2\theta - (V_1/V_2\cos\theta)^2\}^{\frac{1}{2}} \\ &= 0, \quad \theta = \Theta, \\ &= \pi, \quad \theta = \tfrac{1}{2}\pi.\end{aligned}\right\} \quad (2.139)$$

Fig. 2.30. Normal-mode propagation. (a) Phase and group velocity versus normalized frequency for a liquid layer on an elastic substratum where $\alpha_2/\alpha_1 = 2\sqrt{3}$, $\sigma_1 = 0.5$, $\sigma_2 = 0.25$, $\rho_2/\rho_1 = 2.5$ (from Ewing et al., 1957). (b) First-mode wavetrain from a source 4 km away. (c) The high-frequency portion of (b) called the water wave; its onset is used in refraction work to determine the range (see problem 3.14). (From Clay and Medwin, 1977.)

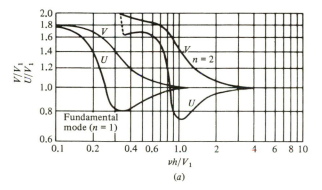

(a)

Typical curves of V versus v for a water layer are shown in fig. 2.30a for $n = 1, 2$.

The group velocity U is given by (2.102), that is,

$$U = V + \omega\frac{dV}{d\omega} = V + v\frac{dV}{dv}.$$

For a water layer, fig. 2.30a shows that the term $v(dV/dv)$ is never positive, hence $U \le V$. Moreover, although $v(dV/dv)$ increases in magnitude at first as v increases from the value v_0, eventually the term approaches zero as v approaches infinity (because the derivative goes to zero faster than v goes to infinity). As a result of these factors, U has the value V_2 at the cutoff frequency v_0, then decreases to a minimum U_m at some frequency v_m after which it increases asymptotically to the value V_1 at $v = \infty$.

A normal-mode wavetrain for a water layer is shown in fig. 2.30b. The first arrival is a wave of frequency v_0 which has traveled with the maximum group velocity V_2; this is followed by waves of increasing v and decreasing U until U reaches the value V_1 at which time a very high-frequency wave, which also has traveled with velocity V_1, is superimposed on the first wave. Following this, the frequencies and group velocities of the two waves approach v_m and U_m respectively. The energy traveling at U_m and the often-abrupt end of the normal-mode wavetrain is called the Airy phase.

Problems

2.1. Consider a cube of elastic isotropic material subjected to a normal stress, σ_{xx}. Equation (2.6) indicates that the modulus of rigidity enters into the calculation of the strain. Explain why shear is involved.

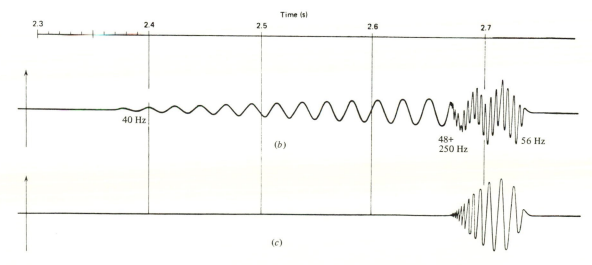

2.2. (a) By substituting $\sigma_{xx} > 0$, $\sigma_{yy} = \sigma_{zz} = 0$ in (2.6), show that $\varepsilon_{yy} = \varepsilon_{zz}$ and verify (2.12). (b) By adding the three equations in (a) for σ_{xx}, σ_{yy}, σ_{zz}, derive (2.11). (c) Substituting $\sigma_{xx} = \sigma_{yy} = \sigma_{zz} = -\mathscr{P}$ in (2.6), derive (2.13).

2.3. Starting with the last row in the table 2.4 (which expresses (2.11), (2.12) and (2.13)) derive the relations between the elastic constants given in the table. Each expression relates the constant at the head of the column to the two constants at the left end of the row.

2.4. (a) Firing an airgun (§5.5.3b) in water creates a pressure transient a small distance away with peak pressure of 5 atmospheres (5×10^5 Pa). If the compressibility of water is 4.5×10^{-10}/Pa, what is the peak energy density? (b) If the same wave is generated in rock with $\lambda = \mu = 3 \times 10^{10}$ Pa, what is the peak energy density? Assume a symmetrical P-wave with $\varepsilon_{xx} = \varepsilon_{yy} = \varepsilon_{zz}$; $\varepsilon_{ij} = 0$ for $i \neq j$.

2.5. To illustrate the interrelationship and magnitude of the elastic constants, complete table 2.5. Note that these values apply to specific specimens; the values for rocks range considerably, especially as porosity and pressure change.

2.6. (a) Verify that ψ in (2.38) and (2.39) are solutions of (2.37). [Hint: Let $\zeta = x - Vt$ and show that

$$\frac{\partial \psi}{\partial x} = \frac{\mathrm{d}f}{\mathrm{d}\zeta} \frac{\partial \zeta}{\partial x} = \frac{\mathrm{d}f}{\mathrm{d}\zeta} = f', \text{ etc.]}$$

(b) Verify that ψ in (2.40) is a solution of (2.23). (c) Using the same technique as in parts (a) and (b), show that ψ in (2.44) satisfies (2.42). This includes both ingoing and outgoing waves.

2.7. (a) Show that the wave equation (2.23) can be written in cylindrical coordinates ($x = r \cos \theta$, $y = r \sin \theta$, $z = z$; see fig. 2.31a) as

Table 2.4. *Relations between elastic constants*

	E	σ	k	μ	λ
(E, σ)			$\dfrac{E}{3(1 - 2\sigma)}$	$\dfrac{E}{2(1 + \sigma)}$	$\dfrac{E\sigma}{(1 + \sigma)(1 - 2\sigma)}$
(E, k)		$\dfrac{3k - E}{6k}$		$\dfrac{3kE}{9k - E}$	$3k\left(\dfrac{3k - E}{9k - E}\right)$
(E, μ)		$\dfrac{E - 2\mu}{2\mu}$	$\dfrac{\mu E}{3(3\mu - E)}$		$\mu\left(\dfrac{E - 2\mu}{3\mu - E}\right)$
(σ, k)	$3k(1 - 2\sigma)$			$\dfrac{3k}{2}\left(\dfrac{1 - 2\sigma}{1 + \sigma}\right)$	$3k\left(\dfrac{\sigma}{1 + \sigma}\right)$
(σ, μ)	$2\mu(1 + \sigma)$		$\dfrac{2\mu(1 + \sigma)}{3(1 - 2\sigma)}$		$\mu\left(\dfrac{2\sigma}{1 - 2\sigma}\right)$
(σ, λ)	$\lambda\dfrac{(1 + \sigma)(1 - 2\sigma)}{\sigma}$		$\lambda\left(\dfrac{1 + \sigma}{3\sigma}\right)$	$\lambda\left(\dfrac{1 - 2\sigma}{2\sigma}\right)$	
(k, μ)	$\dfrac{9k\mu}{3k + \mu}$	$\dfrac{3k - 2\mu}{2(3k + \mu)}$			$k - 2\mu/3$
(k, λ)	$9k\left(\dfrac{k - \lambda}{3k - \lambda}\right)$	$\dfrac{\lambda}{3k - \lambda}$		$\tfrac{3}{2}(k - \lambda)$	
(μ, λ)	$\mu\left(\dfrac{3\lambda + 2\mu}{\lambda + \mu}\right)$	$\dfrac{\lambda}{2(\lambda + \mu)}$	$\lambda + \tfrac{2}{3}\mu$		

Table 2.5. *Example of magnitudes of elastic constants*

	Water	Stiff mud	Sandstone	Limestone	Granite
Young's modulus, E ($\times 10^9$ Pa)			16	54	50
Bulk modulus, k ($\times 10^9$ Pa)	2.1				
Rigidity modulus, μ ($\times 10^9$ Pa)					
Lamé's λ constant ($\times 10^9$ Pa)					
Poisson's ratio, σ	0.5	0.43	0.34	0.25	0.20
Density, ρ (g/cm³)	1.0	1.5	1.9	2.5	2.7
P-wave velocity, α (km/s)	1.5	1.6			
S-wave velocity, β (km/s)					

$$\frac{\partial^2 \psi}{\partial r^2} + \frac{1}{r}\frac{\partial \psi}{\partial r} + \frac{1}{r^2}\frac{\partial^2 \psi}{\partial \theta^2} + \frac{\partial^2 \psi}{\partial z^2} = \frac{1}{V^2}\frac{\partial^2 \psi}{\partial t^2}.$$

(b) Verify that (2.41) is the wave equation by substituting the following coordinate transformation (see fig. 2.31b) into (2.23):

$$x = r \sin\theta \cos\phi,$$

$$y = r \sin\theta \sin\phi,$$

$$z = r \cos\theta.$$

(For an easier solution, see problem 10.6.)

2.8. The magnitudes of T, v, λ, κ are important in practical situations. Calculate them for the situations shown in table 2.6.

2.9. Show that ϕ and χ in (2.56) satisfy (2.21) and (2.26) respectively. [Hint: Calculate $\mathbf{V}\cdot\boldsymbol{\zeta}$ and $\mathbf{V}\times\boldsymbol{\zeta}$.]

2.10. Justify on physical grounds the following boundary conditions for different combinations of media in contact at an interface: (a) solid–fluid: normal stress and displacement are continuous, tangential stress in the solid vanishes at the interface; (b) solid–vacuum: normal and tangential stresses in the solid vanish at the interface; (c) fluid–fluid: normal stresses and displacements are continuous; (d) fluid–vacuum: normal stress in the fluid vanishes at the interface.

2.11. Using (2.1), (2.2), (2.5), (2.6), (2.7) and (2.57) show that the boundary conditions at the xy-plane separating two semi-infinite solids require for a wave in the xz-plane the continuity of the following functions: $(\phi_z - \chi_x)$, $(\phi_x + \chi_z)$, $\lambda\nabla^2\phi + 2\mu(\phi_{zz} - \chi_{xz})$, $\mu(2\phi_{xz} + \chi_{zz} - \chi_{xx})$, where the subscripts denote partial differentiation (these are respectively the normal and tangential displacements, normal and tangential stresses).

Table 2.6. *Magnitudes of* T, λ, κ.

	α(km/s)	For $v = 15$ Hz			For $v = 60$ Hz		
		T(s)	λ(m)	κ(m^{-1})	T(s)	λ(m)	κ(m^{-1})
Weathering (min.)	0.1						
Weathering (av.)	0.5						
Water	1.5						
Poorly consolidated sands–shales at 0.75 km	2.0						
Tertiary clastics at 3.00 km	3.3						
Porous limestone	4.3						
Dense limestone	5.5						
Salt	4.6						
Anhydrite	6.1						

Fig.2.31. Coordinate systems. (a) Cylindrical coordinates; (b) spherical coordinates.

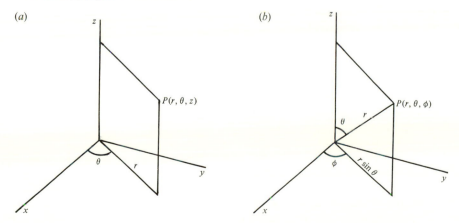

2.12. A source of seismic waves produces a step displacement on a spherical cavity enclosing the source of the form

$$\text{step}_0(t) = 0, \quad t < 0,$$
$$= k, \quad t \geq 0.$$

Show from (2.65) that the displacement is given by

$$u = \frac{r_0^2 k}{r}\left\{\left(\frac{1}{r_0} - \frac{1}{r}\right)e^{-V\zeta/r_0} + \frac{1}{r}\right\}.$$

Is the motion oscillatory? What is the final (permanent) displacement at the distance r?

2.13. Show that for harmonic waves of the form $\phi = (A/r)\cos\omega(r/V - t)$, the two terms in (2.59) are of equal importance at the distance $r = \lambda/2\pi$.

2.14. Equations (2.71) and (2.72) for a Rayleigh wave are valid for $z = 0$ when $\sigma = \frac{1}{4}$. (a) Show that for $z \neq 0$ the expressions for u, w are

$$u = \kappa A(-e^{-0.848\kappa z} + 0.577\,e^{-0.393\kappa z})\sin\kappa(x - V_R t),$$

$$w = \kappa A(-0.848\,e^{-0.848\kappa z} + 1.468\,e^{-0.393\kappa z})\cos\kappa(x - V_R t)$$

(b) What are the values of u, w, θ when $z = 1/2\kappa$? When $z = 1/\kappa$? (c) Is the motion retrograde for all values of z? [Hint: Note that for the motion to change direction, the amplitude of either u or w must change sign, that is, pass through zero.] (d) What are the values of V_R, the Rayleigh-wave velocity, when $\sigma = 0.4$ and when $\sigma = 0.2$? What are the corresponding values of the constants in (a)?

2.15. Assume three geophones so oriented that one records only the vertical component of a seismic wave, another records only the horizontal component in the direction of the shot and the third only the horizontal component at right angles to this. Assume a simple waveshape and draw the response of the three geophones for the following cases: (a) a P-wave traveling directly from the shot to the geophones; (b) a P-wave reflected from a deep horizon; (c) an S-wave generated by reflection of a P-wave at an interface; (d) a Rayleigh wave generated by the shot; (e) a Love wave. Compare the relative magnitudes of the components for short and long offsets.

2.16. (a) A tube wave has a velocity of 1.05 km/s. The fluid in the borehole has a bulk modulus of 2.15×10^9 Pa and density 1.20 g/cm³. The wall rock has $\sigma = 0.25$, $\rho = 2.5$ g/cm³. Calculate μ and α for the wall rock. (b) Repeat for $V_T = 1.20$ km/s and 1.30 km/s. What do you conclude about the accuracy of this method for determining μ?

2.17. The natural logarithm of the ratio of amplitudes is measured in nepers. Show that one neper = 8.686 dB.

2.18. A refraction seismic wavelet assumed to be essentially harmonic with frequency 40 Hz, is found to have amplitudes of 5.00 mm and 4.57 mm on traces 2.50 km

and 3.00 km from the source after correcting for divergence. Assuming a velocity of 3.20 km/s, constant subsurface conditions and ideal recording conditions, what is the ratio of the amplitudes on a given trace of the first and fourth cycles? What percentage of the energy is lost over three cycles? What is the value of h?

2.19. A pulse consists of two frequency components, $\omega_0 \pm \Delta\omega$ of equal amplitudes. We write for the two components

$$A\cos(\kappa_1 x - \omega_1 t), \quad A\cos(\kappa_2 x - \omega_2 t),$$

where $\omega_1 = \omega_0 + \Delta\omega$, $\omega_2 = \omega_0 - \Delta\omega$, $\kappa_0 = 2\pi/\lambda_0 = \omega_0/V$, $\kappa_1 \approx \kappa_0 + \Delta\kappa \approx (\omega_0 + \Delta\omega)/V$, $\kappa_2 \approx \kappa_0 - \Delta\kappa \approx (\omega_0 - \Delta\omega)/V$. (a) Show that the pulse is given approximately by the expression $\mathcal{B}\cos(\kappa_0 x - \omega_0 t)$, where $\mathcal{B} = 2A\cos\Delta\kappa\{x - (\Delta\omega/\Delta\kappa)t\}$. (b) Why do we regard $\mathcal{B}$ as the amplitude? Show that the envelope of the pulse is the graph of $\mathcal{B}$ plus its reflection in the x-axis. (c) Show that the envelope moves with the velocity U where

$$U = \frac{\Delta\omega}{\Delta\kappa} \approx \frac{d\omega}{d\kappa} \approx V - \lambda\frac{dV}{d\lambda} \approx V + \omega\frac{dV}{d\omega}$$

(see fig. 2.18).

2.20. Verify (2.113) using (2.112).

2.21. (a) Starting with either (2.118)–(2.121) or (2.125)–(2.128), show that the displacements at a free surface for an incident P-wave are

$$u/A_0 = 4j\omega pab\,e^{j\omega(px-t)}(c + 2)/(4ab + c^2),$$

$$w/A_0 = -2j\omega pac\,e^{j\omega(px-t)}(c + 2)/(4ab + c^2).$$

Note that a, b, c above have the same meaning as in (2.118)–(2.121). (b) For normal incidence at a free surface, show that at $z = 0$,

$$u/A_0 = 0, \quad w/A_0 = -2j\omega/\alpha.$$

(c) For the free surface of a solid and $\theta = 45°$, $\alpha = 3$ km/s, $\alpha/\beta = \sqrt{2}$, show that

$$u/A_0 = 0.598q, \quad w/A_0 = -0.345q,$$

where

$$q = j\omega\,e^{j\omega(px-t)}.$$

(d) At the surface of the ocean, show that

$$u/A_0 = 0, \quad w/A_0 = -(2j\omega/\alpha)\,e^{j\omega px}\cos\theta.$$

2.22. Deduce Zoeppritz' equations from first principles. [Hint: Express the wave displacements along and perpendicular to the rays in the form $\phi_i = \mathcal{A}_i e^{j\omega\zeta_i}$ and $\chi_i = \mathcal{B}_i e^{j\omega\zeta_i}$ (cf. (2.115) and (2.116)), then apply the boundary conditions (§2.2.8), noting that $u_1 = \phi_0\sin\theta_1 + \phi_1\sin\theta_1$

$+\chi_1 \cos \delta_1, \quad w_1 = -\phi_0 \cos \theta_1 + \phi_1 \cos \theta_1 - \chi_1 \sin \delta_1, \quad u_2 = \phi_2 \sin \theta_2 - \chi_2 \cos \delta_2, \quad w_2 = -\phi_2 \cos \theta_2 - \chi_2 \sin \delta_2.$ [Note: the positive directions of the transverse wave displacements in fig. 2.32 are given by right-hand rotation from the direction of propagation.]

2.23. (*a*) Using (2.50) to represent a plane wave incident on a plane interface show that a complex coefficient of reflection, $R = a + jb$, $a^2 + b^2 < 1$, R being defined by (2.129), corresponds to a reduction in amplitude by the factor $(a^2 + b^2)^{\frac{1}{2}}$ and an advance in phase by $\tan^{-1}(b/a)$. (*b*) Show that an imaginary angle of refraction, θ_2 (see §2.3.4) in (2.125)–(2.128) leads to a complex value of R, and hence to phase shifts.

2.24. Calculate the reflection and transmission coefficients, R and T of (2.129) and (2.130), for a sandstone–shale interface when (*a*) $V_{ss} = 2.43$, $V_{sh} = 2.02$ km/s, $\rho_{ss} = 2.08$ and $\rho_{sh} = 2.23$ g/cm^3; (*b*) $V_{ss} = 3.35$, $V_{sh} = 3.14$ km/s, $\rho_{ss} = 2.21$, $\rho_{sh} = 2.52$ g/cm^3; (*c*) what are the corresponding values in nepers and in dB?

2.25. Assume horizontal layering as follows:

Surface S
  $V = 0.60$ km/s, $\rho = 1.45$ g/cm^3, thickness $= 10$ m

Interface A
  $V = 2.40$ km/s, $\rho = 2.35$ g/cm^3, thickness $= 600$ m

Interface B
  $V = 3.20$ km/s, $\rho = 2.68$ g/cm^3, thickness $= 800$ m

Interface C
  $V = 3.40$ km/s, $\rho = 2.70$ g/cm^3

Assuming a shot at the base of the low-velocity layer, calculate the relative amplitudes and energy densities for (*a*) the primary reflections from B and C; (*b*) the multiples BSA, BAB and BSB; (*c*) compare traveltimes, amplitudes and energy densities of these 5 events.

2.26. Show that when the angles in (2.125)–(2.128) are small so that the squares and products are negligible, (2.129) and (2.130) are still valid, and

$$\frac{\mathcal{B}_1}{\mathcal{A}_0} = \frac{2W_2 q + 4Z_1 r}{(W_1 + W_2)(Z_1 + Z_2)}, \quad \frac{\mathcal{B}_2}{\mathcal{A}_0} = \frac{2W_1 q - 4Z_1 r}{(W_1 + W_2)(Z_1 + Z_2)},$$

$$q = (Z_1 \theta_2 - Z_2 \theta_1), \quad r = (W_1 \delta_1 - W_2 \delta_2).$$

2.27. For an SH-wave (an incident SH-component perpendicular to the paper in fig. 2.25) write the boundary conditions and find the amplitudes of all reflected and refracted waves. The absence of P-waves is important in S-wave studies.

2.28. (*a*) Derive Knott's equations and Zoeppritz' equations for a P-wave incident on a liquid–solid interface

when the incident wave is (i) in the liquid; (ii) in the solid. (*b*) Calculate the amplitude of the reflected and transmitted P- and S-waves where an incident P-wave strikes the interface from a water layer ($\alpha = 1.5$ km/s, $\beta = 0$, $\rho = 1.0$ g/cm^3) at $20°$ when the seafloor is (i) 'soft' ($\alpha = 2.0$ km/s, $\beta = 1.0$ km/s, $\rho = 2.0$ g/cm^3); and (ii) 'hard' ($\alpha = 4.0$ km/s, $\beta = 2.5$ km/s, $\rho = 2.5$ g/cm^3). (*c*) Repeat part (*b*) for an angle of incidence of $30°$.

2.29. Derive the Zoeppritz equations for an incident SV-wave.

Fig.2.32. Relation between amplitudes of wave motion and displacement.

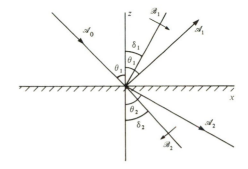

Fig.2.33. Travel paths through a saltdome according to Snell's law. (From Barton, 1929.) (*a*) Vertical section; (*b*) plan view.

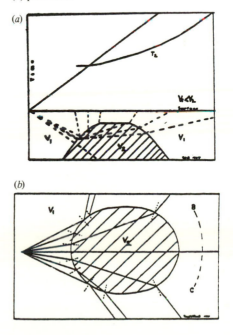

(*a*)

(*b*)

2.30. Show that the maximum amplitude of an incident wave and its reflection at the surface of the ocean occurs at the depth $\lambda/(4\cos\theta)$ where $\theta=$ angle of incidence, by expressing $\mathscr{P}$ in the form used in (2.115) and (2.116) and applying appropriate boundary conditions.

2.31. In early refraction exploration for saltdomes, a 'blind spot' (the region B–C in fig. 2.33) was found when the dome lay directly on the line between the shot and the geophone, that is, the arrivals were often too weak to be detected. This was called 'absorption of the wave' by the saltdome. What is the true explanation for this 'absorption'?

2.32. (*a*) In fig. 2.30*a*, $v_0=40$ Hz; find the water depth using data from the figure; (*b*) find the frequencies which are reinforced when the rays are reflected at angles of 30° and 40° to the vertical; (*c*) calculate θ for successive modes with frequencies 215 and 300 Hz; (*d*) find V for cases (*a*), (*b*) and (*c*). [Ignore the phase change upon reflection at the seafloor.]

3

Geometry of seismic wave paths

Overview

This chapter uses a geometrical-optics approach to derive the basic relationships between traveltime and the locations of reflecting/refracting interfaces; most structural interpretation relies on such an approach.

The accurate interpretation of reflection data requires a knowledge of the velocity at all points along the reflection paths. However, even if we had such a detailed knowledge of the velocity, the calculations would be tedious and often we assume a simple distribution of velocity which is close enough to give useable results. The simplest assumption, which is made in §3.1, is that the velocity is constant between the surface and the reflecting bed. Although this assumption is rarely even approximately true, it leads to simple formulae which give answers which are within the required accuracy in many instances.

The basic problem in reflection seismic surveying is to determine the position of a bed which gives rise to a reflection on a seismic record. In general this is a problem in three dimensions. However, the dip is often very gentle and the direction of profiling is frequently nearly along either the direction of dip or the direction of strike. In such cases a two-dimensional solution is generally used. The arrival time versus offset relation for a plane reflector and constant velocity is hyperbolic. The distance to the reflector can be found from the reflection arrival time at the shotpoint if the velocity is known. The variation of arrival time as a geophone is moved away from the shotpoint, called normal moveout, provides the most important criterion for identifying reflections and a method of determining velocity. The dip is found from differences in arrival times of a reflection at different locations after correction for normal moveout; dip moveout is related to dip and also to the angle of approach of wavefronts at the surface and to apparent velocity. Reflector dip and strike can be found from the components of dip moveout seen at the intersection of seismic lines.

Section 3.2 deals with reflection raypaths where the velocity changes vertically; this results in changes in raypath direction. One solution is to use equivalent average velocity. For parallel velocity layers the effect on the traveltime curve is to give a root-mean-square (rms) velocity. Vertical velocity is often expressed as a function of arrival time or depth. Where velocity is linear with depth, wavefronts are spherical and raypaths are arcs of circles, facts which can be used in graphical plotting of depth sections.

Section 3.3 concerns the geometry of head-wave paths as used in refraction exploration. In most cases we assume a series of beds with the same strike and each with a constant velocity, the velocity increasing as we go to deeper beds, and then we derive formulae relating traveltime, offset, depth, dip and velocities. The cases considered include a single horizontal refractor, several horizontal refractors, a single dipping refractor, and several dipping refractors. Velocities can be found from slopes of the traveltime versus offset curves, depths from the intercepts of projections to the shotpoint, and dip by differences in depth at adjacent shotpoints. Refraction paths in the case of a linear increase in overburden velocity are also considered.

3.1   Geometry of reflection paths for constant velocity
3.1.1   *Horizontal reflector; normal moveout*

The simplest two-dimensional problem is that of zero dip illustrated in the lower part of fig. 3.1. The reflecting bed, AB, is at a depth h below the shotpoint S. Energy leaving S along the direction SC will be reflected in such a direction that the angle of reflection equals the angle of incidence.

Although the reflected ray CR can be determined by laying off an angle equal to α at C, it is easier to make use of the *image point* I which is located on the same normal to the reflector as S and as far below the bed as S is above. If we join I to C and prolong the straight line to R, CR is the reflected ray (since CD is parallel to SI, making all the angles marked α equal).

Denoting the average velocity by V, the traveltime t for the reflected wave is $(SC + CR)/V$. However, $SC = CI$ so that IR is equal in length to the actual path, SCR. Therefore, $t = IR/V$ and in terms of x, the shotpoint-to-

geophone distance (*offset*), we can write

$$V^2 t^2 = x^2 + 4h^2, \tag{3.1}$$

or

$$(V^2 t^2/4h^2) - (x^2/4h^2) = 1. \tag{3.2}$$

Thus the traveltime curve is a hyperbola as shown in the upper part of fig. 3.1.

The geophone at R will also record the *direct wave* which travels along the path SR. Since SR is always less than $(SC + CR)$, the direct wave arrives first. The traveltime is $t_D = x/V$ and the traveltime curves are the straight lines OM and ON passing through the origin with slopes of $\pm 1/V$.

When the distance x becomes very large the difference between SR and $(SC + CR)$ becomes small and the reflection traveltime approaches the direct wave traveltime asymptotically.

The location of the reflecting bed is determined by measuring t_0, the traveltime for a geophone at the shot-

Fig.3.1. Traveltime curve for horizontal reflector.

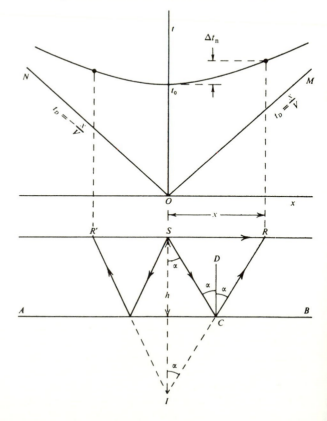

point. Setting $x = 0$ in (3.1) we see that

$$h = \tfrac{1}{2}Vt_0. \tag{3.3}$$

Equation (3.1) can be written

$$t^2 = (x^2/V^2) + (4h^2/V^2) = (x^2/V^2) + t_0^2. \tag{3.4}$$

If we plot t^2 against x^2 (instead of t versus x as in fig. 3.1), we obtain a straight line of slope $(1/V^2)$ and intercept t_0^2. This forms the basis of a well-known scheme for determining V, the 'X^2–T^2 method'; this will be described in §7.3.3a.

We can solve (3.1) for t, the traveltime measured on the seismic record. Generally $2h$ is appreciable larger than x so that we can use a binomial expansion as follows:

$$t = (2h/V)\{1 + (x/2h)^2\}^{\frac{1}{2}} = t_0\{1 + (x/Vt_0)^2\}^{\frac{1}{2}}$$
$$= t_0\{1 + \tfrac{1}{2}(x/Vt_0)^2 - \tfrac{1}{8}(x/Vt_0)^4 + \dots\}. \tag{3.5}$$

If t_1, t_2, x_1 and x_2 are two traveltimes and offsets, we have to the first approximation

$$\Delta t = t_2 - t_1 \approx (x_2^2 - x_1^2)/2V^2 t_0. \tag{3.6}$$

In the special case where one geophone is at the shotpoint, Δt is known as the *normal moveout* (NMO) which we shall denote by Δt_n. Then,

$$\Delta t_n \approx x^2/2V^2 t_0. \tag{3.7}$$

At times we retain another term in the expansion (see also problem 3.1c):

$$\Delta t_n \approx (x^2/2V^2 t_0) - (x^4/8V^4 t_0^3)$$
$$= (x^2/2V^2 t_0)\{1 - (x/4h)^2\}. \tag{3.8}$$

From (3.7) we note that the normal moveout increases as the square of the offset x, inversely as the square of the velocity, inversely as the first power of the traveltime (or depth – see (3.3)). Thus reflection curvature increases rapidly as we go to more distant geophones, at the same time the curvature becoming progressively less with increasing record time.

The concept of normal moveout is extremely important. It is the principal criterion by which we decide whether an event observed on a seismic record is a reflection or not. If the normal moveout differs from the value given by (3.7) by more than the allowable experimental error, we are not justified in treating the event as a reflection. One of the most important quantities in seismic interpretation is the change in arrival time caused by dip; to find this quantity we must eliminate normal moveout. Normal moveout must also be eliminated before 'stacking' (adding together) common-depth-point records (see §5.3.1). Finally, (3.7) can be used to find V by measuring

x, t_0 and Δt_n; this forms the basis of the T–ΔT method of finding velocity (see §7.3.3b) and also of velocity analysis (§8.2.3). Brown (1969) discusses refinements to handle dip and long offset.

3.1.2 *Dipping reflector ; dip moveout*

When the bed is dipping in the direction of the profile, we have the situation shown in fig. 3.2, ξ being the dip and h the distance normal to the bed. To draw the raypath for the reflection arriving at the geophone R, we join the image point I to R by a straight line cutting the bed at C. The path is then SCR and t is equal to $(SC + CR)/V$; since $(SC + CR) = IR$, application of the cosine law to the triangle SIR gives

$$V^2 t^2 = IR^2$$
$$= x^2 + 4h^2 - 4hx\cos(\tfrac{1}{2}\pi + \xi)$$
$$= x^2 + 4h^2 + 4hx\sin\xi. \tag{3.9}$$

On completing the squares, we obtain

$$\frac{V^2 t^2}{(2h\cos\xi)^2} - \frac{(x + 2h\sin\xi)^2}{(2h\cos\xi)^2} = 1.$$

Thus, as before the traveltime curve is a hyperbola but the axis of symmetry is now the line $x = -2h\sin\xi$ instead of the t-axis. This means that t has different values for geophones symmetrically placed on opposite sides of the shotpoint, unlike the case for zero dip.

Fig. 3.2. Traveltime curve for dipping reflector.

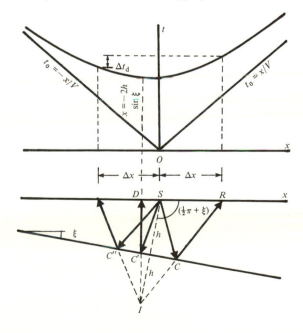

Setting x equal to 0 in (3.9) gives the same value for h as in (3.3); note, however, that h is not measured vertically as it was in the earlier result. We call the points C, C', C'' in fig. 3.2, where the angles of incidence and reflection are equal, *reflecting points*. (These are sometimes called 'depth points' but this term is also used for the point on the surface midway between source and receiver; we call the latter a *midpoint* and to avoid confusion we shall avoid the term 'depth point'.) The updip displacement of reflecting points compared to midpoints for dipping reflectors is important in migrating data (§5.6.3) and in the common-depth-point method (§5.3.1).

To obtain the dip ξ, we solve for t in (3.9) by assuming that $2h$ is greater than x and expanding as in the derivation of (3.5). Then

$$t = \frac{2h}{V}\left(1 + \frac{x^2 + 4hx\sin\xi}{4h^2}\right)^{\frac{1}{2}}$$

$$\approx t_0\left(1 + \frac{x^2 + 4hx\sin\xi}{8h^2}\right) \qquad (3.10)$$

using only the first term of the expansion. The simplest method of finding ξ is from the difference in traveltimes for two geophones equally distant from, and on opposite sides of, the shotpoint. Letting x have the values $+\Delta x$ for the downdip geophone and $-\Delta x$ for the updip geophone and denoting the equivalent traveltimes by t_1 and t_2, we get

$$t_1 \approx t_0\left(1 + \frac{(\Delta x)^2 + 4h\,\Delta x\sin\xi}{8h^2}\right),$$

$$t_2 \approx t_0\left(1 + \frac{(\Delta x)^2 - 4h\,\Delta x\sin\xi}{8h^2}\right),$$

$$\Delta t_d = t_1 - t_2 \approx t_0\left(\frac{\Delta x\sin\xi}{h}\right) \approx \frac{2\Delta x}{V}\sin\xi.$$

The dip ξ is given by

$$\sin\xi \approx \tfrac{1}{2}V\left(\frac{\Delta t_d}{\Delta x}\right). \qquad (3.11)$$

The quantity $\Delta t_d/\Delta x$ is called the *dip moveout*. (Note that dimensionally dip moveout is time/distance whereas normal moveout is time.) For small angles, ξ is approximately equal to $\sin\xi$ so that the dip is directly proportional to Δt_d under these circumstances. To obtain the dip as accurately as possible, we use as large a value of Δx as the data quality permits; for symmetrical spreads (§5.3.2), we measure dip moveout between the geophone groups at the opposite ends of the spread, Δx then being half the spread length (fig. 3.2).

Dip moveout can also be measured by the time

difference between t_0 at different shotpoints. As shown in fig. 3.3, $\Delta t_d = t_{01} - t_{02}$ and

$$\sin\xi = \tfrac{1}{2}V\left(\frac{\Delta t_d}{\Delta x}\right), \qquad (3.12)$$

where Δx is the distance between shotpoints. When we measure dip on a record section (§5.6.3), Δx is the distance between any two convenient points.

It should be noted that normal moveout was eliminated in the derivation of (3.11). The terms in $(\Delta x)^2$ which disappeared in the subtraction represent the normal moveout.

Fig. 3.4 illustrates diagrammatically the relation between normal moveout and dip moveout. The diagram at the left represents a reflection from a dipping bed; the alignment is curved and unsymmetrical about the shotpoint. Diagram (B) shows what would have been observed if the bed had been horizontal; the alignment is curved symmetrically about the shotpoint position owing to the normal moveout. The latter ranges from 0 to 13 ms (1 millisecond $= 10^{-3}$ s $=$ 1 ms, the unit of time commonly used in seismic work) at an offset of 400 m. Diagram (C) was obtained by subtracting the normal moveouts shown in (B) from the arrival times in (A). The resulting alignment shows the effect of dip alone; it is straight and has a time difference between the outside curves of 10 ms, that is, $\Delta t_d = 10$ ms when $\Delta x = 400$ m. Thus, we find that the dip is $2500(10 \times 10^{-3}/800) = 0.031$ rad $= 1.8°$.

The method of normal-moveout removal illustrated in fig. 3.4c was used to demonstrate the difference between normal moveout and dip moveout. If we require only the

Fig.3.3. Geometry involved in dip moveout measured between shotpoints or on record sections.

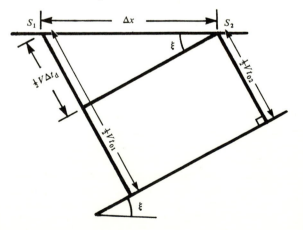

dip moveout Δt_d we merely subtract the traveltimes for the two outside geophones in (A).

Frequently we do not have a symmetrical spread and we find the dip moveout by removing the effect of normal moveout. As an example, refer to fig. 3.4 curve (D), which shows a reflection observed on a spread extending from $x = -133$ m to $x = +400$ m. Let $t_0 = 1.225$ s, $t_1 = 1.223$ sec, $t_2 = 1.242$ s, $V = 2800$ m/s. From (3.7) we get for Δt_n at offsets of 133 and 400 m respectively the values 1 ms and 8 ms (rounded off to the nearest millisecond since this is usually the precision of measurement on seismic records). Subtracting these values we obtain for the corrected arrival times $t_1 = 1.222$, $t_2 = 1.234$; hence the dip moveout is 12 ms/(533 m/2). The corresponding dip is $\xi = 2800(12 \times 10^{-3}/533) = 0.063$ rad $= 3.6°$.

An alternative to the above method is to use the arrival times at $x = -133$ m and $x = +133$ m, thus obtaining a symmetrical spread and eliminating the need for calculating normal moveout. However, doing this would decrease the effective spread length from 533 m to 266 m and thereby reduce the accuracy of the ratio ($\Delta t_d/\Delta x$).

The *apparent velocity* V_a of a wavefront is the ratio of the distance (Δx) between two points on a surface (usually the surface of the ground) to the difference in arrival times (Δt) for the same event at the two points. It is given by

$$V_a = \Delta x/\Delta t = V/\sin \alpha, \qquad (3.13)$$

where α is the *angle of approach* (fig. 3.5); α is sometimes called *apparent dip*. This equation is somewhat similar to (3.11) and (3.12) but it has different significance, since it gives the direction of travel of a plane wave as it reaches the spread, V being the average velocity between C and the surface. In (3.11) and (3.12) V is the equivalent velocity down to the reflector and ξ is the angle of dip. Because $\sin \alpha$ can be very small, V_a can be very large and for energy approaching vertically, $V_a = \infty$.

3.1.3 *Cross-dip*

When the profile is at an appreciable angle to the direction of dip, the determination of the latter becomes a three-dimensional problem and we use the methods of solid analytical geometry. In fig. 3.6 we take the xy-plane

Fig.3.4. Relation between normal moveout and dip moveout. For curves (A), (B) and (C), $t_0 = 1.000$ s, $V = 2500$ m/s. For curve (D), $t_0 = 1.225$ s, $t_1 = 1.223$ s, $t_2 = 1.242$ s, $V = 2800$ m/s.

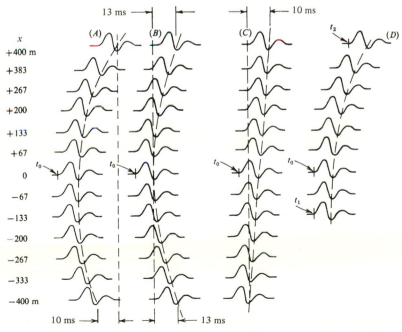

Fig.3.5. Finding the angle of approach of a wave.

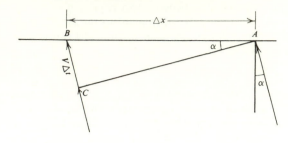

as horizontal with the z-axis extending vertically downward. The line OP of length h is perpendicular to a dipping plane bed which outcrops (that is, intersects the xy-plane) along the line MN if extended sufficiently.

We write θ_1, θ_2, θ_3 for the angles between OP and x-, y- and z-axes, l, m, n for the direction cosines of OP. The angle Ξ between MN and the x-axis is the direction of strike of the bed while $\theta_3 = \xi$, the angle of dip.

The path of a reflected wave arriving at a geophone R on the x-axis can be found using the image point I. The line joining I to R cuts the reflector at Q, hence OQR is the path. Since $OQ = QI$, the line IR is equal to Vt, t being the traveltime for the geophone at R. The coordinates of I and R are respectively $(2hl, 2hm, 2hn)$ and $(x, 0, 0)$, hence we have

$$V^2t^2 = (IR)^2$$
$$= (x - 2hl)^2 + (0 - 2hm)^2 + (0 - 2hn)^2$$
$$= x^2 + 4h^2(l^2 + m^2 + n^2) - 4hlx$$
$$= x^2 + 4h^2 - 4hlx,$$

Fig.3.6. Three-dimensional view of a reflection path for a dipping bed.

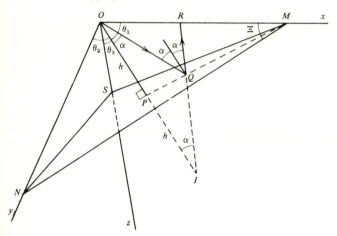

since $l^2 + m^2 + n^2 = 1$.

When $x = 0$, we obtain the same relation between h and t_0 as in (3.3). Proceeding as in the derivation of (3.10) we get for the approximate value of t,

$$t \approx t_0\left(1 + \frac{x^2 - 4hlx}{8h^2}\right).$$

By subtracting the arrival times at two geophones located on the x-axis at $x = \pm\Delta x$, we find

$$\Delta t_x \approx t_0(l\Delta x/h)$$
$$\approx 2l\Delta x/V,$$
$$l = \cos\theta_1 \approx \tfrac{1}{2}V\left(\frac{\Delta t_x}{\Delta x}\right). \tag{3.14}$$

If we also have a spread along the y-axis (*cross-spread*), we get

$$m = \cos\theta_2 \approx \tfrac{1}{2}V\left(\frac{\Delta t_y}{\Delta y}\right), \tag{3.15}$$

where Δt_y is the time difference ('cross-dip') between geophones a distance $2\Delta y$ apart and symmetrical about the shotpoint. Since

$$n = \cos\xi = \{1 - (l^2 + m^2)\}^{\frac{1}{2}},$$
$$\sin\xi = (1 - n^2)^{\frac{1}{2}} = (l^2 + m^2)^{\frac{1}{2}}$$
$$= \tfrac{1}{2}V\left\{\left(\frac{\Delta t_x}{\Delta x}\right)^2 + \left(\frac{\Delta t_y}{\Delta y}\right)^2\right\}^{\frac{1}{2}}. \tag{3.16}$$

Fig.3.7. Determination of strike.

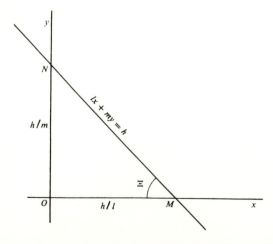

The components of dip moveout, $\Delta t_x/\Delta x$ and $\Delta t_y/\Delta y$, are also called *apparent dips*.

   To find the strike Ξ, we start from the equation of a plane (that is, the reflector) which has a perpendicular from the origin of length h and direction cosines (l, m, n), namely

$$lx + my + nz = h.$$

Setting $z = 0$ gives the equation of the line of intersection of the reflector and the surface; this strike line has the equation

$$lx + my = h.$$

The intercepts of this line on the x- and y-axes are h/l and h/m. Referring to fig. 3.7, we find that

$$\tan \Xi = \frac{h/m}{h/l} = \frac{l}{m}$$

$$= \frac{(\Delta t_x/\Delta x)}{(\Delta t_y/\Delta y)}. \qquad (3.17)$$

   Consider the case where the profile lines are not perpendicular, for example, where they are in the $\mathbf{r}_1$ and $\mathbf{r}_2$ directions of fig. 3.8a and the dip is in the $\mathbf{r}_0$ direction. We express the dip moveout as the vector $(\mathrm{d}t/\mathrm{d}x)\mathbf{r}_0 = \mathbf{AO}$; the component of dip moveout on the line in the $\mathbf{r}_2$ direction is thus $(\mathrm{d}t/\mathrm{d}x)\mathbf{r}_0 \cdot \mathbf{r}_2 = (\mathrm{d}t/\mathrm{d}x)\cos\beta = |\mathbf{OB}|$ (see problem 3.2a). The converse problem of finding the total dip moveout from measurements of the components of dip moveout OB and OC can be done graphically as shown in fig. 3.8b (see also problem 3.2b), or mathematically as follows. We take one profile along the x-axis and the other along the y'-axis at an angle α to the x-axis. Taking the length of a symmetrical spread along the y'-axis as $2\Delta y'$, the coordinates of the ends of the spread (relative to the x-, y-axes) are $\pm\Delta y'\cos\alpha$, $\pm\Delta y'\sin\alpha$. Then

$$V^2 t_{\pm}^2 = (2hl \pm \Delta y'\cos\alpha)^2 + (2hm \pm \Delta y'\sin\alpha)^2 + (2hn)^2$$

$$= (\Delta y')^2 + 4h^2 \pm 4h\Delta y'(l\cos\alpha + m\sin\alpha).$$

The dip moveout along this line, $\Delta t'/\Delta y'$, is

$$\Delta t'/\Delta y' = 2(l\cos\alpha + m\sin\alpha)/V. \qquad (3.18)$$

Since α is known, l can be found from $\Delta t/\Delta x$ and m from (3.18).

3.2   Vertical velocity gradient and raypath curvature
3.2.1   *Effect of velocity variation*

   The assumption of constant velocity is not valid in general, the velocity usually changing as we go from one point to another. In petroleum exploration we are usually dealing with more-or-less flat-lying bedding and the

changes in seismic velocity as we move horizontally are for the most part small, being the result of slow changes in density and elastic properties of the beds. These horizontal variations are generally much less rapid than the variations in the vertical direction where we are going from bed to bed with consequent lithological changes and increasing pressure with increasing depth. Because the horizontal changes are gradual they can often be taken into account by dividing the survey area into smaller areas within each of which the horizontal variations can be ignored and the same vertical velocity distribution used. Such areas are often large enough to include several structures of the size of interest in oil exploration so that

Fig.3.8. Determining dip and strike from non-perpendicular observations. (*a*) Relation between the point of observation O and the reflecting point A (A is always updip from O); (*b*) example of graphical solution.

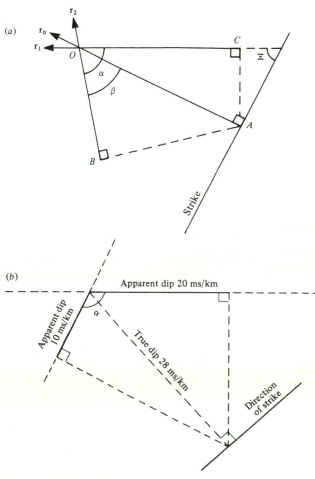

changes from one velocity function to another do not necessarily impose a serious burden upon the interpreter.

3.2.2 *Equivalent average velocity*

Vertical variations in velocity can be taken into account in various ways . One of the simplest is to use a modification of the constant velocity model. We assume that the actual section existing between the surface and a certain reflecting horizon can be replaced with an equivalent single layer of constant velocity $\overline{V}$ equal to the average velocity between the surface and the reflecting horizon; $\overline{V}$ is the *equivalent average velocity*. This velocity is usually given as a function of depth (or of t_0, which is nearly the same except when the dip is large). Thus the section is assigned a different constant velocity for each of the reflectors below it. Despite this inconsistency the method is useful and is extensively applied. The variation of the average velocity with t_0 is found using one of the methods described in §7.3. For the observed values of the arrival time t_0 we select the average velocity $\overline{V}$ corresponding to

Fig.3.9. Derivation of formula for rms velocity $\overline{\overline{V}}$ in two-layer medium. (*a*) Reflection path; (*b*) $X^2 - T^2$ curve.

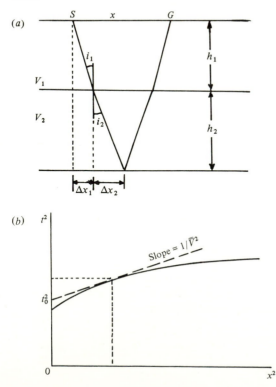

(*a*)

(*b*)

this reflector; using the values of t_0, the dip moveout $\Delta t_{\rm d}$ and $\overline{V}$, we calculate the depth h and the dip ξ using (3.3) and (3.11).

3.2.3 *Velocity layering*

A method which is commonly used to take into account velocity variations is to replace the actual velocity distribution with an approximate one corresponding to a number of horizontal layers of different velocities, the velocity being constant within each layer. Simple equations such as (3.3) and (3.11) are no longer appropriate because rays are bent at each interface. A graphical method can be used to find the depth and dip. The method uses a 'wavefront chart'; the preparation and use of these charts will be described in §5.6.3. In effect this method replaces the actual raypaths with a series of line segments which are straight within each layer but undergo abrupt changes in direction at the boundaries between layers. Dix (1955) shows that the effect of this on (3.4) is to replace the average velocity $\overline{V}$ by its rms (root-mean-square) value, $\overline{\overline{V}}$. His derivation is as follows. Referring to fig. 3.9*b*, the $X^2 - T^2$ curve for the reflection shown in fig. 3.9*a* is curved and for a given offset x we write the equation of the tangent as

$$t^2 = x^2/\overline{\overline{V}}^2 + \overline{\overline{t}}_0^2 ;$$

hence

$$\mathrm{d}t/\mathrm{d}x = x/\overline{\overline{V}}^2 t. \tag{3.19}$$

The angle of approach, i_1, is given by

$$\sin i_1 = V_1 \frac{\mathrm{d}t}{\mathrm{d}x} = \frac{V_1 x}{\overline{\overline{V}}^2 t}, \tag{3.20}$$

using (3.19). Also, writing Δt_i for the one-way vertical traveltime through the ith bed, and keeping x small, we have

$$\begin{aligned}
\tfrac{1}{2}x &= \Delta x_1 + \Delta x_2 = h_1 \tan i_1 + h_2 \tan i_2 \\
&\approx V_1 \Delta t_1 \sin i_1 + V_2 \Delta t_2 \sin i_2 \\
&\approx (V_1^2 \Delta t_1 + V_2^2 \Delta t_2) \sin i_1 / V_1 \\
&\approx (V_1^2 \Delta t_1 + V_2^2 \Delta t_2)(x/\overline{\overline{V}}^2 t)
\end{aligned}$$

from (3.20). Since $t \approx 2(\Delta t_1 + \Delta t_2)$, we get

$$\overline{\overline{V}}^2 \approx \sum_{i=1}^{2} V_i^2 \Delta t_i \bigg/ \sum_{i=1}^{2} \Delta t_i .$$

This equation can be generalized for n horizontal beds, giving

$$t^2 \approx x^2/\overline{\overline{V}}^2 + \overline{\overline{t}}_0^2, \tag{3.21}$$

$$\overline{V^2} \approx \sum_{i=1}^{n} V_i^2 \Delta t_i \Big/ \sum_{i=1}^{n} \Delta t_i. \tag{3.22}$$

(See Shah and Levin, 1973, for higher-order approximations necessary to give more accuracy for large values of x.)

3.2.4   Velocity functions

At times the assumption is made that the velocity varies in a systematic continuous manner and therefore can be represented by a velocity function. The actual velocity usually varies extremely rapidly over short intervals, as shown by sonic logs (see §7.3.2); however, if we integrate these changes over distances of a wavelength or so (30–100 m), we obtain a function which is generally smooth except for discontinuities at marked lithological changes. If the velocity discontinuities are small, we are often able to represent the velocity distribution with sufficient accuracy by a smooth velocity function. The path of a wave traveling in such a medium is then determined by two integral equations.

To derive the equations, we assume that the medium is divided into a large number of thin beds in each of which the velocity is constant; on letting the number of beds go to infinity, the thickness of each bed becomes infinitesimal and the velocity distribution becomes a continuous function of depth. Referring to fig. 3.10, we have for the nth bed

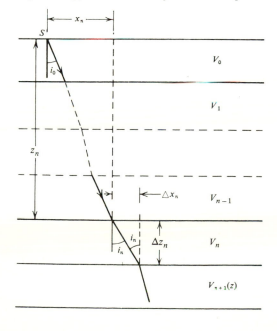

Fig.3.10. Raypath where velocity varies with depth.

$$\frac{\sin i_n}{V_n} = \frac{\sin i_0}{V_0} = p,$$

$$V_n = V_n(z),$$

$$\Delta x_n = \Delta z_n \tan i_n,$$

$$\Delta t_n = \frac{\Delta z_n}{V_n \cos i_n}.$$

The raypath parameter p (see (2.103)) is a constant which depends upon the direction in which the ray left the shotpoint, that is, upon i_0.

In the limit when n becomes infinite, we get

$$\frac{\sin i}{V} = \frac{\sin i_0}{V_0} = p, \quad V = V(z), \tag{3.23}$$

$$\frac{dx}{dz} = \tan i, \quad \frac{dt}{dz} = \frac{1}{V \cos i},$$

$$x = \int_0^z \tan i \, dz, \quad t = \int_0^z \frac{dz}{V \cos i};$$

hence

$$\left. \begin{aligned} x &= \int_0^z \frac{pV \, dz}{\{1 - (pV)^2\}^{\frac{1}{2}}}, \\ t &= \int_0^z \frac{dz}{V\{1 - (pV)^2\}^{\frac{1}{2}}}. \end{aligned} \right\} \tag{3.24}$$

Since V is a function of z, (3.24) furnishes two integral equations relating x and t to the depth z. These equations can be solved by numerical methods when we have a table of values of V at various depths.

3.2.5   Linear increase of velocity with depth

Sometimes we can express V as a continuous function of z and integrate (3.24). One case of considerable importance is that of a linear increase of velocity with depth, namely

$$V = V_0 + az,$$

where V_0 is the velocity at the horizontal datum plane, V the velocity at a depth z below the datum plane and a is a constant whose value is generally between 0.3 and 1.3/s.

If we introduce a new variable $u = pV = \sin i$, then $du = p \, dV = pa \, dz$, and we can solve for x and t as follows (p is the raypath parameter):

$$x = \frac{1}{pa} \int_{u_0}^{u} \frac{u \, du}{(1 - u^2)^{\frac{1}{2}}} = \frac{1}{pa} (1 - u^2)^{\frac{1}{2}} \Big|_{u}^{u_0} = \frac{1}{pa} \cos i \Big|_{i}^{i_0}$$

$$= \frac{1}{pa} (\cos i_0 - \cos i), \tag{3.25}$$

$$t = \frac{1}{a}\int_{u_0}^{u}\frac{du}{u(1-u^2)^{\frac{1}{2}}} = \frac{1}{a}\ln\left\{\frac{u}{1+(1-u^2)^{\frac{1}{2}}}\right\}\bigg|_{u_0}^{u}$$

$$= \frac{1}{a}\ln\left[\frac{\sin i}{\sin i_0}\left(\frac{1+\cos i_0}{1+\cos i}\right)\right] = \frac{1}{a}\ln\left(\frac{\tan\frac{1}{2}i}{\tan\frac{1}{2}i_0}\right); \quad (3.26)$$

hence,

$$i = 2\tan^{-1}(e^{at}\tan\tfrac{1}{2}i_0), \tag{3.27}$$

$$z = (V - V_0)/a = (\sin i - \sin i_0)/pa. \tag{3.28}$$

The parametric equations (3.25) and (3.28) give the coordinates x and z, the parameter i being related to the one-way traveltime t by (3.26) or (3.27).

The raypath given by (3.25) and (3.28) is a circle; this can be shown by calculating the radius of curvature ρ which turns out to be a constant:

$$\rho = (1 + x'^2)^{\frac{3}{2}}/x'',$$

where

$$x' = \frac{dx}{dz} = \tan i, \text{ using (3.25), (3.28),}$$

$$x'' = \frac{d^2x}{dz^2} = \frac{d}{di}(\tan i)\frac{di}{dz} = \sec^2 i\frac{di}{dz}$$

$$= pa\sec^3 i, \text{ using (3.28).}$$

Hence

$$\rho = \frac{(1 + \tan^2 i)^{\frac{3}{2}}}{pa\sec^3 i} = \frac{1}{pa} = \left(\frac{V_0}{a}\right)\frac{1}{\sin i_0} = \text{constant.}$$

Fig. 3.11 shows a ray leaving the shotpoint at the angle i_0. The center, O, of the circular ray lies above the surface a distance $\rho\sin i_0$, that is, V_0/a. Since this is independent of i_0, the centers of all rays lie on the same horizontal line. This line is located where the velocity would be zero if the velocity function were extrapolated up into the air (since $z = -V_0/a$ at this elevation).

To determine the shape of the wavefront, we make use of fig. 3.12. The raypaths SA and SB are circular arcs with centers O_1 and O_2 respectively. If we continue the arcs upwards to meet the vertical through S at the point S', the line O_1O_2 bisects $S'S$ at right angles. Next we select any point C on the downward extension of $S'S$ and draw the tangents to the two arcs, CA and CB. From plane geometry we know that the square of the length of a tangent to a circle from an external point (for example, CA^2) is equal to the product of the two segments of any chord drawn from the same point ($CS \cdot CS'$ in fig. 3.12). Using both circles we see that

$$CS \cdot CS' = CA^2 = CB^2,$$

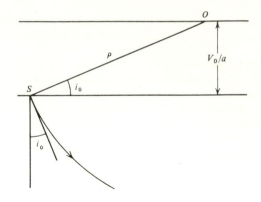

Fig.3.11. Circular ray leaving the shotpoint at the angle i_0.

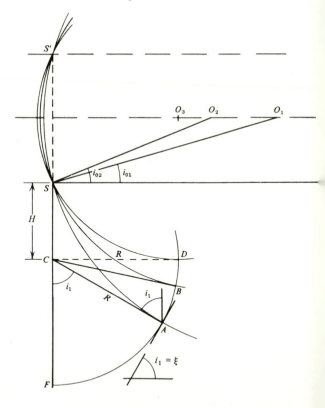

Fig.3.12. Construction of wavefronts for linear increase of velocity.

hence $CA = CB$. Thus a circle with center C and radius $R = CA$ cuts the two raypaths at right angles. Since SA and SB can be any raypaths and a wavefront is a surface which meets all rays at right angles, the circle with center C must be the wavefront which passes through A and B. Even though the arc SA is longer than SB, the greater path length is exactly compensated for by the higher velocity at the greater depth of the raypath SA.

We can draw the wavefront for any value of t if we can obtain the values of H and R in fig. 3.12. Thus, the quantities H and R are equal to the values of z and x for a ray which has $i = \frac{1}{2}\pi$ at the time t, that is, SD in the diagram. Substitution of $i = \frac{1}{2}\pi$ in (3.25), (3.27) and (3.28) yields

$$\tan \tfrac{1}{2} i_0 = e^{-at}, \ \sin i_0 = \operatorname{sech} at, \ \cos i_0 = \tanh at,$$

$$\left.\begin{aligned}
H &= (1/pa)(1 - \sin i_0) \\
&= (V_0/a)\{(1/\sin i_0) - 1\} \\
&= (V_0/a)(\cosh at - 1), \\
R &= (1/pa)\cos i_0 = (V_0/a)\cot i_0 \\
&= (V_0/a)\sinh at.
\end{aligned}\right\} \tag{3.29}$$

Equation (3.29) shows that the center of the wavefront moves downward and the radius becomes larger as time increases.

Field measurements yield values of the arrival time at the shotpoint t_0 and angle of approach $\Delta t/\Delta x$. Since the ray which returns to the shotpoint must have encountered a reflecting horizon normal to the raypath and retraced its path back to the point of origin, the dip is equal to the angle i_1 at the time $t = \frac{1}{2}t_0$. Thus, to locate the segment of reflecting horizon corresponding to a set of values of t_0 and $\Delta t/\Delta x$, we make the following calculations:

$(a)\ t = \tfrac{1}{2}t_0,\quad (b)\ i_0 = \sin^{-1}\left(V_0 \dfrac{\Delta t}{\Delta x}\right),$

$(c)\ i_1 = 2\tan^{-1}(e^{at}\tan\tfrac{1}{2}i_0),$

$(d)\ H = (V_0/a)(\cosh at - 1),$

$(e)\ R = (V_0/a)\sinh at.$

With these values we find C, lay off the radius R at the angle i_1, and draw the reflecting segment perpendicular to the radius as shown at the point A in fig. 3.12. This method is easily adapted to a simple plotting machine (Daly, 1948) or to wavefront charts (Agocs, 1950).

Refraction studies involving linear increase of overburden velocity are discussed in §3.3.5.

3.3   Geometry of refraction paths
3.3.1   *Single horizontal refractor*

Refraction seismology involves the study of head waves, which have been introduced in §2.4.7. In the case of a single horizontal refracting horizon, we can readily derive a formula expressing the arrival time in terms of the offset, the depth and the velocities. In fig. 3.13 the lower part shows a horizontal plane refractor separating two beds of velocities V_1 and V_2, where $V_2 > V_1$. For a geophone at R, the path of the refracted wave is $OMPR$, Θ being the critical angle. The traveltime t can be written

$$\begin{aligned}
t &= \frac{OM}{V_1} + \frac{MP}{V_2} + \frac{PR}{V_1} = \frac{MP}{V_2} + 2\frac{OM}{V_1} \\
&= \frac{x - 2h\tan\Theta}{V_2} + \frac{2h}{V_1\cos\Theta} \\
&= \frac{x}{V_2} + \frac{2h}{V_1\cos\Theta}\left(1 - \frac{V_1}{V_2}\sin\Theta\right) \\
&= \frac{x}{V_2} + \frac{2h\cos\Theta}{V_1},
\end{aligned} \tag{3.30}$$

where we have used the relation $\sin\Theta = V_1/V_2$ in the last step. This equation can also be written

$$t = (x/V_2) + t_1, \tag{3.31}$$

Fig. 3.13. Relation between reflection and refraction raypaths and traveltime curves.

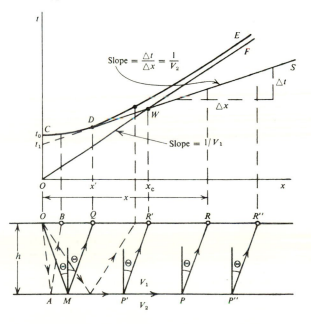

where

$$t_1 = (2h\cos\Theta)/V_1,$$

or

$$h = \tfrac{1}{2}V_1 t_1/\cos\Theta. \qquad (3.32)$$

Obviously the head wave will not be observed at offsets less than the *critical distance*, OQ in fig. 3.13; writing x' for the critical distance,

$$x' = OQ = 2h\tan\Theta = 2h\tan\{\sin^{-1}(V_1/V_2)\}$$
$$= 2h\{(V_2/V_1)^2 - 1\}^{-\frac{1}{2}}. \qquad (3.33)$$

The relation between x'/h and V_2/V_1 is shown in fig. 3.14. As the ratio V_2/V_1 increases, x' decreases. When V_2/V_1 equals 1.4, x' is equal to $2h$. As a rule of thumb, offsets should be greater than twice the depth to the refractor to observe refractions without undue interference from shallower head waves.

Equations (3.30) and (3.31) represent a straight line of slope $1/V_2$ and *intercept time* t_1. This is illustrated in fig. 3.13 where OMQ, $OMP'R'$, $OMPR$ and $OMP''R''$ are a series of refraction paths and DWS the corresponding time–distance curve. Note that this straight-line equation does not have physical meaning for offsets less than x' since the refracted wave does not exist for such values of x; nevertheless we can project the line back to the time axis to find t_1.

The problem to be solved usually is to find the depth h and the two velocities V_1 and V_2. The slope of the direct-wave time–distance curve is the reciprocal of

Fig. 3.14. Relation between critical distance x', crossover distance x_c and velocity contrast.

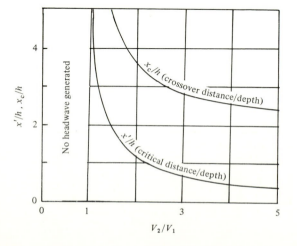

V_1 and the same measurement for the refraction event gives V_2. We can then calculate the critical angle Θ from the relation $\Theta = \sin^{-1}(V_1/V_2)$, and use the intercept time, t_1, to calculate h from (3.32).

In fig. 3.13 the time–distance curves for the reflection from the interface AP'' and for the direct path are represented by the hyperbola CDE and the straight line OF, respectively. Since the path OMQ can be regarded either as a reflection or as the beginning of the refracted wave, the reflection and refraction time–distance curves must coincide at $x = x'$, that is, at the point D. Moreover, differentiating (3.1) to obtain the slope of the reflection time–distance curve at $x = x'$, we find

$$\left[\frac{dt}{dx}\right]_{x=x'} = \left[\frac{x}{V_1^2 t}\right]_{x=x'} = \frac{1}{V_1}\left[\frac{OQ}{OM+MQ}\right]$$
$$= \frac{1}{V_1}\left(\frac{\tfrac{1}{2}OQ}{OM}\right) = \frac{1}{V_1}\sin\Theta = \frac{1}{V_2}.$$

We see therefore that the reflection and refraction curves have the same slope at D, and consequently the refraction curve is tangent to the reflection curve at $x = x'$.

Comparing reflected and refracted waves from the same horizon and arriving at the same geophone, we note that the refraction arrival time is always less than the reflection arrival time (except at D). The intercept time t_1 for the refraction is less than the arrival time t_0 for the reflection at the shotpoint, since

$$t_1 = (2h/V_1)\cos\Theta, \quad t_0 = (2h/V_1); \quad \text{hence } t_1 < t_0.$$

Starting at the point Q, we see that the direct wave arrives ahead of the reflected and refracted waves since its path is the shortest of the three. However, part of the refraction path is traversed at velocity V_2, so that as x increases, eventually the refraction wave will overtake the direct wave. In fig. 3.13 these two traveltimes are equal at the point W. If the offset corresponding to W is x_c, we have

$$\frac{x_c}{V_1} = \frac{x_c}{V_2} + \frac{2h}{V_1}\cos\Theta$$

$$\therefore h = \frac{x_c}{2}\left(1 - \frac{V_1}{V_2}\right)\bigg/\cos\Theta$$

$$= \frac{x_c}{2}\left(\frac{V_2-V_1}{V_2}\right)\frac{V_2}{(V_2^2-V_1^2)^{\frac{1}{2}}}$$

$$h = \frac{x_c}{2}\left(\frac{V_2-V_1}{V_2+V_1}\right)^{\frac{1}{2}}. \qquad (3.34)$$

This relation is sometimes used to find h from measurements of the velocities and the *crossover distance* x_c. However, usually we can determine t_1 more accurately than x_c

and hence (3.32) provides a better method of determining h. The relation between x_c/h and V_2/V_1 is shown in fig. 3.14.

3.3.2   *Several horizontal refractors*

Where all layers are horizontal, (3.30) can be generalized to cover the case of more than one refracting horizon. Consider the situation in fig. 3.15 where we have three layers of velocities, V_1, V_2 and V_3. Whenever $V_2 > V_1$, we have the refraction path $OMPR$ and corresponding time–distance curve WS, just as we had in fig. 3.13. If $V_3 > V_2 > V_1$, travel by a refraction path in V_3 will eventually overtake the refraction in V_2. The refraction paths such as $OM'M''P''P'R'$ are fixed by Snell's law:

$$\frac{\sin\theta_1}{V_1} = \frac{\sin\Theta_2}{V_2} = \frac{1}{V_3},$$

where Θ_2 is the critical angle for the lower horizon while θ_1 is less than the critical angle for the upper horizon. The expression for the traveltime curve ST is obtained as before:

$$
\begin{aligned}
t &= \frac{OM' + R'P'}{V_1} + \frac{M'M'' + P'P''}{V_2} + \frac{M''P''}{V_3} \\
&= \frac{2h_1}{V_1 \cos\theta_1} + \frac{2h_2}{V_2 \cos\Theta_2} + \frac{x - 2h_1 \tan\theta_1 - 2h_2 \tan\Theta_2}{V_3} \\
&= \frac{x}{V_3} + \frac{2h_2}{V_2 \cos\Theta_2}\left(1 - \frac{V_2}{V_3}\sin\Theta_2\right) \\
&\quad + \frac{2h_1}{V_1 \cos\theta_1}\left(1 - \frac{V_1}{V_3}\sin\theta_1\right) \\
&= \frac{x}{V_3} + \frac{2h_2}{V_2}\cos\Theta_2 + \frac{2h_1}{V_1}\cos\theta_1 \\
&= \frac{x}{V_3} + t_2.
\end{aligned}
\left.\rule{0pt}{7.5em}\right\}
\quad (3.35)
$$

Thus the time–distance curve for this refraction is also a straight line whose slope is the reciprocal of the velocity just below the refracting horizon and whose intercept is the sum of terms of the form $(2h_i \cos\theta_i/V_i)$, each layer above the refracting horizon contributing one term. We can generalize for n layers:

$$t = \frac{x}{V_n} + \sum_i \frac{2h_i}{V_i}\cos\theta_i, \quad (3.36)$$

where $\theta_i = \sin^{-1}(V_i/V_n)$. This equation can be used to find the velocities and thicknesses of each of a series of horizontal refracting layers, each of constant velocity higher than any of the layers above it, provided each layer contributes enough of the time–distance curve to permit it to be analyzed correctly. We can find all of the velocities (hence

the angles θ_i also) by measuring the slopes of the various sections of the time–distance curve and then get the thicknesses of the layers from the intercepts:

$$h_n = \frac{V_n}{2\cos\Theta_n}\left(t_n - \sum_{i=1}^{n-1}\frac{2h_i \cos\theta_i}{V_i}\right). \quad (3.37)$$

3.3.3   *Effect of refractor dip*

The simple situations on which (3.30)–(3.37) are based are frequently not valid. One of the most serious defects is the neglect of dip since dip changes the refraction time–distance curve drastically. The lower part of fig. 3.16 shows a vertical dip-section through a refracting horizon. Let t be the traveltime for the refraction path $OMPO'$. Then, we have

$$
\begin{aligned}
t &= \frac{OM + O'P}{V_1} + \frac{MP}{V_2} \\
&= \frac{h_d + h_u}{V_1 \cos\Theta} + \frac{OQ - (h_d + h_u)\tan\Theta}{V_2} \\
&= \frac{x\cos\xi}{V_2} + \frac{h_d + h_u}{V_1}\cos\Theta. \quad (3.38)
\end{aligned}
$$

If we place the shotpoint at O and a detector at O', we are 'shooting downdip'. In this case it is convenient to

Fig. 3.15. Raypaths and traveltime curves for two-refractor case.

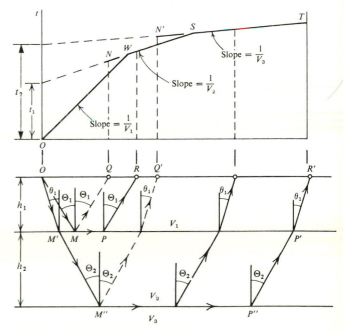

have t in terms of the distance from the shotpoint to the refractor h_d; hence we eliminate h_u using the relation

$$h_u = h_d + x \sin \xi.$$

Writing t_d for the downdip traveltime, we obtain

$$
\left.
\begin{aligned}
t_d &= (x/V_2)\cos \xi + (x/V_1)\cos \Theta \sin \xi + (2h_d/V_1)\cos \Phi \\
&= (x/V_1)\sin(\Theta + \xi) + (2h_d/V_1)\cos \Theta \\
&= (x/V_1)\sin(\Theta + \xi) + t_{1d},
\end{aligned}
\right\}
$$

where $\qquad\qquad\qquad\qquad\qquad\qquad\qquad$ (3.39)

$$t_{1d} = (2h_d/V_1)\cos \Theta.$$

The result for shooting in the updip direction is similarly obtained by eliminating h_d:

$$
\left.
\begin{aligned}
t_u &= (x/V_1)\sin(\Theta - \xi) + t_{1u}, \\
\text{where} & \\
t_{1u} &= (2h_u/V_1)\cos \Theta.
\end{aligned}
\right\}
$$
$\qquad\qquad\qquad\qquad\qquad\qquad\qquad$ (3.40)

Note that the downdip traveltime from O to O' is equal to the updip traveltime from O' to O; this shotpoint-to-shotpoint traveltime is called the *reciprocal time* and is denoted by t_r. The concept that traveltime along a path is the same regardless of the direction of travel is an example of the *principle of reciprocity*.

These equations can be expressed in the same form as (3.31):

$$t_d = (x/V_d) + t_{1d}, \qquad (3.41)$$

$$t_u = (x/V_u) + t_{1u}, \qquad (3.42)$$

where

$$V_d = V_1/\sin(\Theta + \xi), \quad V_u = V_1/\sin(\Theta - \xi). \quad (3.43)$$

V_d and V_u are apparent velocities and are given by the reciprocals of the slopes of the time–distance curves.

For reversed profiles such as shown in fig. 3.16, (3.43) can be solved for the dip ξ and the critical angle Θ (and hence for the refractor velocity V_2):

$$
\left.
\begin{aligned}
\Theta &= \tfrac{1}{2}\{\sin^{-1}(V_1/V_d) + \sin^{-1}(V_1/V_u)\}, \\
\xi &= \tfrac{1}{2}\{\sin^{-1}(V_1/V_d) - \sin^{-1}(V_1/V_u)\}.
\end{aligned}
\right\}
\qquad (3.44)
$$

The distances to the refractor, h_d and h_u, can then be found from the intercepts using (3.39) and (3.40).

Equation (3.43) can be simplified where ξ is small

Fig.3.16. Raypaths and traveltime curves for a dipping refractor.

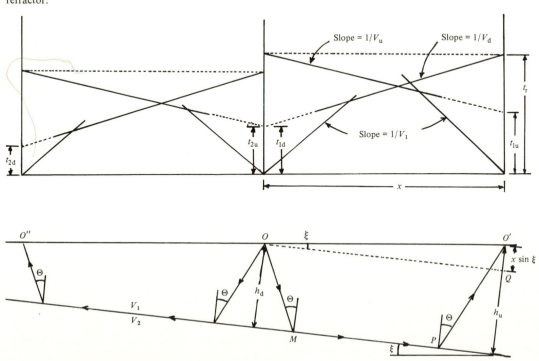

enough that we can approximate by letting $\cos \xi \approx 1$ and $\sin \xi \approx \xi$. With this simplification, (3.43) becomes

$$V_1/V_d = \sin(\Theta + \xi) \approx \sin\Theta + \xi\cos\Theta;$$

$$V_1/V_u = \sin(\Theta - \xi) \approx \sin\Theta - \xi\cos\Theta;$$

hence

$$\sin\Theta = (V_1/V_2) \approx \tfrac{1}{2}V_1\{(1/V_d) + (1/V_u)\},$$

so that

$$1/V_2 \approx \tfrac{1}{2}\{(1/V_d) + (1/V_u)\}. \tag{3.45}$$

An even simpler approximate formula for V_2 (although slightly less accurate) can be obtained by applying the binomial theorem to (3.43) and assuming that ξ is small enough that higher powers of ξ are negligible:

$$V_d = (V_1/\sin\Theta)(\cos\xi + \cot\Theta\sin\xi)^{-1}$$

$$\approx V_2(1 - \xi\cot\Theta),$$

$$V_u \approx V_2(1 + \xi\cot\Theta),$$

hence

$$V_2 \approx \tfrac{1}{2}(V_d + V_u). \tag{3.46}$$

3.3.4   *Several dipping refractors with the same strike*

Equations similar to (3.36) have been derived for the case of several beds with the same strike and different dips. An interesting formula due to Adachi (1954) departs from the usual parameters and uses vertical thicknesses and angles of incidence and refraction measured with respect to the vertical (see fig. 3.17). The derivation of Adachi's formula is straightforward but involves lengthy trigonometric manipulation (see Johnson, 1976), and we merely quote the result:

$$t_n = \frac{x\sin\beta_1}{V_1} + \sum_{i-1}^{n-1}\frac{h_i}{V_i}(\cos\alpha_i + \cos\beta_i), \tag{3.47}$$

where t_n is the traveltime of the refraction at the nth interface (separating layers of velocities V_n and V_{n+1}), α_i and β_i are the angles between the vertical and the downgoing and upgoing rays in the ith layer, h_i the vertical thickness of the ith layer. We define the angles a_i, b_i (see fig. 3.17) as angles of incidence, a_i', b_i' angles of refraction, all measured relative to the normal, and $\xi_{i+1} = $ dip of the ith interface. Then

$$a_i' = \sin^{-1}\{(V_{i+1}/V_i)\sin a_i\},$$

Fig.3.17. Notation used in Adachi's formula.

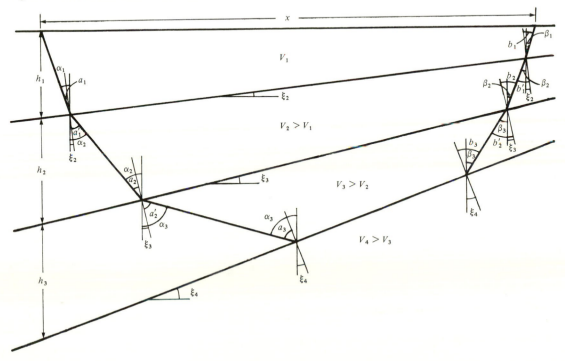

$$b'_i = \sin^{-1}\{(V_{i+1}/V_i)\sin b_i\},$$

$$\alpha_i = a_i + \xi_{i+1}, \quad \beta_i = a_i - \xi_{i+1}$$

$$\alpha_{i+1} = a'_i + \xi_{i+1}, \quad \beta_{i+1} = b'_i - \xi_{i+1}.$$

For the refraction along the *n*th interface, $a_n = b_n = \Theta_n$, the critical angle.

Assuming reversed profiles, we measure V_1, the apparent velocities, V_{2u} and V_{2d}, and the intercepts, t_{1u} and t_{1d}, as usual. For the first interface,

$$\alpha_1 = \sin^{-1}(V_1/V_{2d}), \quad \beta_1 = \sin^{-1}(V_1/V_{2u})$$

$$\Theta_1 = a_1 = b_1 = \tfrac{1}{2}(\alpha_1 + \beta_1),$$

$$\xi_2 = \tfrac{1}{2}(\alpha_1 - \beta_1), \quad (\text{cf. } (3.44)),$$

$$V_2 = V_1/\sin\Theta_1, \quad h_1 = V_1 t_{1u}/(\cos\alpha_1 + \cos\beta_1).$$

To solve for the second interface, we calculate new values of α_1, β_1 and then find the other angles (note that ξ_2 is now known):

$$\alpha_1 = \sin^{-1}(V_1/V_{3d}), \quad \beta_1 = \sin^{-1}(V_1/V_{3u}),$$

$$a_1 = \alpha_1 - \xi_2, \quad b_1 = \beta_1 + \xi_2,$$

$$a'_2 = \sin^{-1}\{(V_2/V_1)\sin a_1\},$$

$$b'_2 = \sin^{-1}\{(V_2/V_1)\sin b_1\},$$

$$\alpha_2 = a'_2 + \xi_2, \quad \beta_2 = b'_2 - \xi_2,$$

$$a_2 = b_2 = \Theta_2 = \tfrac{1}{2}(\alpha_2 + \beta_2) = \tfrac{1}{2}(a'_2 + b'_2),$$

$$V_3 = V_2/\sin\Theta_2, \quad \xi_3 = \tfrac{1}{2}(\alpha_2 - \beta_2),$$

$$t_{2u} = (h_1/V_1)(\cos\alpha_1 + \cos\beta_1)$$
$$+ (h_2/V_2)(\cos\alpha_2 + \cos\beta_2),$$

h_2 being found from the last relation. In principle this iterative procedure can be continued indefinitely but in practice, as with all refraction schemes, the errors and difficulties mount rapidly as the number of layers increase.

Adachi's formula is best suited to simple cases where the refractors are plane, no velocity or structural problems exist and the refractors are shallow. When these conditions are not met, the formula, in common with other similar ones, may be of limited value. Often one is not sure that formulae are applicable to a specific real situation. Where there are more than two refracting horizons it is often difficult to identify equivalent updip and downdip segments, especially if the refractors are not plane or if the dip and strike change.

3.3.5 *Linear increase of overburden velocity*

The concepts of a linear continuous increase of velocity have been discussed in §3.2.5. It is obvious from fig 3.12 that the circular rays will eventually reach the surface again, as shown in fig. 3.18. For this case (3.25) and (3.26) give (see problem 3.18):

$$\left.\begin{aligned} x &= (2V_0/a)\cot i_0 = (2V_0/a)\sinh\tfrac{1}{2}at, \\ t &= (2/a)\ln(\cot\tfrac{1}{2}i_0). \end{aligned}\right\} \quad (3.48)$$

The maximum depth of penetration, h_m, is reached when i equals $\tfrac{1}{2}\pi$, hence

$$h_m = (V_0/a)(\cosh\tfrac{1}{2}at - 1) \quad (3.49)$$

using (3.28) and (3.48).

The case of a high-velocity layer overlain by a layer in which the velocity increases linearly with depth (fig. 3.19) is of considerable practical importance. The relation between t and x for a horizontal refractor can be found as follows:

$$t = t_{SN} + t_{NP} + t_{PR}$$
$$= 2t_{SN} + (x - 2MN)/V_m.$$

Fig.3.18. Linear increase in velocity applied to refraction interpretation.

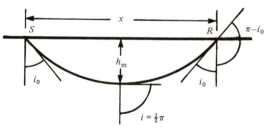

Fig.3.19. Refraction with linear increase in velocity in upper layer.

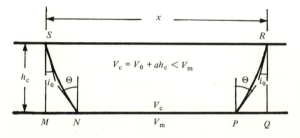

Noting that $\sin i_0/V_0 = (\sin \Theta)/V_c = 1/V_m$, we find from (3.26)

$$t_{SN} = \frac{1}{a}\ln\left(\frac{\tan\frac{1}{2}\Theta}{\tan\frac{1}{2}i_0}\right)$$

$$= \frac{1}{a}\ln\left\{\left(\frac{V_c}{V_m + (V_m^2 - V_c^2)^{\frac{1}{2}}}\right)\left(\frac{V_m + (V_m^2 - V_0^2)^{\frac{1}{2}}}{V_0}\right)\right\}$$

$$- \frac{1}{a}\left\{\cosh^{-1}\left(\frac{V_m}{V_0}\right) - \cosh^{-1}\left(\frac{V_m}{V_c}\right)\right\},$$

where use has been made of the identity $\cosh^{-1} x = \ln\{x + (x^2 - 1)^{\frac{1}{2}}\}$. From (3.25), we get

$$MN = (1/pa)(\cos i_0 - \cos\Theta)$$

$$= (1/pa)[\{1 - (V_0/V_m)^2\}^{\frac{1}{2}} - \{1 - (V_c/V_m)^2\}^{\frac{1}{2}}].$$

Substituting in the first expression for t gives

$$t = (x/V_m) + t_0, \tag{3.50}$$

where $t_0 =$ intercept time:

$$t_0 = (2/a)[\{\cosh^{-1}(V_m/V_0) - \cosh^{-1}(V_m/V_c)\}$$
$$- \{1 - (V_0/V_m)^2\}^{\frac{1}{2}} + \{1 - (V_c/V_m)^2\}^{\frac{1}{2}}]. \tag{3.51}$$

The slope of the traveltime curve gives V_m. A curve is plotted of t_0 against h_c (or V_c) for given values of V_0 and a, and h_c and V_c are read from this curve for particular measurements of t_0.

Problems

3.1. (a) Calculate Δt_n for a geophone 600 m from the shotpoint for a reflection at $t_0 = 2.358$ s, given that $\bar{V} = 2.90$ km/s. (b) Typical errors in x, $\bar{V}$, t_0 might be 0.6 m, 0.2 km/s, 5 ms; calculate the corresponding errors in Δt_n approximately. What do you conclude about the accuracy of Δt_n calculations? (c) Show that (3.8) can be written

$$\Delta t_{nn} = \Delta t_n(1 - \Delta t_n/2t_0)$$

where Δt_n is given by (3.7) and Δt_{nn} is the second approximation. Taking into account the errors in x, $\bar{V}$, t_0, when is this equation useful?

3.2. (a) Show that the quantity dt/dx can be considered as a vector or component of a vector according as dt corresponds to the total dip or component of dip. (b) Using fig. 3.20, verify that the construction of fig. 3.8 gives same results as (3.18). [Hint: Express l, m and OC in terms of OA.]

3.3. (a) Calculate $\bar{V}$ and $\bar{\bar{V}}$ down to each of the beds in the following table. Why do they differ (give a geometrical explanation)?

Depth	Velocity
0–1.00 km	2.00 km/s
1.00–2.50	3.00
2.50–2.80	6.00
2.80–4.80	4.00

(b) A reflection from the 2.5 km horizon has a dip moveout of 104 ms/km. Calculate the dip using $\bar{V}$ and $\bar{\bar{V}}$ and compare with the angle of approach on a split spread. Why are the three values different?

3.4. Show that (3.9) becomes

$$(Vt)^2 = (2x \cos\xi)^2 + 4h_c^2$$

(Gardner, 1947) where h is replaced by h_c, the slant depth at the midpoint between the shotpoint and receiver (see fig. 3.21).

3.5. (a) Using (3.11) and the results of problem 3.4,

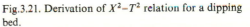
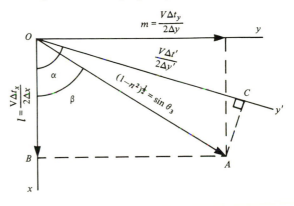

Fig.3.20. Combining dip components.

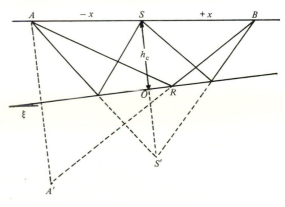

Fig.3.21. Derivation of $X^2 - T^2$ relation for a dipping bed.

verify the following result (due to Favre, according to Dix, 1955):

$$\tan \xi \approx t/(t_{AB}^2 - t_0^2)^{\frac{1}{2}},$$

where $\xi =$ dip, $t = t_{SA} - t_{SB}$, $t_{AB} =$ traveltime between shotpoint A and receiver B, $t_0 =$ traveltime at shotpoint S (see fig. 3.21). (b) Using (3.9), show that

$$\sin \xi = V^2(t_{SA}^2 - t_{SB}^2)/8h_c x.$$

(c) Under what condition is the result in (b) the same as (3.11) and also consistent with (a)?

3.6. The expression for dip in terms of dip moveout, (3.11), involves the approximation of dropping higher-order terms in the quadratic expansion used to get (3.10). What is the effect on (3.11) if an additional term is carried in this expansion? What is the percentage change in dip?

3.7. (a) Shotpoints B and C are respectively 600 m north and 500 m east of shotpoint A. Traveltimes at A, B and C for a certain reflection are $t_0 = 1.750$, 1.825 and 1.796 s. What are the dip and strike of the horizon, $\bar{V}$ being 3.25 km/s? (b) What are the changes in dip and strike if the line AC has the bearing N80°E?

3.8. (a) Two intersecting seismic spreads have bearings N10°E and N140°E. If the first spread shows an event at $t_0 = 1.760$ s with dip moveout of 56 ms/km while the same event on the second spread has a dip moveout of 32 ms/km, find the true dip, depth and the strike, assuming that (i) both dips are down to the south and west, (ii) dip on the first spread is down to the south while the other is down to the southeast. Take the average velocity as 3 km/s. (b) Calculate the position of the reflecting point (migrated position) for each spread in (i) as if the cross information had not been available and each had been assumed to be indicating total moveout; compare with the results of part (a). Would the errors be more serious or less serious if the calculations were made for the usual situation where the velocity increases with depth?

3.9. Using the data in problem 3.3a, plot $\bar{V}$ and $\bar{\bar{V}}$ versus depth and versus traveltime and determine the best-fit straight lines in the four cases. What are the main problems in approximating data with functional fits?

3.10. (a) Assuming flat bedding, calculate depths corresponding to $t_0 = 1.0, 2.0, 2.1$ and 3.1 s using the following velocity functions: (i) $\bar{V}$ from problem 3.3a, (ii) $\bar{\bar{V}}$ from problem 3.3a, (iii) the best-fit functions determined in problem 3.9. (b) What errors are introduced in the above relative to the depths given in problem 3.3a.

3.11. (a) Repeat the calculations of problem 3.10a for flat velocity layering but dip moveout of 104 ms/km and find the dip in each case. (b) Using the velocity data tabulated in problem 3.3a trace a ray down through the various layers and find the arrival time, reflecting points and dips of reflectors located at the depths of each velocity boundary.

3.12. Shotpoints A and B are located at the ends of a 225 m spread of 16 geophones. Using the data below, find the velocities, the dip and depth to the refractor.

x_A	t_A	t_B	x_B	x_A	t_A	t_B	x_B
0 m	0 ms	98 ms	225 m	120 m	70 ms	52 ms	105 m
15	10	92	210	135	73	46	90
30	21	87	195	150	78	43	75
45	30	81	180	165	81	37	60
60	41	75	165	180	85	31	45
75	50	71	150	195	89	21	30
90	59	65	135	210	94	10	15
105	65	60	120	225	98	0	0

3.13. (a) Show that the two geological sections illustrated in fig. 3.22 produce the same time–distance curves. (b) What would be the apparent depth to the lower interface in fig 3.22a, b if $V_3 = 3.15$ km/s instead of 6 km/s?

3.14. Fig. 3.23 shows a refraction profile recorded as a ship firing an airgun steamed away from a sonobuoy.

Fig.3.22. Two different geologic sections which give the same refraction time–distance curve.

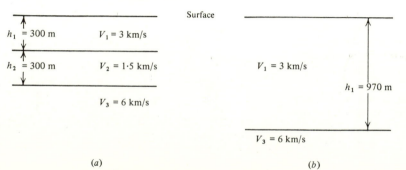

$h_1 = 300$ m $V_1 = 3$ km/s

$h_2 = 300$ m $V_2 = 1.5$ km/s

$V_3 = 6$ km/s

$V_1 = 3$ km/s

$h_1 = 970$ m

$V_3 = 6$ km/s

(a)   (b)

Identify the direct wave through the water and use its slope to give the source-to-sonobuoy distances (assume 1.5 km/s as the velocity in water). (*a*) Identify distinctive head-wave arrivals, determine their velocities, intercept times and depths of the refractors assuming flat bedding and no velocity inversions. (*b*) What is the water depth?

Identify multiples and explain their probable travel paths. (The data in the upper right corner result from paging and actually belong below the bottom of the record.)

3.15. The time–distance observations in fig. 3.24 constitute an engineering refraction problem. (*a*) Solve for the first layer for both sets of reversed profiles and show

Fig.3.23. Sonobuoy refraction profile in Baffin Bay shot with 1000 in³ airgun. (Courtesy Fairfield Industries.)

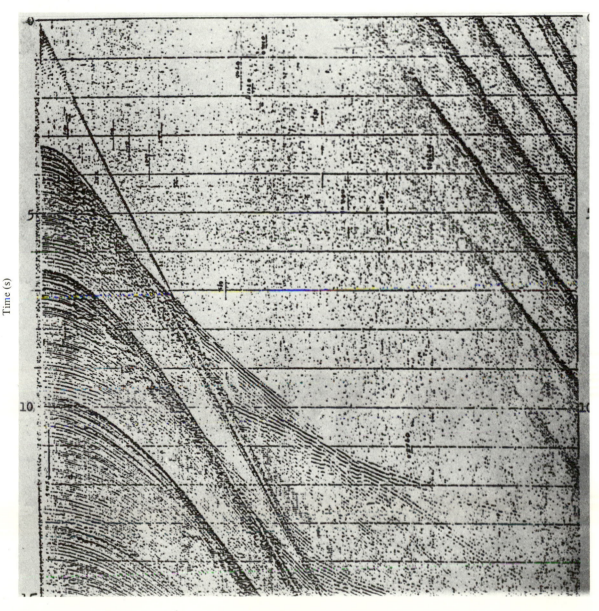

that the layer has a thickness of about 2.9 m and 3.8 m at shotpoints A and C (dip about 0.3°). (b) Apply (3.35) to get approximate thicknesses of the second layer. (c) What is the dip of the deeper interface? (d) Why are the answers in (b) and (c) approximate?

3.16. To find the depth to bedrock in a damsite survey, traveltimes were measured from the shotpoint to 12 geophones laid out at 15 m intervals along a straight line through the shotpoint. The offsets x range from 15 to 180 m. Determine the depth of overburden from the data in the following table. By how much does the depth differ if a two-layer solution is calculated instead of a three-layer solution?

x	t	x	t	x	t
15 m	19 ms	75 m	59 ms	135 m	72 ms
30	29	90	62	150	76
45	39	105	65	165	78
60	50	120	68	180	83

3.17. Given the following data for a reversed refraction profile with shotpoints A, B, use Adachi's method to find velocities, depths and dips.

x	0	0.5	1.0	1.5	2.0	2.5	3.0	3.5	4.0	4.5	5.0 km
t_A	0.00	0.25	0.50	0.74	0.98	1.24	1.50	1.70	1.81	1.91	2.02 s
t_B	3.00	2.90	2.80	2.68	2.52	2.41	2.31	2.20	2.07	1.91	1.80 s

x	5.5	6.0	6.5	7.0	7.5	8.0	8.5	9.0	9.5	10.0 km
t_A	2.16	2.28	2.38	2.44	2.56	2.64	2.72	2.80	2.89	3.00 s
t_B	1.65	1.50	1.40	1.25	1.12	1.00	0.75	0.49	0.23	0.00 s

3.18. Prove (3.48) and (3.49) from (3.25)–(3.28).

3.19. (a) Given the velocity function $V = 1.60 + 0.60 z$ km/s (z in km), find the dip of the reflector and the depth and offset of the point of reflection when $t_0 = 4.420$ s and $\Delta t/\Delta x = 0.155$ s/km. (b) What interpretation would you give of the result in part (a)? If this reflector did not exist and the ray continued without reflection, when and where would it emerge? What moveout would be observed at the recording spread? (c) If you were given the results of part (b) instead of the values of t_0 and $\Delta t/\Delta x$ in (a), calculate the maximum depth of penetration.

3.20. If the velocity function in problem 3.19 applies above a horizontal refractor at a depth of 2.40 km where the refractor velocity is 4.25 km/s, plot the traveltime–distance curve.

3.21. Given that situations (a) through (h) in fig. 3.25 involve the same two rock types, draw the appropriate time–distance curves. Diagram (c) shows two cases for dip in opposite directions. In figs (i) and (j), the velocity in the lower medium varies laterally according to the density of the shading.

3.22. Barton (1929) discusses shooting into a geophone placed in a borehole (fig. 3.26) as a means of determining where the bottom of the borehole is located. (a) Given that A, B, D, E are equidistant from W in the cardinal directions and assuming straight-line travel paths at the velocity V and that the traveltimes from shotpoints D and E in fig. 3.26a are equal, derive expressions for CC' and CW in fig. 3.26b in terms of the traveltimes from shotholes A and B, $t_{AC'}$ and $t_{BC'}$. (b) What are the values of $t_{AC'}$ and $t_{BC'}$ for $V = 2.500$ km/s, $AW = BW = CC' = 1000$ m, $CW = 200$ m? (c) How sensitive is the method, that is, what are $\Delta(CC')/\Delta t_{AC'}$ and $\Delta(CW)/\Delta t_{AC'}$? For the specific

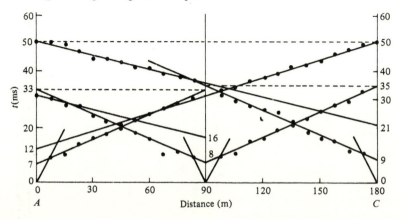

Fig.3.24. Engineering refraction profile.

Fig.3.25. Time–distance curves for various two-layer configurations. This figure is adapted from figures in Donald C. Barton's paper 'The seismic method of mapping geologic structure' (1929), the first publication in English on the seismic method. The upper part (above O) of each diagram provides space for a curve of arrival time versus distance for the model shown in cross-section below (below O). Part (*a*) has been completed to show what is expected. In (*c*), two alternatives are given so two sets of curves should be drawn. In (*i*) and (*j*), velocities are assumed to vary horizontally proportional to the density of the shading. In each case the velocity in the cross-hatched portion is higher than that above.

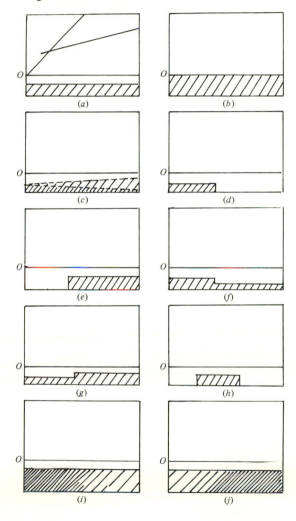

(*a*)   (*b*)

(*c*)   (*d*)

(*e*)   (*f*)

(*g*)   (*h*)

(*i*)   (*j*)

situation in part (*b*), how much change is made in WC and CC' per millisecond error in $t_{AC'}$? (*d*) Modify the assumptions in part (*b*) by taking the velocity as 1.5 km/s for the first 500 m and 3.5 for the lower 500 m; what are the actual traveltimes now and how would these be interpreted assuming the straight-path assumption in part (*a*)?

Fig.3.26. Mapping a crooked borehole by measuring traveltime to a geophone at C' in the borehole. (From Barton, 1929.) (*a*) Plan view; (*b*) vertical section AWB.

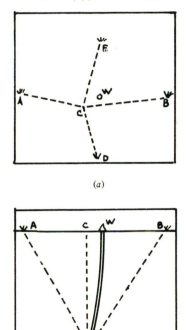

(*a*)

(*b*)

4

Characteristics of seismic events

Overview

The basic task of interpreting seismic records is that of selecting those events on the record which represent primary reflections (or refractions), translating the arrival times for these into depths and dips, and mapping the reflecting (refracting) horizons. In addition, the interpreter must be alert to other features of events such as changes in amplitude or character, and to other types of events which may yield valuable information, such as multiple reflections and diffractions. The characteristics of events is the subject of this chapter.

The features which allow one to recognize and identify an event – coherence, amplitude standout, character, dip and normal moveout – are discussed in §4.1. Differences in arrival time because of offset provide an especially useful method of distinguishing between reflections, refractions, multiples and other types of events.

Section 4.2 takes up the characteristics of several classes of non-primary reflection events. Diffractions have more normal moveout than reflections and are curved events on stacked unmigrated sections. They bear certain relationships to the reflections from reflectors whose termination often generates them. The crest of a diffraction gives the location of the diffracting point for simple velocity situations; this point is useful in locating bedding terminations such as occur at faults and saltdome flanks. Diffractions also provide a mechanism for getting seismic energy into regions which cannot be reached on a geometrical-optics basis.

Multiples are classified as long-path if they show as separate events or short-path if their effect is merely that of changing reflection waveshape. Peg-leg multiples are important factors in changing the waveshape and removing high frequencies with increasing traveltime. Long-path multiples confuse interpretation unless they are recognized for what they are. The characteristics of surface waves conclude §4.2.

Section 4.3 deals with characteristics of reflections. Reflector curvature affects reflection amplitude. Where the center of curvature of a synclinal reflector is below the observing level, a buried-focus occurs and a reverse branch appears. The reverse branch has convex-upward curvature and a reversed sense of traverse and involves a phase shift.

Resolution refers to the ability to distinguish between adjacent features. Vertical resolution concerns the minimum separation between interfaces for them to show as separate reflectors; the resolvable limit is about a quarter-wavelength. Horizontal resolution, the minimum distance separating features for them to be seen as separate, is about the width of the first Fresnel zone.

The shape of the seismic wavelet changes with traveltime because of absorption and peg-leg multiples, and also because of filtering actions in recording and processing. Minimum-phase wavelets are distinguished from zero-phase ones, the Ricker wavelet being the most commonly assumed zero-phase wavelet.

Distinction is made between coherent and incoherent noise, and repeatable and ambient noise. A discussion of the attenuation of noise concludes this chapter.

Fig.4.1. Characteristics of seismic events.

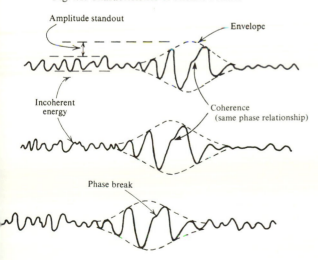

4.1   Distinguishing features of events

Recognition and identification of seismic events are based upon five characteristics: (*a*) coherence, (*b*) amplitude standout, (*c*) character, (*d*) dip moveout and (*e*) normal moveout. The first of these is by far the most important in recognizing an event. Whenever a wave recognizable as such reaches a spread, it produces approximately the same effect on each geophone. If the wave is strong enough to override other energy arriving at the same time, the traces will look more-or-less alike during the interval in which this wave is arriving; this similarity in appearance from trace to trace is called *coherence* (see fig. 4.1) and is a necessary condition for the recognition of any event. *Amplitude standout* refers to an increase of amplitude such as results from the arrival of coherent energy; it is not always marked, especially when *AGC* (see §5.4.5) is used in recording. *Character* refers to a distinctive appearance of the waveform which identifies a particular event; it involves primarily the shape of the envelope, the number of cycles which show amplitude standout and irregularities in phase resulting from interference between components of the event. *Moveout*, which refers to a systematic difference from trace to trace in the arrival time of an event, has been discussed in §3.1.1 and §3.1.2.

Coherence and amplitude standout tell us whether or not a strong seismic event is present, but they say nothing about the type of event. The most distinctive criterion for identifying the nature of events is moveout. Character is also often very useful, especially the frequency content and the number of cycles observed. Reflections exhibit normal moveouts which must fall within certain limits set by the velocity distribution. The dip moveout is usually small but occasionally reflections have large dip moveouts (as with fault-plane reflections). Reflection events rarely involve more than two or three cycles and are often rich in frequency components in the range 15–60 Hz; deep reflections at times have considerable energy even below this range.

One very powerful technique for distinguishing between reflections, diffractions, reflected refractions and multiples is to display the data after correcting for (*a*) weathering and elevation (*static corrections*, since the correction is the same for all arrival times on a given trace; see §5.6.2) and (*b*) normal moveout (*dynamic corrections*, since the amount of the correction decreases with increasing arrival time). Such corrected records can be made in data processing. Provided the correct normal moveout was removed, reflections appear (fig. 4.2) as straight lines while diffractions and multiples still have some curvature (since their normal moveouts are larger than those of primary reflections) and refractions and other formerly straight alignments have inverse curvature.

4.2 Events other than primary reflections
4.2.1 *Diffractions*

Diffraction phenomona were discussed in §2.3.5. It was shown that the reflection from a half-plane and the diffraction from its edge are continuous and indistinguishable on the basis of character. Diffractions, however, usually exhibit distinctive moveout. In fig. 4.3*a* for all shotpoint and geophone positions such that the point of reflection R is to the left of the diffraction source A, the reflection traveltime curve is given by (3.4), that is

$$t_r = (1/V)(x^2 + 4h^2)^{\frac{1}{2}} \approx (2h/V) + (x^2/4Vh)$$
$$= t_0 + \Delta t_n,$$

assuming that x is smaller than h. The reflection traveltime curve is a hyperbola as shown in fig. 3.1. For the case where the shotpoint is directly above the diffraction source A (S_2 in fig. 4.3*b*), the diffraction traveltime curve is given by the equation

$$t_d = (1/V)\{h + (x^2 + h^2)^{\frac{1}{2}}\} \approx (2h/V) + (x^2/2Vh)$$
$$= t_0 + 2\Delta t_n. \tag{4.1}$$

Thus, the normal moveout for the diffraction shown in fig. 4.3*b* is twice that of a reflection at the same offset;

Fig.4.2. Types of events on a seismic record. Identities of events are: a = direct wave, V = 650 m/s; b = refraction at base of weathering, V_H = 1640 m/s; c = refraction from flat refractor, V_R = 4920 m/s; d = reflection from refractor in c, $\overline{V}$ = 1640 m/s; e = reflection from flat reflector, $\overline{V}$ = 1970 m/s; f = reflection from flat reflector, $\overline{V}$ = 2300 m/s; g = reflection from dipping reflector, $\overline{V}$ = 2630 m/s; h = multiple of d; i = multiple of e; j = ground roll, V_R = 575 m/s; k = airwave, V = 330 m/s; l = reflected refraction from in-line disruption of refractor in c; m = reflected refraction from broadside disruption of refractor in c. After proper normal-moveout correction the primary reflections are straight. In processing, data are usually *muted*, with a 'front-end mute'; this involves setting to zero all values earlier than some 'mute schedule', here indicated by the dotted line. Consequently the data which might otherwise appear in the upper right triangles are usually not seen.

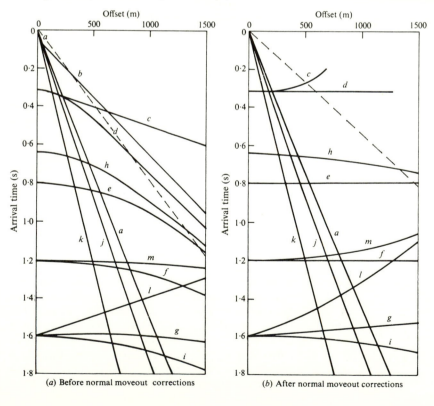

(a) Before normal moveout corrections   (b) After normal moveout corrections

the reflection corresponds to a virtual source at a depth $2h$ while the diffraction comes from a source at depth h. The earliest arrival time on a diffraction curve is for the trace which is recorded directly over the diffracting point (except for unusual velocity-distribution situations) but the diffraction will not necessarily have its peak amplitude on this trace. The diffraction moveout term in (4.1), $x^2/2Vh$, decreases with the depth of the diffracting point, h, so that diffraction curvature decreases with depth. If we consider a coincident source–geophone section as shown in fig. 4.3c, the diffraction traveltime curve will be given by

$$t'_d = 2(x^2 + h^2)^{\frac{1}{2}}/V \approx (2h/V) + (x^2/Vh)$$
$$= t_0 + 4\Delta t_n. \qquad (4.2)$$

This is the sort of section which common-depth-point stacking seeks to simulate. A diffraction on such a section will be a curved event approaching the slope $\pm 2/V$ for large x (see problem 4.1).

Consider three half-planes at the same depth but with different dips (fig. 4.4). The diffractions for each of these crest at the location of the edge of the half-plane, have the same curvature, are tangent to the reflection, and the maximum amplitude of the diffraction occurs at this point of tangency. Thus the diffraction crest locates the diffracting point, the diffraction curvature depends on the depth and the velocity above the diffracting point, and

the amplitude distribution along the diffraction depends on the attitude of the half-plane.

A reflector which is bent sharply as shown in fig. 4.5 could be thought of as the superposition of two dipping half-planes, each terminating at the bend point. Thus the two diffraction curves would coincide in arrival time and would add together constructively in the region between the respective reflections. In fig. 4.5b, the reflector to the right of $x = 2.1$ km gives rise to $P'B$, the reflector to the left of $x = 2.1$ km gives AP; diffraction fills in the gap PP' and makes the seismic event continuous without a sharp break in slope.

As another example of diffraction effects, consider the reflection from a reflector with a hole in it, as shown in fig. 4.6. Diffraction tends to fill in the hole.

Fig. 4.7 shows the location and amplitude of wave motion a short time after a plane wavefront has passed by the point of a wedge which is a perfect reflector. The reflected wavefront has an associated diffraction BAC and the portion of the wavefront which missed the reflecting wedge also has an associated diffraction FDE. The portion of the diffraction DE represents energy reaching into the shadow zone hidden from the incident wavefront by the reflector. Diffraction provides a mechanism for getting seismic energy into regions which cannot be reached on the basis of geometrical optics.

The downgoing diffraction FDE in fig. 4.7 of course

Fig. 4.3. Diffraction traveltime curves. (a), (b) single source and many geophones; (c) source and geophone coincident at each location.

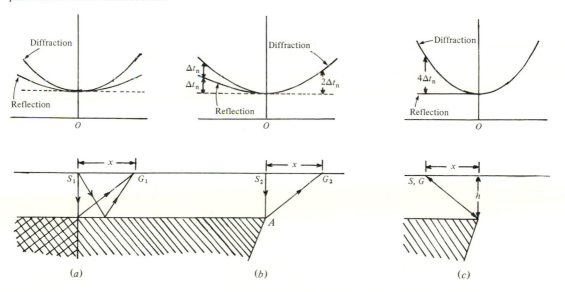

(a)   (b)   (c)

might be subsequently reflected by another reflector to give a reflected diffraction, or an upcoming reflection could be diffracted to give a diffracted reflection, etc. The diffraction curvature for such compound events will differ from that of simple diffraction events which have the same arrival time. Additional compound events are shown in fig. 4.31.

Fig. 4.8a shows a vertical step. The top of the step A will generate diffraction, D_1, in fig. 4.8b. The bottom of the step B will also generate a diffraction, D_1', but the portion of this diffraction to the right of B will travel partially at a lower velocity than the reflection to the left of the step.

The reflection from the left portion of the base of the model will arrive earlier than the reflection from the right portion because more of its travel path is at the higher velocity. Furthermore, each of these reflection segments will have an associated diffraction appearing to come from C, even though the base of the model is continuous; such diffractions are called *phantom diffractions* and they result from lateral velocity changes in the section above the reflector. Consider a geophone just to the right of a point directly over the step in fig. 4.8a. The travel path for a reflection from the base of the model is shown by the dashed path, but some energy will travel the dotted path and arrive earlier than the reflection; the latter

Fig.4.4. Reflections and diffractions from half-planes terminating at the arrows (coincident sources and receivers). (Courtesy Chevron.) (a) Termination at downdip end of half-plane; (b) flat half-plane; (c) termination at updip end of half-plane.

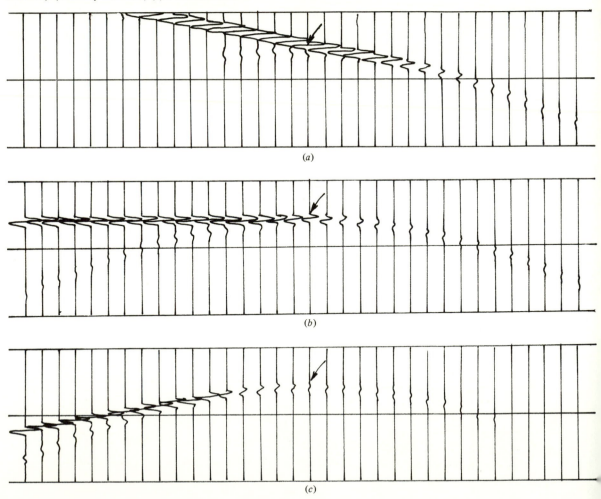

(a)

(b)

(c)

produces the phantom diffraction (diffraction energy does not necessarily obey the rules of geometrical optics). Phantom diffractions are occasionally seen on seismic records, especially in areas of thrusting where some travel paths pass through thrust plates and others do not.

4.2.2 *Multiples*

(*a*) *Distinction between types. Multiples* are events which have undergone more than one reflection. Since the amplitude of multiples is proportional to the product of the reflection coefficients for each of the reflectors involved and since R is very small for most interfaces, only the largest impedance contrasts will generate multiples strong enough to be recognized as distinctive events.

We distinguish between two classes of multiples which we call long-path and short-path. A *long-path multiple* is one whose travel path is long compared with primary reflections from the same deep interfaces and hence long-path multiples appear as separate events on a seismic record. A *short-path multiple*, on the other hand, arrives so soon after the associated primary reflection from the same deep interface that it interferes with and adds tail to the primary reflection; hence its effect is that of changing waveshape rather than producing a separate event. Possible raypaths for these two classes are shown in fig. 4.9.

Fig.4.5. Reflections and diffractions from a sharply bent reflector. Dips are 31° and 11° to the left and right of $x = 2.1$ km. (Courtesy Chevron.) (*a*) Model; (*b*) reflections and diffraction (dashed curve).

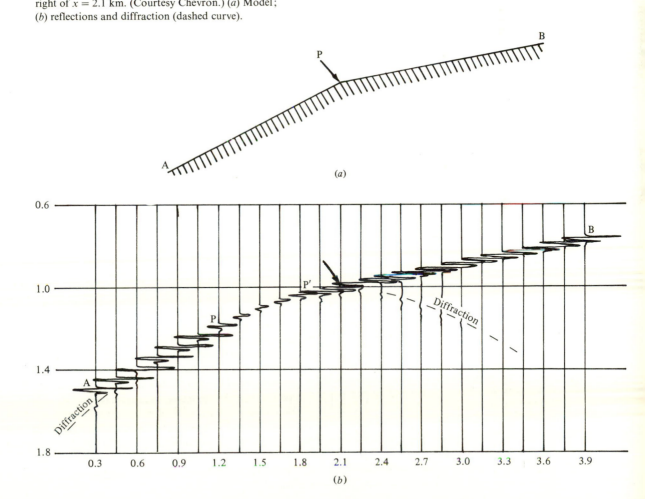

(b) *Short-path multiples.* Short-path multiples which have been reflected successively from the top and base of thin reflectors (fig. 4.10a) on their way to or from the principal reflecting interface with which they are associated (often called *peg-leg multiples*) are important in determining the waveforms of the events recorded on a seismogram. These peg-leg multiples have the effect of delaying part of the energy and therefore lengthening the wavelet. The stronger peg-leg multiples often have the same sign as the primary since successive large impedance contrasts tend to be in opposite directions (otherwise the successive large changes in velocity would cause the velocity to exceed its allowable range). This effectively lowers the signal frequency as time increases. Fig. 4.10b shows how a simple impulse (such as an explosion might generate) becomes modified as a result of it passing through a sequence of interfaces. Frequency spectra of these wavelets in fig. 4.10c show the loss of high frequencies with time. Fig. 4.10d shows that the attenuation ($\eta\lambda$; see (2.101)) because of peg-leg multiples is 0.085 ± 0.055 dB.

Ghosts are the special type of multiple illustrated in fig. 4.9. The energy traveling downward from the shot has superimposed upon it energy which initially traveled upward and was then reflected downward at the base of the low-velocity layer (LVL) (in land surveys) or at the surface of the water (in marine surveys). A 180° phase shift, equivalent to half a wavelength, occurs at the additional reflection and hence the effective path difference between the direct wave and the ghost is $(\frac{1}{2}\lambda + 2D_s)$, where D_s is the depth of the shot below the reflector producing the ghost. The interference between the ghost and the primary depends on the fraction of a wavelength repre-

sented by the difference in effective path length; since the seismic wavelet is made up of a range of frequencies, the interference effect will vary for the different components. Thus, the overall effect on the wavelet shape will vary as D_s is varied. Relatively small changes in shot depth can result in large variations in reflection character, creating serious problems for the interpreter. Therefore the depth of the shot below the base of the weathering or the surface of the water is maintained as nearly constant as possible.

Ghosts affect directivity as well as waveshape. Fig. 4.11 shows a point source at a depth $D_s = c\lambda$; if the source emits the wave $A \cos(\kappa r - \omega t)$, then the effect at P is

$$\psi_P = A \cos(\kappa r_1 - \omega t) - A \cos(\kappa r_2 - \omega t)$$
$$= A \cos(\kappa r - \omega t - \kappa c\lambda \cos\theta)$$
$$\qquad - A \cos(\kappa r - \omega t + \kappa c\lambda \cos\theta).$$

Expanding and noting that $\kappa\lambda = 2\pi$, we get approximately

$$\psi_P = 2A \sin(\kappa r - \omega t)\sin(2\pi c \cos\theta)$$
$$= 2A \sin(2\pi c \cos\theta)\cos(\kappa r - \omega t - \tfrac{1}{2}\pi). \qquad (4.3)$$

Thus, for r large, the total wave motion lags the original wave motion by 90° and has an amplitude which depends on θ. Since a wavelet contains a spectrum of frequencies, different components will add differently at various angles θ, resulting in changes in waveshape.

Ghosts are especially important in marine surveys because the surface of the water is almost a perfect reflector and consequently the ghost interference will be strong. If D_s is small in comparison with the dominant wavelengths, appreciable signal cancellation will occur. At depths of 10 to 15 m interference is constructive for frequencies of 40 to 25 Hz which is in the usual seismic range. The same effect occurs with the upcoming signal from the reflectors we wish to map. Hence marine sources and marine detectors are often operated at such depths.

A particularly troublesome type of multiple produces the coherent noise known as *singing* (also called *ringing* or *water reverberation*) which is frequently encountered in marine work (and occasionally on land). This is due to multiple reflections in the water layer. The large reflection coefficients at the top and bottom of this layer result in considerable energy being reflected back and forth repeatedly, the reverberating energy being reinforced periodically by reflected energy. Depending upon the water depth, certain frequencies are enhanced and as a result the record looks very sinusoidal (see fig. 4.12). Not only is the picking of reflections difficult but measured traveltimes and dip moveouts will probably be in error. This type of noise and its attenuation are discussed in §8.1.2d.

Fig.4.6. Effect of a hole in a reflector. (Courtesy Chevron.)

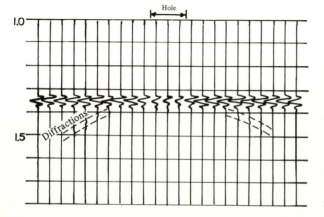

(c) *Long-path multiples*. The strongest long-path multiples involve reflections at the surface, the seafloor or (on land) at the base of the low-velocity layer (LVL, also called weathered layer; see table 2.3, §2.4.5) where the reflection coefficient is very large because of the large acoustic-impedance contrast. Since this type of multiple involves at least two reflections at depth, its amplitude depends mainly on the magnitude of reflection coefficients at depth and multiples of this type will be observed as distinctive events when these coefficients are abnormally

high. Since R in (2.129) may be as large as 0.7 at the base of the LVL and perhaps 0.2 for the strongest interfaces at depth, the maximum effective R for such multiples will be of the order of $0.2 \times 0.7 \times 0.2 = 0.03$. This value is in the range of typical reflection coefficients so that such multiples may have sufficient energy to be confused with primary events. The principal situation where weaker long-path multiples may be observable is where primary energy is nearly absent at the time of arrival of the multiple energy so that the gain of the recording system is very high. It

Fig.4.7. Diffraction from a perfectly reflecting wedge. (Courtesy Chevron.)

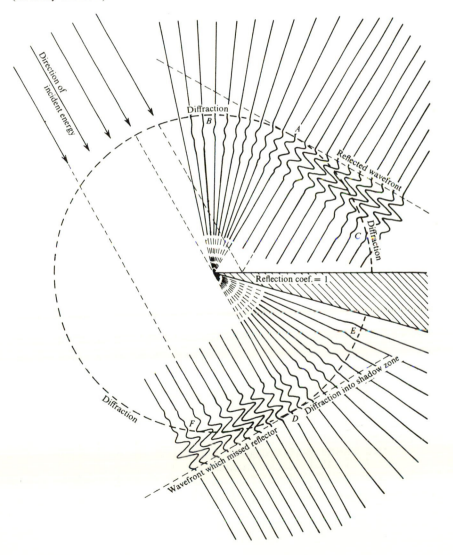

is important that multiples be recognized as such so that they will not be interpreted as reflections from deeper horizons.

Because velocity generally increases with depth, multiples usually exhibit more normal moveout than primary reflections with the same traveltime. This is the basis of the attenuation of multiples in common-depth-point processing which will be discussed in §8.2.5. However, the difference in normal moveout is often not large enough to identify multiples. The attenuation of multiples is also the primary objective of predictive deconvolution (§8.2.1e).

The effect of dip on multiples which involve the surface or the base of the LVL can be seen by tracing rays using the method of images. In fig. 4.13 we trace a multiple arriving at symmetrically disposed geophones, G_1 and

G_{24}. The first image point, I_1, is on the perpendicular from S to AB as far below AB as S is above. We next draw the perpendicular from I_1 to the surface of the ground where the second reflection occurs and place I_2 as far above the surface as I_1 is below. Finally, we locate I_3 on the perpendicular to AB as far below as I_2 is above. We can now draw the rays from the source S to the geophones (working backwards from the geophones). The dip moveout is the difference between the path lengths $I_3 G_{24}$ and $I_3 G_1$; it is about double that of the primary ($I_1 G_{24} - I_1 G_1$). The multiple at the shotpoint will appear to come from I_3 which is updip from I_1, the image point for the primary, and $I_3 S$ is slightly less than twice $I_1 S$. Hence we can see that if the reflector dips, the multiple involves a slightly different portion of the reflector than the primary and has a traveltime slightly less than double the traveltime of the primary. The latter fact makes identifying multiples by merely doubling the arrival time of the primary imprecise whenever appreciable dip is present. The arrival time of the multiple will be approximately equal to that of a primary reflection from a bed at the depth of I_1. If the actual dip at I_1 is not double that at AB (and one would not in general expect such a dip), then the multiple will appear to have anomalous dip. If the multiple should be misidentified as a primary, one might incorrectly postulate an unconformity or updip thinning which might lead to erroneous geologic conclusions.

In deep-water marine surveys multiples of the sea-floor may be so strong as to virtually obliterate primary reflections. Furthermore, the range of reflection angles may be so great that the effective reflection coefficient varies widely for different offset distances. Fig. 4.14 shows the build-up of amplitude observed near the critical angle

Fig.4.8. Effects of a step. (From Angona, 1960.) (a) Model; (b) split-spread record with source over step: P = direct wave, S = surface wave, R_r and R_L = reflections from upper and lower part of step, R_b = reflections from base of model, D_1 and D_1' = diffractions from top and base of step, M = multiple, $(R)_{PS}$ = converted wave. Note phantom diffractions (e.g., D_1'') which continue R_b beyond the center.

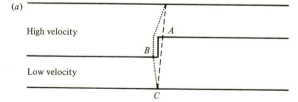

(a)

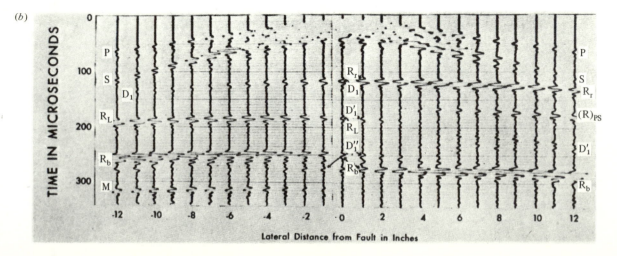

(b)

Fig.4.9. Types of multiples.

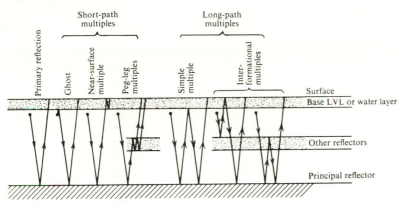

Fig.4.10. Changes in waveshape resulting from passage through a layered sequence. (*a*) Schematic diagram showing peg-leg multiples adding to the wavetrain; (*b*) waveshapes after different traveltimes (after O'Doherty and Anstey, 1971); (*c*) frequency spectra of wavetrains shown in (*b*); (*d*) histogram of attenuation coefficients due to peg-leg multiples, based on data from 31 wells in different basins; η and λ have the same meaning here as in §2.3.2*b*, except for the attenuation mechanism; (after Schoenberger and Levin, 1978).

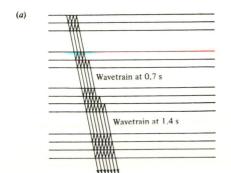

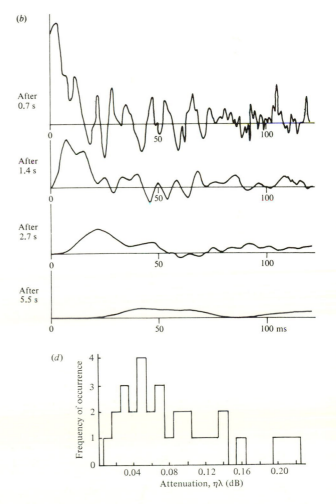

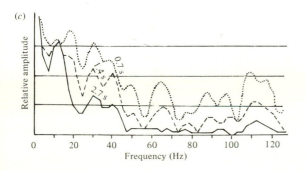

and how this occurs on different traces for successive multiples. Predictive deconvolution is generally ineffective in attenuating such multiples because it assumes that for any given trace the reflection coefficients at the reflectors involved are constant. However, if seafloor dip is small, the reflection angles for the simple reflection at offset x will be the same as for the first multiple at offset $2x$ (fig. 4.15), and for the next multiple at offset $3x$, etc., so that the primary on one trace can be used to predict and compensate for the multiple seen on another trace at another offset. This provides the basis for radial multiple suppression.

4.2.3 *Refractions*

Refractions (head waves) are relatively low frequency and usually include several cycles. They are often followed by a number of parallel alignments, that is, they seem to involve a long wavetrain consisting of several cycles, and the number of cycles usually increases with distance, an effect called *shingling*. Multiple reflections as shown in fig. 4.16*a* probably play a large part in this process of adding legs. Some of the energy which peels off the refractor will be returned to the refractor at the critical angle by reflection at beds parallel to the refractor to form a delayed head wave. Such multiple bounces can occur anywhere and they will add to the same head wave (when the reflector to refractor distance is small) and add additional head waves (when the distance is great enough that the reflected refraction is separated from the primary refraction). These effects will increase as the offset distance increases, so that the net effect will be to shift the energy later in the head wave and to make more legs appear.

Fig.4.11. Directivity of source plus ghost. S = source; I = image (effective source of the ghost); P = observing point.

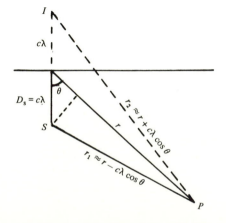

Head waves are not usually considered to be a problem on reflection records. Head waves from shallow refractors may be observed wherever the offset distance exceeds the critical distance and, as shown in fig. 3.14 and by (3.33), the critical distance is less than the refractor depth for $V_2/V_1 > 2.24$. Velocity contrasts of this magnitude are possible, for example, where carbonates or evaporites are overlain by sands or shales. Where long spreads are used, as in much common-depth-point recording, refractors as deep as 1 or 2 km may produce head waves on the long-offset traces. While head waves generally have straight alignments (prior to normal-moveout correction) in contrast to reflections and diffractions, they often do not appear on enough traces to make this feature useful in identifying them. On stacked records they usually disappear in the muting of the first-break region (the upper right triangular region of fig. 4.2).

4.2.4 *Reflected refractions*

Where a refractor is terminated, such as shown in figs. 4.16*b* or *c*, the head wave will be reflected backward. It may appear on the later portion of a reflection record some distance from the actual refractor termination. Where the refractor termination is nearly perpendicular to the seismic line, the reflected head wave will have a nearly straight alignment with an apparent velocity approximately the negative of the refractor velocity. The head wave will be reflected even though Snell's law is not satisfied at the refractor termination. The refractor termination may be either against lower or higher impedance material (though the former is more common) so the reflected head wave may have polarity either opposite to or the same as the head wave. Where the refractor is massive, reflections as in fig. 4.16*d* may appear much like reflected head waves (fig. 4.16*c*). Where the refractor termination is off to the side of the line (fig. 4.16*e*), the event may have some curvature (pseudo-normal-moveout) across the record (see problem 4.5).

4.2.5 *Surface waves*

Surface waves (often called *ground roll*) are usually present on reflection records. For the most part these are Rayleigh waves with velocities ranging from 100 to 1000 m/s or so. Ground-roll frequencies are usually lower than those of reflections and refractions, often with the energy concentrated below 10 Hz. Ground-roll alignments are straight, just as in the case of refractions, but they represent much lower velocity. The envelope of ground roll builds up and decays very slowly and often includes many cycles. Surface-wave energy generally is high enough even in the reflection band to override all but the strongest

Fig.4.12. Seismic record showing singing. (Courtesy Petty-Ray Geophysical.) (*a*) Field record; (*b*) same after singing has been removed by deconvolution processing (§8.1.2*d*).

Time (s)

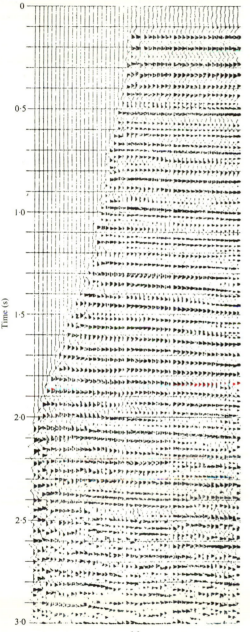

(*a*)

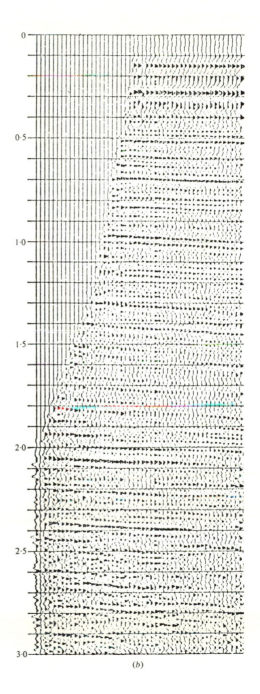

(*b*)

reflections; however, because of the low velocity, different geophone groups are affected at different times so that only a few groups are affected at any one time. Sometimes there is more than one ground-roll wavetrain, each with different velocities. Occasionally where surface waves are exceptionally strong, in-line offsets are used to permit recording the desired reflections before the surface waves reach the spread.

Surface-wave effects can be attenuated by the use of arrays (§5.3.3 and problem 5.5), by frequency filtering (ground roll can be seen on the 0–6 Hz and slightly on the 6–12 Hz panels of fig. 8.11 and by apparent-velocity filtering (see fig. 8.23).

Fig.4.13. Raypath of a multiple from a dipping bed.

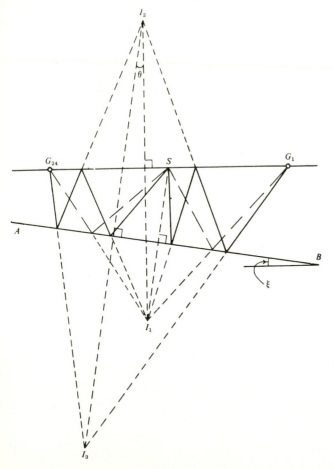

4.3   Characteristics of reflections
4.3.1   *Effects of reflector curvature*

Geometrical focusing as a result of curvature of a reflector affects the amplitude of a reflection.

In a constant-velocity medium, the wave generated by a point source is spherical with radius of curvature equal to the source-to-wavefront distance. We shall assume that such a wave encounters a spherical reflecting horizon centered directly below the shotpoint and with radius $\rho_{\mathscr{S}}$. We can then apply the well-known formula of geometrical optics for reflections from curved mirrors,

$$1/u + 1/v = 2/R,$$

where u and v are the object and image distances and R is the radius of the mirror; positive R corresponds to a concave mirror, negative v to a virtual image. The object and image distances are equivalent to the radii of curvature of the incident and reflected waves ρ_i and ρ_r, the radius of the curvature of the reflecting surface $\rho_{\mathscr{S}}$ being positive for a syncline, negative for an anticline. The formula now becomes

$$1/\rho_i + 1/\rho_r = 2/\rho_{\mathscr{S}}. \tag{4.4}$$

If the distance to the reflector is h, $\rho_i = h$ for a point source and

$$\rho_r = h\rho_{\mathscr{S}}/(2h - \rho_{\mathscr{S}}). \tag{4.5}$$

When the reflected wave reaches the surface, its radius of curvature ρ is

$$\rho = \rho_r - h = 2h\{(\rho_{\mathscr{S}} - h)/(2h - \rho_{\mathscr{S}})\}. \tag{4.6}$$

We can express this in terms of curvature, the reciprocal of radius of curvature, in a normalized form as

$$\frac{h}{\rho} = -\left(\frac{(h/\rho_{\mathscr{S}}) - \frac{1}{2}}{(h/\rho_{\mathscr{S}}) - 1}\right). \tag{4.7}$$

Fig. 4.17 is a graph of (4.7). Reflector curvature of $\pm\infty$ corresponds to a point diffractor, while zero curvature ($h/\rho_s = 0$) corresponds to a plane reflector for which the convex-upward curvature of the wavefront produces normal moveout. When the reflector center of curvature is at the surface ($\rho_{\mathscr{S}} = h$), the reflected energy concentrates to a point. Curvature greater than this produces a buried-focus as discussed below. Fig. 4.18a shows the image points for a diffracting point, anticline, plane and syncline, while fig. 4.18b shows the buried-focus case.

Consider a cone of energy from the source which is reflected from a spherical cap, *MN* (fig. 4.18a); the reflected energy is spread over a larger area at the surface for the anticline than for the plane, and over a smaller area for the gentle syncline. Thus reflections should appear stronger over gentle synclines and weaker over anticlines.

When a syncline has a radius of curvature less than its depth, ρ_r is positive while ρ is negative and the energy passes through a focus below the surface (see fig. 4.18b); this is a *buried-focus* situation. Obviously the likelihood of a buried-focus increases with reflector depth. A reflection involving a buried-focus is called a *reverse branch*. The sense of traverse of the reverse branch is reversed from the usual, i.e., as the shotpoint travels from left to right, the reflecting point travels from right to left.

Where the shotpoint and geophone are not coincident (that is, for offset traces), the reflected wave may focus even where the reflector's center of curvature is not below the surface, as in fig. 4.18c. Thus, long-offset traces may involve buried-focus effects even where short-offset

traces do not. Common-depth-point stacking where short- and long-offset traces are combined after normal-moveout correction usually does not allow for this situation correctly.

When the curvature of a syncline is not constant, as in fig. 4.18d reflections may be obtained from more than one part of the reflector and the reflected energy has *multiple branches*, most commonly three. The two deeper reflections in fig. 4.19 involve multiple branches; each shows branches from each flank of the syncline plus a reverse branch from the curved bottom of the syncline. Fig. 8.26 shows a number of buried-focus effects. Three-dimensional multiple-branch effects are discussed in §4.3.3.

Just as light can be focused by passing through a lens, seismic waves can also be focused by curved velocity

Fig.4.14. Change of amplitude with offset for seafloor multiples, offshore Eastern Canada. Trace spacing, 100 m, offset of first trace = 425 m. The amplitude build-up occurs near the critical angle. (See problem 4.12.) (Courtesy Chevron.)

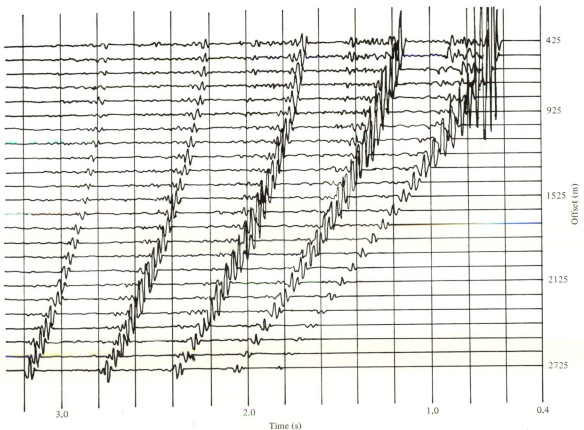

Fig.4.15. Relation between offset and angle of reflection for primary and multiple reflections from a flat reflector.

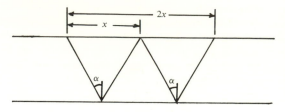

Fig.4.16. Reflected refractions. (*a*) Multiply-reflected refraction; (*b*), (*c*), (*d*) refractions reflected from faults or saltdomes; (*e*) isometric drawing of refractions reflected from termination of refractor to side of spread; paths are shown from source *S* to geophones G_1 and G_2; the lines shown by long dashes indicate head wave travel in the refractor.

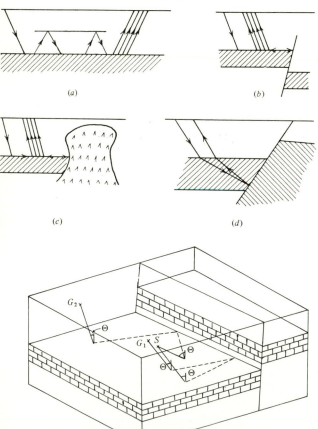

surfaces which result in seismic rays being bent by refraction; such situations are often very complex. Curvature at the base of the weathering can be especially important because of the large velocity contrast usually associated with this surface. Variations in permafrost thickness and gas accumulations can also cause focusing effects (fig. 4.20).

Fermat's principle explains that a wave will take that raypath for which the traveltime is stationary with respect to minor variations of the raypath, that is, for which the change in the traveltime for an incremental change in raypath is zero. For most situations the raypath involves the minimum traveltime between the points, that is, travel over any neighboring path will take longer; hence Fermat's principle is often called the *principle of least time* or the *brachistochrone principle*. Snell's law, Huygens' principle and many other laws of geometrical optics can be derived from this principle.

An incident wavefront approaching the reflector in a buried-focus situation (fig. 4.18*e*) encounters the reflector before the wave reaches the reflecting point *R* which satisfies Snell's law. Contributions to the reflection from the region surrounding *R* will thus arrive earlier than the reflection from *R* itself, that is, the reflection point involves a maximum in a Fermat's principle sense, thus contrasting with the more usual situation where the

Fig.4.17. Normalized curvature of wavefront at surface, h/ρ, as a function of reflector curvature, $h/\rho_\mathscr{S}$, for a point source. The letters *d*, *a*, *r*, *s* refer to curves in fig. 4.18*a*, the letter *b* to fig. 4.18*b*.

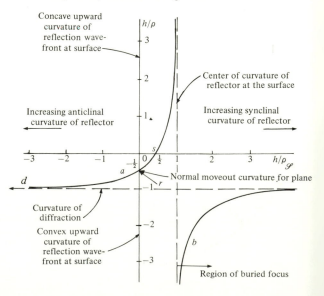

Fig.4.18. Effects of reflector curvature. (*a*) Effect on wavefront curvature as reflector changes from anticline to syncline; *d*, diffracting point; *a*, anticline; *r*, plane reflector; *s*, syncline. Respective wavefronts *D*, *A*, *R*, *S*; image points I_D, I_A, I_R, I_S; and radii of curvature, ρ_D, ρ_A, ρ_R, ρ_S. (*b*) Raypaths passing through focal region when center of curvature *C* is below surface. (*c*) Long-offset traces S_1 and S_2 pass through focus even where short-offset traces S_0 do not. (*d*) Reflections from several points on reflector when curvature changes. (*e*) Incident wavefront of radius ρ_i impinging on reflector of radius $\rho_{\mathscr{S}}$ when $\rho_i > \rho_{\mathscr{S}}$.

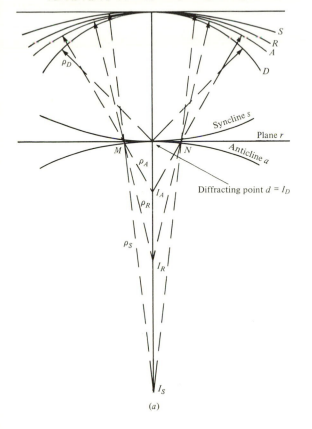

(*a*)

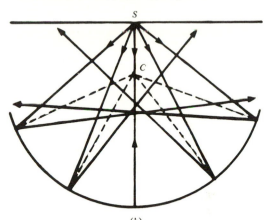

(*b*)

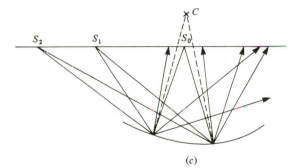

(*c*)

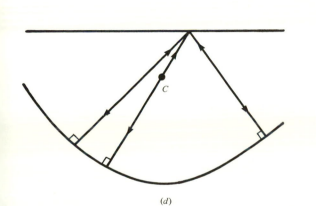

(*d*)

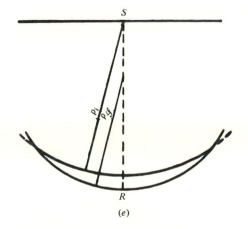

(*e*)

reflection point involves minimum traveltime. The fact that reflection contributions from the region surrounding the reflection point arrive earlier manifests itself as a change in the waveshape of the reverse branch reflection compared with normal branches.

Equation (2.43) for a harmonic spherical wave is

$$\psi = (A/r)\,e^{j\omega(r/V - t)}.$$

Neglecting the time-variant portion, we can write

$$\psi = \frac{A\omega}{V}\left\{ \frac{\cos{(\omega r/V)}}{\omega r/V} + j\frac{\sin{(\omega r/V)}}{\omega r/V} \right\}. \qquad (4.8)$$

These two terms are plotted in fig. 4.21 along with the corresponding terms for a plane wave (which does not involve the r in the denominator). The phase of a wave passing through a focus behaves differently from that of a plane wave passing the point. The phase, $\gamma = \tan^{-1}$ (imaginary part/real part), approaches π as r approaches zero

Fig.4.19. Reflections from cylindrically curved reflectors. For all three reflectors, radius of curvature = 1000 m, V = 2000 m/s. Depths to the bottom of the syncline are 800, 1200 and 1600 m respectively. The traces are 100 m apart with sources and receivers coincident. (Courtesy Chevron.)

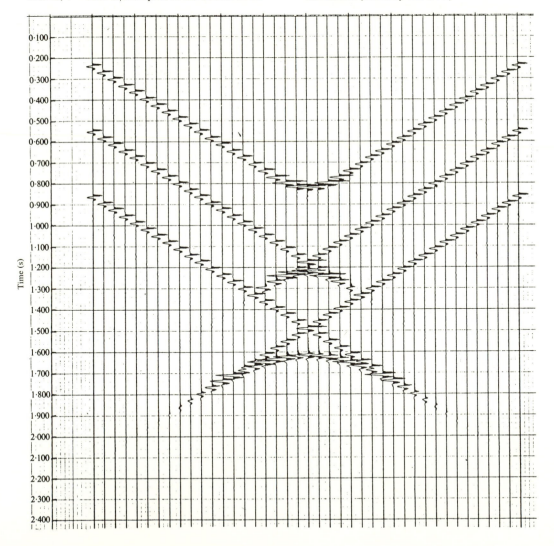

Fig.4.20. Focusing produced by velocity variations. (*a*) Low velocity in a gas accumulation producing buried-focus effects on deeper reflections. (*b*) High-velocity wedge producing buried-focus effects.

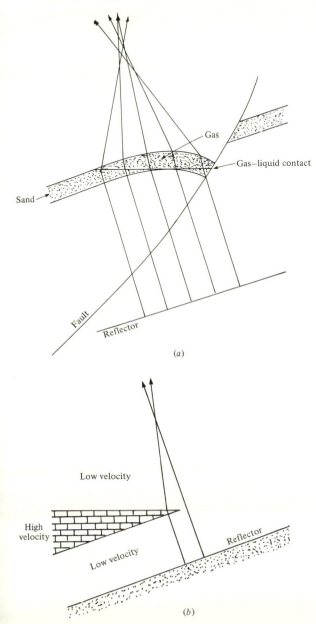

(a)

(b)

for negative *r* and 0 as *r* approaches zero for positive *r*. With wavefronts which pass through a focus (as for the reverse branch), this phase shift is π if the wavefront is spherical or $\frac{1}{2}\pi$ if it is cylindrical. A $\frac{1}{2}\pi$ phase shift can be seen by comparing the waveshape of the reverse branch for the lower event in fig. 4.19 with other events; the reflectors here are cylindrical. Such a phase shift is rarely useful in identifying buried-focus events, but it will affect calculations of reflector depth where picking is done systematically on the same phase, for example, always picking troughs, and it affects common-depth-point stacking.

4.3.2 *Resolution*

(*a*) *Vertical resolution.* Resolution refers to the minimum separation between two features such that we can tell that there are two separate features rather than only one. With respect to seismic waves we may think of (i) how far apart (in space or time) two interfaces must be to show as separate reflectors or (ii) how far two features involving a single interface must be separated to show as separate features.

Fig.4.21. Behavior of a spherical wave (solid curve) passing through a focal point compared to a plane wave (dashed curve). (*a*) The $\cos(\omega r/V)/(\omega r/V)$ term in (4.8); (*b*) the $\sin(\omega r/V)/(\omega r/V)$ term.

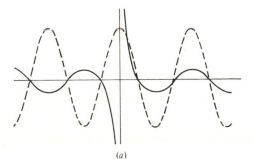

(a)

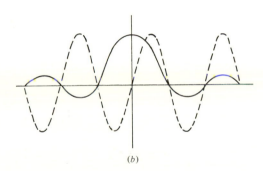

(b)

Fig. 4.22 (caption on facing page)

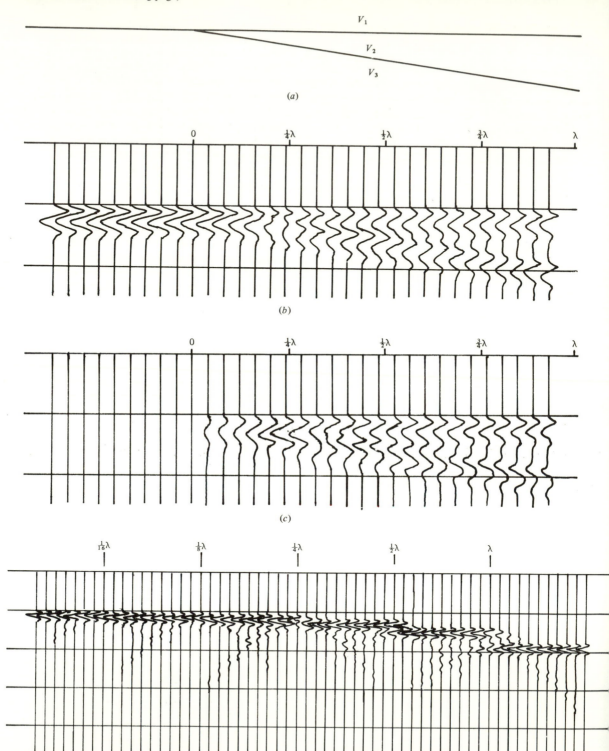

V_1

V_2

V_3

(a)

0    $\frac{1}{4}\lambda$ $\frac{1}{2}\lambda$ $\frac{3}{4}\lambda$ λ

(b)

0    $\frac{1}{4}\lambda$ $\frac{1}{2}\lambda$ $\frac{3}{4}\lambda$ λ

(c)

$\frac{1}{16}\lambda$ $\frac{1}{8}\lambda$ $\frac{1}{4}\lambda$ $\frac{1}{2}\lambda$ λ

(d)

If seismic wavelets were extremely sharp, resolution would not be a problem. However real seismic wavelets involve a limited range of frequencies and hence have appreciable breadth.

Let us first consider resolution in the direction of wave propagation. For two horizontal reflectors a distance Δz apart, the deeper reflection lags behind the shallower by the fraction $2\Delta z/\lambda$ of a wavelength. We can tell that there are two waves when the arrival of the second wave causes a perceptible change in the appearance of the first wave.

Rayleigh (Jenkins and White, 1957, p. 300) defined the *resolvable limit* as being when the two events are separated by a half-cycle, so that interference effects are maximized; the interference may be constructive (for example, fig. 4.22c) or destructive (fig. 4.22b). Ricker (1953b) used a slightly different criterion which resulted in a slightly smaller resolvable limit; Widess (1973) also used a different criterion.

For a boxcar frequency spectrum (see (10.130)), the wavelet shape is that of a sinc function. The Rayleigh criterion is equivalent to a width of approximately $2/(3v_u)$, where v_u is the upper frequency limit of the boxcar (see problem 4.9). Thus we must record higher frequencies if we are to achieve higher resolution (Sheriff, 1977).

As an illustration of vertical resolution, fig. 4.22b shows the effect of a wedge whose velocity is intermediate between that above and below it. The thickness of the wedge is indicated as fractions of a wavelength; the waveshape clearly indicates more than one reflector when the wedge thickness exceeds $\frac{1}{4}\lambda$. Fig. 4.22c shows a wedge with a velocity different from that of the surrounding material. Again the resolvable limit is about $\frac{1}{4}\lambda$, where the amplitude is at a maximum because of constructive interference. Note that the wedge still produces a significant reflection when it is appreciably thinner than the resolvable limit, and a bed only $\frac{1}{20}\lambda$ to $\frac{1}{30}\lambda$ in thickness is detectable although its thickness cannot be determined from the waveshape. Similar resolution considerations

apply to structural features; fig. 4.22d shows a series of faults with varying amounts of throw, the resolution being about $\frac{1}{4}\lambda$. These three examples suggest that the Rayleigh definition of resolvable limit is reasonable.

(b) *Horizontal resolution.* To discuss horizontal resolution, we introduce the concept of Fresnel zones used in optics (Burnett *et al.*, 1958). In fig. 4.23, S is a source and coincident detector, SP_0 is perpendicular to a reflecting plane and $R_1, R_2, \ldots$ are such that the distances SP_0, $SP_1, SP_2, \ldots$ differ by $\frac{1}{4}\lambda$; thus $h_{n+1} - h_n = \frac{1}{4}\lambda$. Generally $h_n \gg R_n \gg \lambda$; hence $R_n \approx (\frac{1}{2}n\lambda h_0)^{\frac{1}{2}}$ and $\Delta\mathscr{S} \approx \pi\lambda h_0(n - \frac{1}{2})$ where $\Delta\mathscr{S}$ is the area of each of the zones.

We shall calculate the energy returning to S from the $(n + 1)$th zone. If we apply (2.110) to a circle with the origin over the center, the integration with respect to θ merely multiplies by 2π. If we apply this result to fig. 4.23, h becomes h_0 and ξ becomes h_n. If we now calculate the Laplace transform $\Phi(s)$ for two circles of radii R_n and R_{n+1} and subtract the second from the first, we obtain the effect of the $(n + 1)$th zone:

$$\Phi(s) = \tfrac{1}{2}ch_0\{(1/h_n^2)\,\mathrm{e}^{-2sh_n/V} - (1/h_{n+1}^2)\,\mathrm{e}^{-2sh_{n+1}/V}\}. \tag{4.9}$$

This solution corresponds to a unit impulse source, $\delta(t)$ (see §2.3.5b). Taking the inverse transform (see (10.159) and (10.166)), we have in the time domain

$$\phi(t) = \tfrac{1}{2}ch_0\{(1/h_n^2)\delta(t - t_n) - (1/h_{n+1}^2)\delta(t - t_n - \tfrac{1}{2}T)\}, \tag{4.10}$$

where $t_n = 2h_n/V$, $t_{n+1} = (2h_n + \tfrac{1}{2}\lambda)/V = t_n + \tfrac{1}{2}T$, T being the period.

If the input at the source had been $A\cos\omega t$ instead

Fig. 4.23. Geometry of Fresnel zones.

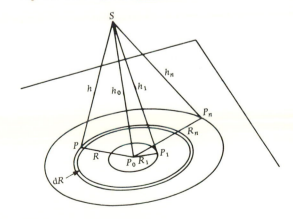

◄   Fig.4.22. Reflections illustrating vertical resolution. (After Sheriff, 1976.) (a) Model of a wedge; (b) reflection from the wedge where $V_3 > V_2 > V_1$; wedge thickness is indicated in fractions of the dominant wavelength. (c) Reflection from a thin wedge embedded in a medium of different velocity ($V_3 = V_1 \neq V_2$). (d) Reflections from a faulted reflector, with the fault throw indicated as fractions of the dominant wavelength.

of an impulse, (4.9) would have the additional factor $As/(s^2 + \omega^2)$ (see (10.162)) and upon taking the inverse transform, (10.166) would give

$$\phi(t) = \tfrac{1}{2}ch_0 A\{(1/h_n^2)\, \text{step}\,[t - t_n]\cos\omega(t - t_n)$$
$$- (1/h_{n+1}^2)\, \text{step}\,[t - t_n - \tfrac{1}{2}T]\cos\omega(t - t_n - \tfrac{1}{2}T)\}$$
$$= \tfrac{1}{2}ch_0 A\{(1/h_n^2)\, \text{step}\,[t - t_n]$$
$$+ 1/(h_n + \tfrac{1}{4}\lambda)^2\, \text{step}\,[t - t_n - \tfrac{1}{2}T]\}\cos\omega(t - t_n)$$
$$\approx (ch_0/2h_n^2)A\{\text{step}\,[t - t_n]$$
$$+ (1 - \lambda/2h_n)\, \text{step}\,[t - t_n - \tfrac{1}{2}T]\}\cos\omega(t - t_n).$$

When $t > (t_n + \tfrac{1}{2}T)$,

$$\phi(t) \approx A\{(ch_0/h_n^2)(1 - \lambda/4h_n)\}\cos\omega(t - t_n)$$
$$\approx A\{ch_0/h_n^2\}\cos\omega(t - t_n). \qquad (4.11)$$

Since t_n and t_{n+1} differ by $\tfrac{1}{2}T$, contributions from successive zones are alternately plus and minus. Thus the effect at S is an alternating series, $\phi_T = S_1 - S_2 + S_3 - S_4 + \dots$, where S_{n+1} is a positive quantity given by A times the factor in the curly brackets in (4.11). Since S_n decreases as n increases, the series converges and we may write

Fig.4.24. Nomogram for determining Fresnel-zone radii. A straight line connecting the two-way time and the frequency intersects the central line at the same point as a straight line connecting the average velocity and the radius of the zone. For example, a 20 Hz reflection at 2.0 s and a velocity of 3.0 km/s has a Fresnel-zone radius of 470 m.

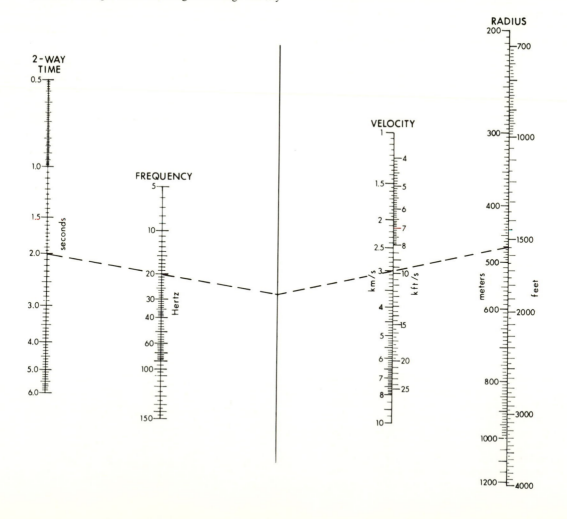

$$\phi_T = \tfrac{1}{2}S_1 + (\tfrac{1}{2}S_1 - S_2 + \tfrac{1}{2}S_3) + (\tfrac{1}{2}S_3 - S_4 + \tfrac{1}{2}S_5) + \cdots$$

The terms in parentheses are approximately zero, hence

$$\Phi_T \approx \tfrac{1}{2}S_1, \qquad (4.12)$$

that is, the major contribution to the reflected signal comes from the first Fresnel zone. Therefore the radius of this zone,

$$R_1 = (\tfrac{1}{2}\lambda h_0)^{\frac{1}{2}} = \tfrac{1}{2}V(t/v)^{\frac{1}{2}}, \qquad (4.13)$$

where h_0 is the depth, t the arrival time, V the average velocity and v the frequency, can be taken as a measure of the horizontal resolution. For a depth of 3 km and velocity of 3 km/s ($t = 2$ s), the Fresnel-zone radius (the adjective 'first' is often dropped for the first zone) ranges from 300 m to 470 m for frequencies of 50 to 20 Hz (see fig. 4.24). Fig. 4.25 shows the response of small segments of a reflecting surface at a depth of 1500 m for a 30 m dominant wavelength for which the Fresnel-zone radius is 150 m. When the reflector dimensions are somewhat smaller than the Fresnel zone, the response is essentially that of a diffracting point.

The foregoing discussion assumes a point source, for which the travel paths from source back to detector differ by a half-cycle for successive Fresnel zones. Fresnel zones are sometimes specified with respect to a plane incident wave rather than a spherical wave, in which case the half-cycle differences between successive Fresnel

zones have to be accommodated entirely in the reflector-to-detector portion of the travel path. This results in an enlargement of the Fresnel zone, the radius in this case being

$$R_1 = \tfrac{1}{2}(\lambda h_0)^{\frac{1}{2}} = \tfrac{1}{2}V(2t/v)^{\frac{1}{2}}. \qquad (4.14)$$

The Fresnel zone concept can be thought of as involving the volume which contains trajectories longer than the minimum (Fermat principle) path by $\tfrac{1}{2}\lambda$ (see fig. 4.26). This concept is useful for certain types of problems.

(c) Resolution of migrated sections. One of the concepts in migrating seismic data (§8.3.4) is that of downward-continuation of the reflected wave field, that is, effectively lowering the geophones into the Earth so that they approach the reflectors. This procedure shrinks the Fresnel zone so that the limitations are the migration process, noise and the spatial sampling interval.

As actually applied, migration is performed on sampled data (sampled spatially, that is, at discrete geophone locations as well as at discrete time intervals). Spatial aliasing considerations (§8.1.2b) limit the angle of approach which in turn limits the amount of dip which can be migrated.

Horizontal uncertainty almost always exceeds vertical uncertainty. Schneider (1978) gives an example showing that 5% velocity error smears the position of a

Fig.4.25. Reflections from strips of varying widths (expressed in terms of the Fresnel-zone width). (After Neidell and Poggiagliolmi, 1977.) (a) Cross-section of model; spacing of vertical lines equals the Fresnel-zone diameter; (b) seismic section.

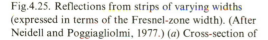

(a)

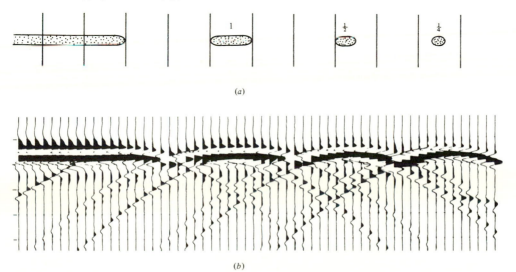

(b)

discontinuity over a horizontal distance equal to 5% of its depth, and local velocities are usually not known better than this. Thus the aggregate effect of migration on horizontal resolution is difficult to quantify because it depends on so many factors, especially the presence of noise.

4.3.3 *Three-dimensional effects*

In §3.1.2 we defined reflecting point as the point at which the angle of incidence equaled the angle of reflection. Seismic data are usually mapped at reflecting points. A line connecting reflecting points is called the *subsurface trace*. There is a subsurface trace for each reflector. Where there is a component of dip perpendicular to the seismic line, the reflecting point lies to the side of the seismic line rather than below it. Such cross-dip effects are often ignored, usually because the cross-dip is not measured, and sometimes such neglect leads to serious errors. In §4.2.1, for example, we assumed that the diffracting point was in the vertical plane containing the seismic line; if instead it is from the truncation of a reflector by a fault which is not perpendicular to the line, then the diffracting point may move along the fault as

source and/or receivers move and the curvature of the diffraction on the seismic record will be less than that given by (4.1) or (4.2).

In §4.3.1 we examined reflector curvature effects and in fig. 4.19 showed multiple-branch effects. If a seismic line crosses a syncline other than at right angles (line *BB'* in fig. 4.27*b*), the reflection branches may come from opposite sides of the seismic line and the length of the reverse branch may be stretched out and thus show smaller curvature (compare fig. 4.27*d* with fig. 4.27*c*). In the extreme situation where the seismic line is parallel to the axis of the syncline, the multiple branches appear as parallel horizontal reflectors (fig. 4.27*e*). Where the syncline is plunging, the different branches will not be parallel.

The Fresnel-zone concept of §4.3.2*b* replaces a reflecting 'point' with a reflecting 'area', the area of the first Fresnel zone. Features to the side of the reflecting point but within the reflecting area will produce effects on the seismic line, as shown in fig. 4.28. Hilterman (1970) showed that such effects can make structures appear to be appreciably larger in area than they actually are.

Fig.4.26. Volume involving all raypaths within the first Fresnel zone. Energy from the source *S* traveling within the shaded zone (for example, by the dotted paths) can reach the geophone *G* within a half-cycle of the first arrival (heavy line) which travels by the geometrical raypath. Note that a portion of the shaded area penetrates below the reflector. Circles indicate wavefronts. (After Hagedoorn, 1954.)

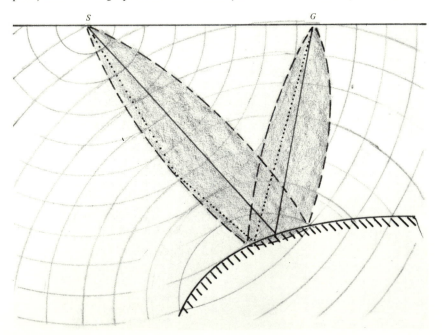

4.3.4 *Shape of the seismic wavelet*

It has been shown that resolution depends on the highest frequency components present. However, we usually do not record appreciable energy above 60 Hz or so. Two possible reasons for the loss of high frequencies, absorption and peg-leg multiples, have been discussed. If there are small time-shift errors, stacking tends to combine high-frequency components out-of-phase while still combining low-frequency components in-phase (see fig. 5.19). Other filtering actions may also attenuate high frequencies. One objective of data processing is to restore

high frequencies as much as possible and to obtain the effect of a short wavelet of specified shape.

Most of the natural mechanisms which affect wavelet shape are minimum-phase (§8.1.4 and §10.6.6) or nearly so (see Sherwood and Trorey, 1965). A minimum-phase wavelet has the energy concentrated in the early part of the wavelet. Real wavelets are causal (§10.6.6*a*) and the first detectable peak or trough is always delayed from the onset of the wavelet so that picking and timing of arrival times are always late. Furthermore, since the high frequencies are attenuated more rapidly than low fre-

Fig.4.27. Buried-focus effects on lines at different angles to strike. (*a*) Cross-section of syncline; (*b*) contour map of reflector showing subsurface traces for

BB′ (——) and *CC′* (-------); (*c*) arrival times along line *AA′*; (*d*) arrival times along line *BB′*; (*e*) arrival times along *CC′*.

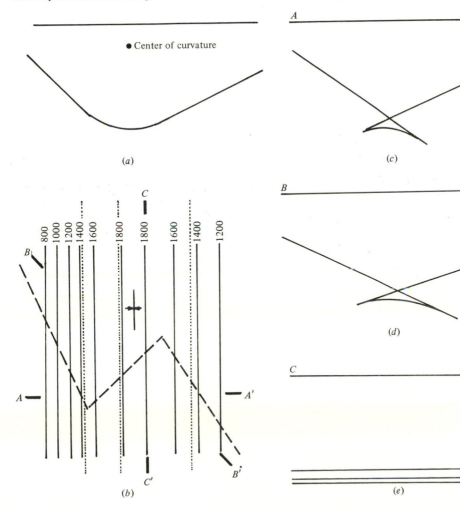

quencies, as arrival times increase the frequency spectrum shifts towards the low frequencies, the wavelet builds up more slowly and the delay between a reflection's onset and its detection increases as arrival times increase. Sometimes compensation for delays is made in data processing but the compensation for time-variable effects is very difficult to achieve correctly.

The effective wavelet is sometimes converted to a zero-phase equivalent in data processing (§8.1.4 and §10.6.6*d*). Zero-phase wavelets are symmetrical and the time scale is shifted (not always correctly) so that the center of the wavelet indicates the arrival time. Conversion to a zero-phase equivalent does not necessarily solve the time-variable filtering effects referred to above.

Fig.4.28. Line across box structure. Trace spacing, 85 m, Fresnel-zone radius, 280 m for 30 Hz. (Courtesy Geoquest.) (*a*) Model; (*b*) line *A* across top of box; (*c*) line B which does not cross box.

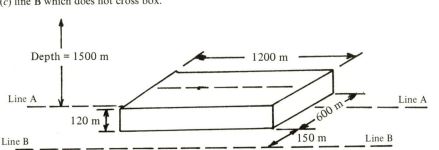

(*a*)

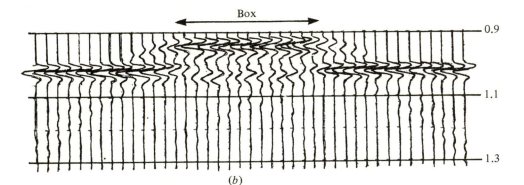

(*b*)

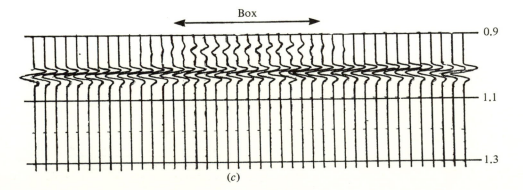

(*c*)

The most common zero-phase wavelet is the *Ricker wavelet* (Ricker, 1940, 1944, 1953a), expressed in the time-domain (fig. 4.29a) as

$$f(t) = (1 - 2\pi^2 v_{\mathrm{M}}^2 t^2) e^{-\pi^2 v_{\mathrm{M}}^2 t^2}, \qquad (4.15)$$

or in the frequency domain (fig. 4.29b) as

$$F(v) = (2/\sqrt{\pi})(v^2/v_{\mathrm{M}}^3) e^{-v^2/v_{\mathrm{M}}^2}, \quad \gamma(v) = 0, \qquad (4.16)$$

where $f(t) \leftrightarrow F(v)$ and v_{M} is the peak frequency (see problem 4.11). The distance between flanking side lobes in the time-domain, T_{D} (fig. 4.29a), is

$$T_{\mathrm{D}} = (\sqrt{6}/\pi)/v_{\mathrm{M}}.$$

4.4 Noise
4.4.1 *Types of seismic noise*

The reliability of seismic mapping is strongly dependent upon the quality of the records. However, the quality of seismic data varies tremendously. At one extreme we have areas where excellent reflections (or refractions) are obtained without any special measures being taken; at the other extreme are those areas in which the most modern equipment, extremely complex field techniques and sophisticated data processing methods do not yield usable data (often called NR *areas*, that is,

Fig.4.29. Ricker wavelet. (*a*) Time-domain representation; (*b*) frequency-domain representation.

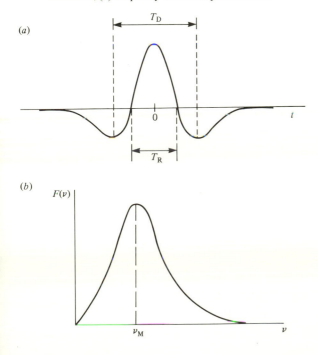

(*a*)

(*b*)

areas of 'no reflections'). In between these extremes lie the vast majority of areas in which useful results are obtained but the quantity and quality of the data could be improved with beneficial results.

We use the term *signal* to denote any event on the seismic record from which we wish to obtain information. Everything else is *noise*, including coherent events which interfere with the observation and measurement of signals. The *signal-to-noise ratio*, abbreviated S/N, is the ratio of the signal energy in a specified portion of the record to the total noise energy in the same portion. Poor records result whenever the signal-to-noise ratio is small; just how small is to some extent a subjective judgment. Nevertheless, when S/N is less than unity, the record quality is usually marginal and deteriorates rapidly as the ratio decreases further.

Seismic noise may be either (*a*) coherent, or (*b*) incoherent. *Coherent noise* can be followed across at least a few traces; *incoherent noise* is dissimilar on all traces and we cannot predict what a trace will be like from a knowledge of nearby traces. Sometimes the difference between coherent and incoherent noise is merely a matter of scale and if we had geophones more closely spaced the incoherent noise would be seen as coherent noise. Nevertheless, incoherent noise is defined with respect to the records being used without regard for what closer spacing might reveal.

Incoherent noise is often referred to as *random noise* (spatially random) which implies not only non-predictability but also certain statistical properties; more often than not the noise is not truly random. (It should be noted that spatial randomness and time randomness may be independent; the usual seismic trace is apt to be random in time since we do not know when a reflection will occur on the basis of what the trace has shown previously, with the exception of multiples.)

Coherent noise is sometimes subdivided into (*a*) energy which travels essentially horizontally, and (*b*) energy which reaches the spread more-or-less vertically. Another important distinction is between (*a*) noise which is repeatable, and (*b*) noise which is not; in other words, whether the same noise is observed at the same time on the same trace when a shot is repeated. The three properties – coherence, travel direction and repeatability – form the basis of most methods of improving record quality.

Coherent noise includes surface waves, reflections or reflected refractions from near-surface structures such as fault planes or buried stream channels, refractions carried by high-velocity stringers, noise caused by vehicular traffic or farm tractors, multiples, etc. (Olhovich 1964). All of the preceding except multiples travel essentially horizontally and all except vehicular noise are repeatable on successive shots.

Incoherent noise which is spatially random and also repeatable is due to scattering from near-surface irregularities and inhomogeneities such as boulders, small-scale faulting, etc.; such noise sources are so small and so near the spread that the outputs of two geophones will only be the same when the geophones are placed almost side-by-side. Non-repeatable random noise may be due to wind shaking a geophone or causing the roots of trees to move generating seismic waves, stones ejected by the shot and falling back to the Earth near a geophone, a person walking near a geophone and so on.

4.4.2 *Attenuation of noise*

The signal-to-noise ratio generally depends on frequency. If the noise has appreciable energy outside the principal frequency range of the signal, frequency filtering can be used to advantage. Very-low-frequency components (such as high-energy surface waves rich in low frequencies) may be filtered out during the initial recording provided the low frequencies are sufficiently separated from the reflection frequencies. However, the spectrum of the noise often overlaps the signal spectrum and then frequency filtering is of limited value in improving record quality. With modern digital recording, the only low-frequency filtering in the field is often that resulting from the limited low-frequency response of the geophones.

If we add several random noises together, there will be some cancellation because they will be out-of-phase with each other. Assume that we have n geophones, each of which is responding to coherent signal S but has random noise N_i superimposed on it. A measurement x_i will then be

$$x_i = S + N_i.$$

Our best estimate of the signal is the mean of n measurements, $\bar{x} = (1/n)\Sigma_i x_i$ and we identify the standard deviation σ with the rms noise, that is,

$$\sigma^2 = (1/n)\sum_i N_i^2.$$

The signal-to-noise ratio of the sum is thus

$$\left(\sum_i x_i\right) \Big/ \left(\sum_i N_i^2\right)^{\frac{1}{2}} = n\bar{x}/(n\sigma^2)^{\frac{1}{2}} = n^{\frac{1}{2}}\bar{x}/\sigma. \qquad (4.17)$$

As n becomes large, σ approaches a limit which depends on the statistical properties of the noise; hence the signal-to-noise ratio varies as $n^{\frac{1}{2}}$ for n large and random noise.

Cancellation of random noise does not place any restrictions on the geophone locations (except that they cannot be too close together) but the cancellation of

coherent noise requires that geophones be spaced along the direction of the wave travel. If the geophones are distributed uniformly over one wavelength, we may achieve a much larger signal-to-noise improvement than for random noise (see §5.3.3*b* and problem 5.5*b*).

These principles are the bases of the use of multiple geophones or multiple sources (called geophone or source *arrays*; see §5.3.3) to cancel noise. If we connect together, for example, 16 geophones which are spaced far enough apart that the noise is spatially random but still close enough together that reflected energy traveling almost vertically is essentially in-phase at all 16 geophones, the sum of the 16 outputs will have a signal-to-noise ratio four times greater than the output when the 16 geophones are placed side-by-side. If on the other hand we are attenuating coherent noise and the 16 geophones are spread evenly over one wavelength of a coherent noise wavetrain (for example, ground roll), then the coherent noise will be greatly reduced.

Noise can also be attenuated by adding together traces shot at different times or different places or both. This forms the basis of several *stacking* techniques including vertical stacking, common-depth-point stacking, uphole stacking and several more complicated methods. The gain in record quality often is large because of a reduction in the level of both random and coherent noise. Provided the static and dynamic corrections are accurately made, signal-to-noise improvements for random noise should be about 2.5 for 6-fold and about 5 for 24-fold stacking (or 8 and 14 dB respectively).

Vertical stacking involves combining together several records for which both the source and geophone locations remain the same. It is extensively used with weak surface energy sources and many marine sources (see §5.4.3 and §5.5.3). Vertical stacking usually implies that no trace-to-trace corrections are applied but that corresponding traces on separate records are merely added to each other. The effect, therefore, is essentially the same as using multiple shots or multiple source units simultaneously. In difficult areas both multiple source units and vertical stacking may be used. In actual practice the surface source is moved somewhat (3–10 m) between successive recordings. Up to 20 or more separate records may be vertically stacked, but the stacking of many records becomes expensive both in field time and in processing costs whereas the incremental improvement becomes small after the first few. Vertical stacking is often done in the field, sometimes in subsequent processing. Marine vertical stacking rarely involves more than four records because at normal ship speeds the ship moves so far during the shooting that the data are *smeared* when stacked; smearing means that changes in the reflecting points affect the arrival times so much that the signal may be adversely affected by summing (the effect is similar to using a very large geophone or source array).

The *common-depth-point* technique which is almost universally used is very effective in attenuating several kinds of noise. The summation traces comprise energy from several shots using different geophone and shotpoint locations. The field technique will be discussed in §5.3.1 and the processing (which is almost always done in a processing center rather than in the field) in §8.2.5. A number of other noise-attenuating techniques (such as apparent-velocity filtering) are also applied in processing and described in chapter 8.

Problems

4.1. (a) Show that the slope of the diffraction event with shotpoint S_2 in fig. 4.3b approaches $\pm 1/V$ for large x. [Hint: expand the expression in (4.2) for $x \gg h$.] (b) What is the slope of the asymptote for fig. 4.3c?

4.2. (a) Given that $0 < c < +1$ in (4.3), discuss the conditions under which the amplitude of ψ_P is zero. (b) Compare the amplitude and energy of ghosts generated at the base of the low-velocity layer and at the surface of the ground, given that $V_H = 1.9$ km/s, $V_W = 0.40$ km/s, and that the densities just below and within the LVL are 2.0 and 1.6 g/cm³ respectively. (c) Assume that the LVL is $\frac{1}{2}\lambda$ in thickness and that $\eta\lambda = 0.6$ dB for the LVL; now what are the ratios of the ghost amplitudes and energies?

4.3. An airgun is fired at a depth of 10 m. The waveform includes frequencies in the range 10–80 Hz, the amplitudes of the 10 Hz and 80 Hz components being the same near the shot. Compare the amplitudes of these components for the wavelet plus ghost at considerable distance from the shot in the directions 0°, 30°, 60°, and 90° to the vertical.

4.4. A multiple reflection is produced by a horizontal bed at a depth of 1.100 km, the average velocity being 2.95 km/s. A primary reflection from a depth of 3.250 km coincides with the multiple. (a) By how much do arrival times differ at points 200, 400, 800 and 1000 m from the

shotpoint? (b) If the shallow bed dips 10°, how much do the arrival times at 400 and 800 m change? What dip of the deeper horizon would this correspond to?

4.5. (a) A horizontal refractor at a depth of 1.20 km is being mapped along a N–S line. The overburden velocity is 2.50 km/s and the refractor velocity 4.00 km/s. The refractor is terminated by a linear vertical fault 3.50 km from the shotpoint. Determine the traveltime curves when the fault strikes (i) E–W, (ii) N–S, (iii) N30°W. (b) Repeat for the E–W fault for a refractor which dips 10° to the north with the shotpoint to the south. (c) What effects will the manner of terminating the refractor have, that is, how will the amplitude of the reflected refraction depend on the dip of the terminating fault? (d) Most commonly a refractor will terminate against rock of lower acoustic impedance but the opposite situation can also happen; what differences will this make? (e) Extend the profile for case (a) part (i) an appreciable distance beyond the fault so as to plot the diffraction from the refractor termination. Assume uniform 2.50 km/s material beyond the refractor termination.

4.6. (a) Determine the traveltime curves for the refraction $SMNPQR$ and the reflected refraction $SMNTUWPQR$ in fig. 4.30. (b) Determine the traveltime curves when both refractor and reflector dip 8° down to the left, the depths shown in fig. 4.30 now being the slant distances to the interface at S. (c) What happens when the reflector dips 3° to the left and the refractor 5° to the left?

4.7. Redraw fig. 4.18a for a plane wave incident on the reflector, and explain the significance of the changes which this makes.

4.8. Explain why waves in fig. 4.22b interfere destructively and in fig. 4.22c constructively when the wedge thickness is $\frac{1}{4}\lambda$.

4.9. (a) A wavelet has a flat frequency spectrum from 0 to v_u, above which the spectrum is zero; show that the Rayleigh criterion gives a resolvable limit t_r where $t_r = 0.715/v_u$. [Hint: Transform a boxcar spectrum $box_{2v_0}(v)$ to the time-domain and find the location of the first

Fig.4.30. Multiply-reflected refraction.

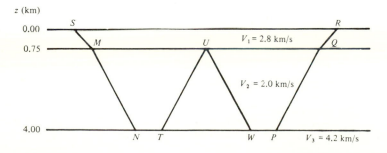

trough.] (b) Show that the value of t_r for a wavelet with a flat spectrum extending from v_L to nv_L (that is, m octaves wide where $n = 2^m$) is given by the solution to the equation

$$nx \cos nx - \sin nx - x \cos x + \sin x = 0,$$

where $x = 2\pi v_L t_r$. (c) Solve the equation in (b) for $m = 3$, 2 and 1.5 and compare the relation between t_r and m. (d) Noting that part (a) involves an infinite number of octaves, how many octaves bandwidth are required to give nearly the same resolution?

4.10. A saltdome is roughly a vertical circular cylinder with a flat top of radius 400 m at a depth of 3.2 km. If the average velocity above the top is 3.8 km/s, what is the minimum frequency which will give a recognizable reflection from the dome?

4.11. (a) Using the result $e^{-at^2} \leftrightarrow (\pi/a)^{\frac{1}{2}} e^{-\omega^2/4a}$, verify that

(4.16) follows from (4.15); (b) show that v_M is the peak of the frequency spectrum; (c) show that $T_D/T_R = \sqrt{3}$ (see fig. 4.29) and that $T_D v_M = \sqrt{6}/\pi$.

4.12. Assume that fig. 4.14 shows relative amplitudes correctly (divergence having been allowed for). The water depth is 420 m and the velocity below the seafloor 2590 m/s. (a) If the reflection coefficient is maximum at the critical angle, on what traces would you expect the maximum amplitude for the first, second, third and fourth multiples? (b) What should be the ratio of the amplitude of the successive multiples on the short-offset trace? How do these calculations compare with observations? What unaccounted-for factors affect this comparison?

4.13. In the table below, classify different types of events and noise on the basis of commonly observed characteristics.

	Incoherent (spacing > 2 m)	Predictable trace-to-trace	Predictable from earlier arrivals	Repeatable on successive shots	Low apparent velocity (<25 km/s)	Distinctive apparent velocity	Moveout linear with offset	Attenuated by frequency filtering	Attenuated by geophone arrays	Attenuated by CDP stacking	Attenuated by app. vel. filtering	Attenuated by front-end muting	Distinguishable by 3-component recording
Primary reflections, dip <10°													
Multiples													
Primary reflections, dip >25°													
Diffractions													
Head waves													
Reflected refractions													
Ground roll													
Wind noise													
Airwave													
SV-waves (reflected)													
SH-waves (reflected)													

Fig.4.31. Reflections and diffractions involving a horst.

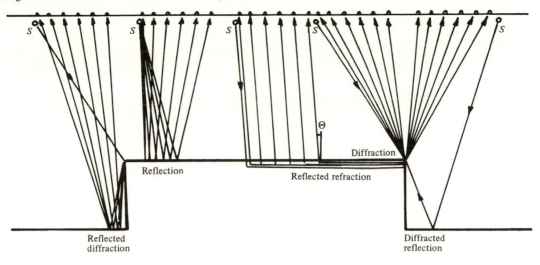

Reflection

Diffraction

Reflected refraction

Reflected
diffraction

Diffracted
reflection

Fig.4.32. Two models which give identical first-arrival
curves. (After Pautsch, 1927.)

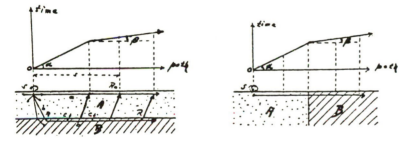

Fig.4.33. Directivity of a harmonic source located at
various depths below a free surface; h = depth
as a fraction of wavelength. (After Waters, 1978.)
(a) Source at $h = 0.1$; (b) $h = 0.5$; (c) $h = 1.0$.

(a)

(b)

(c)

4.14. Draw arrival-time curves for the five events shown in fig. 4.31.

4.15. Pautsch (1927) showed that a horizontal or vertical interface could give identical first-arrival curves (fig. 4.32). Add secondary arrivals and reflections to these diagrams to show how they can distinguish the two cases.

4.16. Show that (4.3) gives the directivity diagrams shown in fig. 4.33.

4.17. Select random numbers between ± 9 to represent noise N_i and add to each a signal $S = 2$. Sum 4 values of $S + N_i$ and determine the mean, the standard deviation σ, and the ratio signal/(signal + noise). Repeat for 8, 16 and 32 values. Note how the mean converges toward S as the number of values increases, how σ approaches a limiting value (which depends on the statistical properties of the noise; see §10.3.10), and how the ratio signal/(signal + noise) converges toward 1. [A table of random numbers is given in appendix C.]

5

Reflection field methods and equipment

Overview

Field methods and equipment for the acquisition of seismic reflection data vary considerably, depending on whether the area is land or marine, on the nature of the geologic problem, and on the accessibility of the area.

The organization of field crews who acquire seismic data and procedures for carrying out a survey on land are described. The common-depth-point (CDP) method is the field method most widely used today. An array of geophones usually feeds each data channel and sometimes source arrays are also used. Arrays have response characteristics which depend on the spectrum, velocity and direction from which a seismic wave comes; this property is used to attenuate certain types of noise. The selection of field parameters depends on both geologic objectives and noise conditions. Methods used for special objectives, such as uphole surveys, three-dimensional methods, extended resolution, vertical profiling and use of channel waves, are discussed.

Explosives are the predominant source of seismic energy on land, but nearly half the land crews use sources on the surface, Vibroseis being the predominant surface source. Geophones convert seismic energy into electrical signals, which are recorded in either analog or digital forms. Newer equipment enables stacking in the field, digitization near the geophones and the use of telemetry.

Marine surveys (§5.5) acquire data at a very fast rate and high hourly cost; these fundamental facts plus special positioning methods distinguish marine operations from those on land. The bubble effect often determines the waveshape generated. The predominant marine energy source is the airgun but explosive, implosive, and other types are sometimes used. Pressure-sensitive hydrophones mounted in streamers are used to detect the returning energy.

Corrections have to be made for elevation and weathering variations to prevent such variations from influencing (and sometimes completely obscuring) the reflection data on which interpretation of geologic features is based. The first, and often the most important, corrections are those calculated by the field crew. Additional (or residual) corrections are subsequently made in data processing.

S-waves have not been generally employed in exploration work although they have potential for giving additional geologic information. Methods of generating S-waves usually also generate P-waves. Appreciable experimentation with S-wave generation and analysis techniques is currently underway.

5.1 Field-crew organization

5.1.1 *Clients and contractors*

Most geophysical work (95% in 1979, based on money spent) is performed by seismic contractors for client companies which utilize the geophysical data obtained in their search for oil. Contracting companies often process data and manufacture geophysical equipment both for their own use and for sale to others. Some client companies operate their own field crews in addition to employing contract crews.

The client company usually designates one of its personnel as *client representative* or 'birddog'. He is the communication channel between his company and the crew supervisor and is responsible for the client's interests, which include checking that sound techniques are employed and that data quality and crew efficiency are maintained.

5.1.2 *Land-crew organization*

Land seismic crews differ greatly in size, ranging from two to three people for a shallow survey for engineering objectives to more than a hundred people for surveys in jungle areas where many men are required to cut trails. Consequently, the organization of the crew varies, but those shown in fig. 5.1 are representative.

The *supervisor* is usually responsible for a single crew, occasionally more. He generally reports to an area manager; the supervisor is usually a professional geophysicist and is responsible for quality and cost control at all levels. In a large operation a *party chief* may be next in

line of responsibility. The *computer* or *seismologist* in the area office prepares the stacking charts (fig. 5.5b) and other documents for use in subsequent data processing; this includes calculating *field corrections* (§5.6.2) based on elevation, uphole and first-break data. Occasionally the seismologist or party chief will make a preliminary interpretation. The supervisor and/or party chief have many non-geophysical responsibilities such as cost control, data security, labor relations, public relations, record keeping, tax calculation, personnel training, safety, etc; on large crews many of these are often handled by an administrator.

The field work is managed by a *party manager*; his primary responsibility is to obtain maximum production and adequate quality at reasonable cost. He hires most of the field helpers and he is also responsible for safety, equipment maintenance, maintaining adequate supplies, paying bills and operation of the field camp.

The *surveyor* has the responsibility of locating the line and the points along the line in their proper places. As the advance man on the ground, he anticipates difficulties and problems which shooting the line will involve and seeks to avoid or resolve them. This often involves investigating alternative line locations so that the objectives of the survey may be achieved at minimum cost. Frequently he is assisted by a *permit man* who contacts land owners or tenants and secures permission to work on their land. The surveyor is also assisted by rodmen who help measure the seismic line and map the area. In areas of difficult access he may also supervise brush cutters or bulldozer operators who clear the way for the seismic line.

The *observer* is usually next after the party manager in authority in the field. His primary responsibility is to operate the instruments but he is also responsible for the actual field layouts and data acquisition. The observer is usually assisted by a junior observer and by a crew of 'jug hustlers' who actually lay out the cable and the geophones.

Other members of a field crew vary in numbers depending on the nature of the survey. A crew may have one to four drillers, occasionally more, plus assistants to help drill and haul water for the drilling operation, or one to perhaps four operators of surface source units (see §5.4.3). A *shooter* is responsible for detonating explosives at the proper time and for cleaning up the shothole area afterwards. Cooks and mechanics may be included where operations are performed out of field camps.

5.1.3 *Marine-crew organization*

A marine seismic crew is usually headed by a party manager who is responsible for the seismic work performed. A marine seismic crew includes several observers and junior observers (who relieve each other during the continuous operations), several technicians who operate the navigation equipment, helpers who handle the streamer (see §5.5.4) when it is being let out or pulled in, and mechanics who maintain the source units.

The seismic ship is under the command of a captain whose authority is final; however, the captain usually follows the instructions of the party manager except where safety is involved. Many supporting personnel are required for a seismic ship: cooks, maintenance engineers, mechanics, etc. Data processing or analysis is usually not done aboard ship.

5.2   Field methods for land surveys
5.2.1   *The program*

Usually the seismic crew receives the *program* from the client in the form of lines on a map which indicate where data are to be obtained. The seismic crew ordinarily is not responsible for laying out the program, a factor

Fig.5.1. Organization of a land seismic crew using explosives in drilled holes. (*a*) Crew in developed accessible area; (*b*) crew in remote area involved in continuous operations; party managers, survey super-

visors, drill supervisors, observers and other field personnel work 14–28 consecutive days before being rotated for time off.

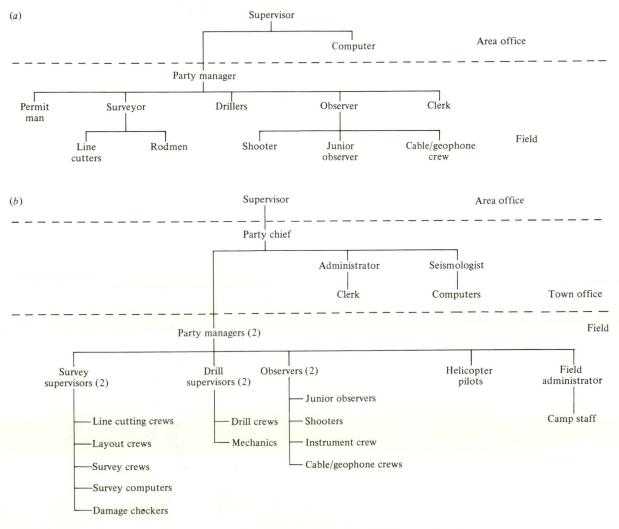

which sometimes contributes to the failure of surveys since it encourages the attitude that the objective is to shoot a certain program rather than to obtain certain information. Without understanding program objectives clearly, the wrong decisions in operating alternatives may be made. Good practice (Agnich and Dunlap, 1959) is to 'shoot the program on paper' before beginning the survey, estimating what the data are likely to show, anticipating problems which may occur, asking what alternatives are available and how data might be obtained which will distinguish between alternative interpretations.

Before beginning a survey the question should be asked, 'Is it probable that the proposed lines will provide the required information?' Data migration (§5.6.3) may require that lines be located elsewhere than directly on top of features in order to measure critical aspects of a structure. Crestal areas may be so extensively faulted that lines across them may be non-definitive. The structures being sought may be beyond seismic resolving power. Near-surface variations along a proposed line may be so large that the data are difficult to interpret whereas moving the seismic line a short distance may improve data quality. Obstructions to shooting along a proposed line may increase difficulties unnecessarily whereas moving the line slightly may achieve the same objectives at reduced cost. Where the dip is considerable, merely running a seismic line to a wellhead may not tie the seismic data to the well data. Lines may not extend sufficiently beyond faults and other features to establish the existence of such features unambiguously or to determine fault displacements. In general, lines should extend with full coverage beyond the area of interest to a distance equal to the target depth. Lines may cross features such as faults so obliquely that their evidences are not readily interpretable. Lack of cross control may result in features located below the seismic line being confused by features to the side of the line.

5.2.2 *Permitting*

Once the seismic program has been decided upon, it is usually desirable (or necessary) to meet with the owners of the land to be traversed. Permission to enter upon lands to carry out a survey may involve a payment, often a fixed sum per shothole, as compensation in advance for 'damages which may be incurred'. Even where the surface owners do not have the right to prevent entry, it is advantageous to explain the nature of the impending operations. Of course, a seismic crew is responsible for damages resulting from their actions whether or not permission is required to carry out the survey.

5.2.3 *Laying out the line*

Once the preliminary operations have been completed, the survey crew lays out the lines to be shot. This is usually done by a transit-and-chain survey which determines the positions and elevations of both the shotpoints and the centers of geophone groups. The *chain* is often a wire equal in length to the geophone group interval. Using this chain the successive group centers are laid out along the line, each center being marked in a conspicuous manner, commonly by means of brightly coloured plastic

Fig.5.2. Laser survey instrument. The top instrument measures the range to the rod at distances up to 3 km with accuracy of ± 6 mm. The lower instrument measures the grade to 3 seconds of arc, calculates the elevation of the rod in terms of the instrument elevation and measures horizontal angles with respect to a reference direction. (Courtesy Keuffel and Esser.)

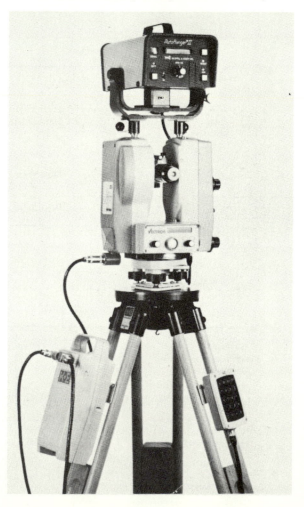

ribbon called *flagging*. The transit is used to keep the line straight and to obtain the elevation of each group center by sighting on a rod carried by the lead chainman.

Electronic surveying instruments (fig. 5.2) which depend on the traveltime of light (a laser beam) from the transit to the rod and back are increasing in use on seismic crews. These instruments operate very quickly and are very accurate; a digital readout gives the distance, elevation difference and direction, thus greatly reducing the likelihood of reading errors. Such instruments may be used to run in a base survey and to tie to benchmarks and wellheads even where not used for all shotpoint and geophone group locations. Radio systems (§5.5.5b) are sometimes used for horizontal control, especially in marsh and shallow-water areas where elevation control can be obtained from the water level.

Small survey minicomputers (fig. 5.3) are used to reduce survey data, adjust loop closures, tie in sun shots and other control, etc. Such a computer in the basecamp can reduce the data and plot an updated map for the next day's use, and also transmit the data to a host computer over a telephone line so that the progress can be followed in a headquarters office.

The surveyor should indicate in his data and on his map the locations of all important features such as streams, buildings, roads, fences, etc. The surveyor also plans access routes so that drills, recording trucks, etc., can get to their required locations most expeditiously.

In areas of difficult terrain or heavy vegetation, trail-building or trail-cutting crews may be required. These often precede the survey crew but usually are under the direct supervision of the surveyor who is therefore responsible for the preparation of a straight trail in the proper location.

Fig.5.3. Micromap computer. Survey data are entered by means of the keyboard and stored on a floppy disk. The computer reduces the survey data and plots a map. (Courtesy Seiscom Delta.)

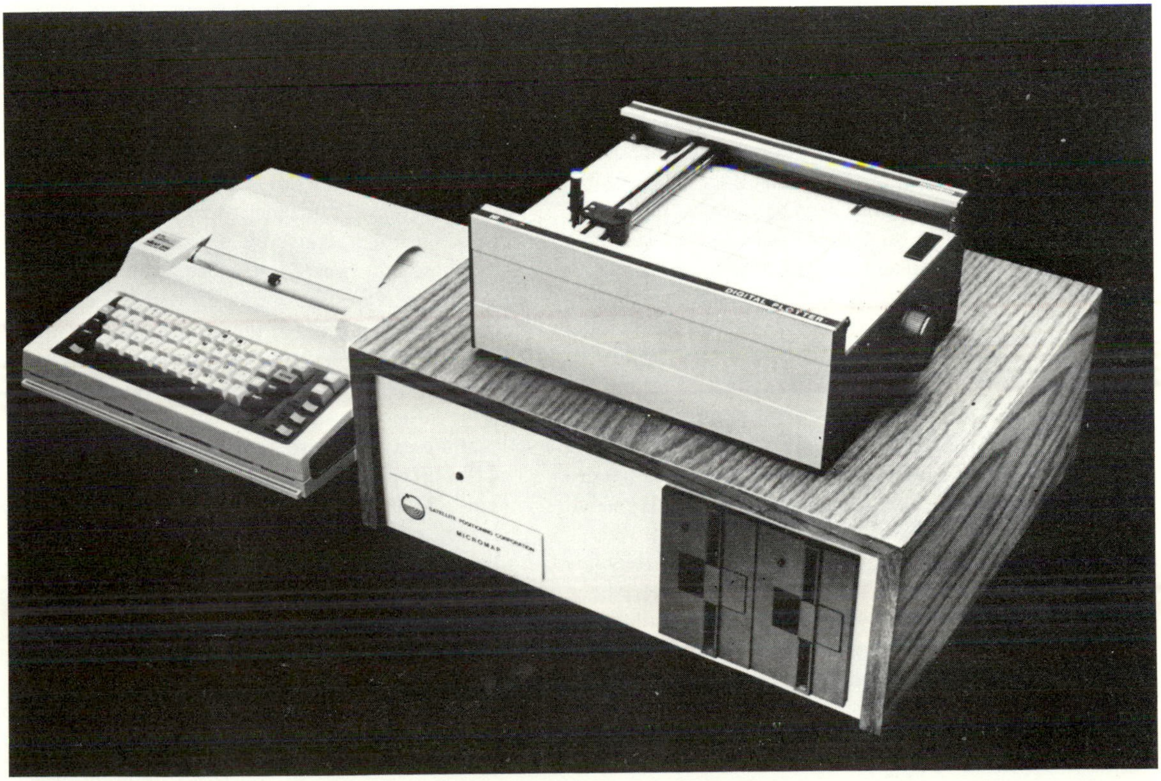

5.2.4 *Shothole drilling*

The next unit on the scene is the drilling crew (when dynamite is used as the energy source); depending upon the number and depth of holes required and the ease of drilling, a seismic crew may have from one to ten drilling crews. Whenever conditions permit, the drills are truck-mounted. Water trucks are often required to supply the drills with water for drilling. In areas of rough terrain the drills may be mounted on tractors or portable drilling equipment may be used. In swampy areas the drills are often mounted on amphibious vehicles. Usually the drilling crew places the dynamite in the holes before leaving the site. Drilling is often a major part of data acquisition costs.

5.2.5 *Recording*

The drilling crews are followed by the recording unit. This unit can be divided into three groups on the basis of primary function: the shooting crew responsible for loading the shotholes (if the drillers have not already loaded them) and for setting off the dynamite; the jug hustlers who lay out the cables, place the geophones in their proper locations and connect them into the cables; and the recording crew which does the actual recording of signals.

The jug hustlers lay out the cables and connect the geophones, the observer tests and adjusts the amplifiers and other units of the recording system and tests the cables for continuity to ensure that all geophones are connected. Finally, he gives the signal to the shooter via telephone or radio to fire the charge (or to start the surface source units). When the shooter receives the signal, he 'arms' his *blaster*, the device used to set off the explosive (see §5.4.2), by a safety switching arrangement and advises the observer that he is ready. The observer then presses an 'arm' button which causes a 'tone' to be transmitted to the shooter and starts the recording system. A coded series of signals sent from the recording equipment actually fires the shot. The blaster then transmits back to the recording equipment the shot instant (*time-break*). The data are recorded and the observer studies a monitor record to see that the record is free of obvious defects, and the equipment is moved to the location for the next shot.

With common-depth-point recording, source points are close together – of the order of a hundred meters apart versus 400 or 500 meters ($\frac{1}{4}$ or $\frac{1}{3}$ mile) with conventional split-dip recording. The high production and high efficiency needed in order to achieve a low cost per kilometer have altered field procedures. At the same time the redundancy of coverage has lessened the dependence on any individual record so that occasional missed records can

be tolerated. Also, the broad dynamic range of digital recording has removed some of the need for filtering in the field and the need to tailor instrument settings to particular local conditions.

Cost considerations dictate that the recording operation must not wait on other units. Field experimentation is minimized and time is not spent repeating shots to improve data quality or on frequent physical moving of the recording truck. Shotholes may be drilled for the entire line before recording even begins so that the recorder never waits on the drills. Extra cables and geophones are laid out and checked out in advance of the recording unit. A *roll-along switch* makes it possible for the recording unit to be located physically at a different place than where it is located electrically. The recording unit connects to the seismic cable at any convenient location, for example, the intersection of a road and the seismic line. The roll-along switch is adjusted so that the proper geophones are connected and the shooters are advised to activate the blaster. Following the shot, the shooters move to the next shothole (which is not very far away) and the observer adjusts the roll-along switch so that the next geophones are connected. The time between shots may be only a few minutes and the recording truck may not move all day long. Holes where misfires occur are not reloaded and reshot. The shooting unit often walks the line since they need no equipment except the blaster and firing line and perhaps shovels to fill in the shothole after the shot. The recording unit does not have to traverse the line and so is subject to less abuse. Damages are reduced because less equipment traverses the line. Thus other benefits accrue besides increased efficiency of recording.

When a seismic crew uses a surface energy source, the source unit moves into place and the signal from the recorder activates the source so that the energy is introduced into the ground at the proper time. Despite the fact that an explosive may not be involved, terms such as 'shot' and 'shotpoint' are still used. The energy from each surface source is usually small compared to the energy from a dynamite explosion so that many records are made for each sourcepoint and vertically stacked to make a single record. Several source units may be used and these may advance a few meters between the component 'subshots' which will be combined to make one profile. It is not uncommon to use three or four source trucks and combine thirty or so component subshots.

A monitor record is usually made in the field to make certain that all of the equipment is functioning properly and also to determine weathering corrections (discussed in §5.6.2). However, monitor records are usually not used for interpretation.

The magnetic tapes are shipped to a data-processing center where corrections are applied and various processing techniques are used, for example, velocity analysis, filtering, stacking (see chapter 8). The end result of the data processing is usually record sections from which an interpretation is made.

5.2.6 *Logistics*

Several points should be noted in the foregoing discussion. Field operations require moving a series of units through the area being surveyed and balance has to be achieved so that the units do not delay each other, especially so that the recording unit is not delayed. Extra drills or layout personnel or overtime are usually added to achieve the required balance. Crews often work irregular hours, working long days sometimes to make up for time lost because of weather. A variety of transport vehicles are used: trucks where possible, marsh and swamp buggies where the ground is soft, tractors in light forests, boats, jack-up barges, air boats, helicopters, etc. Generally the energy source units (drills, vibrators, etc.) are the heaviest units and determine the transport method. In some areas operations are completely portable, everything including small drills being carried on men's backs. Transport often represents an important part of a crew's cost and determines how much production can be achieved.

5.2.7 *Field records*

Complete records should be kept so that several years later it is possible to determine field conditions from the records without ambiguity. The most important records will be those of the surveyor and observer, but drillers and other units should also submit complete reports. All reports should include the date and time of day and should be written as events happen rather than at the end of the day. The observer's daily report should include the tape reel number collated with sourcepoint numbers, specification of source and spread configuration,

notes about any deviations from surveyed positions, information about all recordings including repeats, all recording settings, size of charge and depth to its top and bottom, any facts which affect the validity of data such as electrical leakage, changes in surface material, excessive noise, reasons for delays in the work, etc.

5.3   **Field layouts**

5.3.1   *Split-dip and common-depth-point recording*

Virtually all routine seismic work consists of *continuous profiling*, that is, the cables and shotpoints are arranged so that there are no gaps in the data other than those due to the fact that the geophone groups are spaced at intervals rather than continuously. 'Conventional' coverage implies that each reflecting point is sampled only once except at the ends of each profile; these end points (*tie points*) are sampled again with the adjacent profiles to reduce the likelihood of errors in following an event from one record to the next. This is in contrast to common-depth-point, or *redundant*, *coverage* where each reflecting point is sampled more than once. Areal or cross-coverage indicates that the dip components perpendicular to the seismic line have been measured as well as the dip components along the line. Each of these methods can employ various relationships of sourcepoints to the geophone groups.

Continuous-coverage *split-dip* recording is illustrated by fig. 5.4. Sources are laid out at regular intervals along the line of profiling, often 400 to 540 m apart. A seismic cable which is two shotpoint intervals long is used. Provision is made to connect groups of geophones (for example, 24 groups) at regular intervals along the cable (called the *group interval*). Thus with sourcepoints 400 m apart, 24 groups are distributed along 800 m of cable making the group centers 34 m apart. With the cable stretched from point O_1 to point O_3, sourcepoint O_2 is used; this gives subsurface control (for flat dip) between A and B. The portion of cable between

Fig.5.4. Symmetrical spread with continuous subsurface coverage.

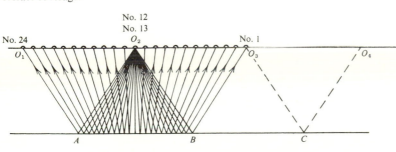

O_1 and O_2 is then moved between O_3 and O_4 and source-point O_3 is used; this gives subsurface coverage between B and C. The travel path for the last group from source-point O_3 is the reversed path for the first group from sourcepoint O_2 so that the subsurface coverage is continuous along the line.

Common-depth-point (CDP) or 'roll-along' recording (Mayne, 1962, 1967) is illustrated in fig. 5.5a. We have evenly-spaced geophone groups which we shall number by their sequence along the seismic line rather than by the trace which they represent on the seismic record. Geophone groups 1 to 24 are connected to the amplifier inputs in the recording truck and source A is used. Assuming a

horizontal reflector this gives subsurface coverage from a to g. Geophone groups 3 to 26 are then connected to the amplifier inputs, the change being made by means of the roll-along switch rather than by physically moving the seismic cable. Source B is then used giving subsurface coverage from b to h. Source C is now used with geophones 5 to 28 giving coverage from c to i, and so on down the seismic line. Note that the reflecting point for the energy from source A into geophone group 21 is point f which is also the reflecting point for the energy from B into geophone group 19, from C into 17, from D into 15, from E into 13 and from F into 11. After removal of normal moveout, these six traces will be combined

Fig.5.5. Common-depth-point profiles. The symbols $\times$ and $\otimes$ represent geophone groups and sourcepoints respectively. (*a*) Vertical section illustrating common-

depth-point recording; (*b*) surface stacking chart for common-depth-point recording.

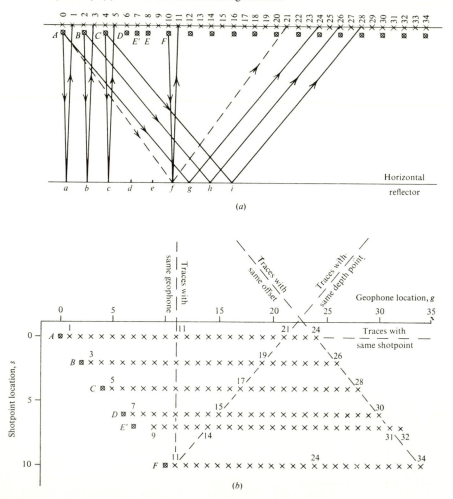

(*stacked*) together in a subsequent data-processing operation. Thus reflecting point f is sampled six times and the coverage is called '600%' or '6-fold' recording. Obviously, the multiplicity tapers off at the end of the line. Most present-day recording uses at least 6-fold multiplicity and 12 and 24 are common, especially in marine recording.

Occasionally one of the regularly-spaced locations will not be a suitable place for a source (perhaps because of risk of damage to nearby buildings) and irregularly-spaced sourcepoints will be used. Thus if point E could not be used, a source might be taken at E' instead and then geophone group 14 (instead of 13) would receive the energy reflected at f. To help keep track of the many traces involved, *stacking charts* are used (Morgan, 1970). A surface stacking chart has geophone location g as one coordinate and source location s as the other, i.e., the trace observed at g from source s is indicated by the location (g, s). (A variation of this chart, a subsurface stacking chart, has the trace plotted at $[\frac{1}{2}(g + s), s]$; see problem 5.1.) Fig. 5.5b shows the surface stacking chart when E' is used instead of E. Note how the six traces which have the common reflecting point f line up along a diagonal; points along the opposite diagonal have a common offset while points on a horizontal line have the same sourcepoint and points on a vertical line represent traces from a common geophone group. Stacking charts are useful in making static and dynamic corrections and to ensure that the traces are stacked properly.

5.3.2 *Spread types*

By *spread* we mean the relative locations of the sourcepoint and the centers of the geophone groups used to record the energy. Several spread types are shown in fig. 5.6. In split-dip recording the source is at the center of a line of regularly-spaced geophone groups. If there are 24 groups, the source is usually midway between groups 12 and 13. Such an arrangement does not give an exact time-tie to the next record (since the sourcepoint does not coincide with a geophone group), and therefore groups 12 and 13 are sometimes located together at the sourcepoint or other minor modifications are made. Placing the source close to a geophone group often results in a noisy trace (because of gases escaping from the shothole and ejection of tamping material, or because the source trucks generate too much noise); hence the source may be moved perpendicular to the seismic line 15–50 m.

In another common arrangement, the source is located at the end of the line of active geophone groups to produce an *end-on spread*. Sometimes in areas of exceptionally heavy ground roll the source is removed (*offset*) an appreciable distance (often 500–700 m) along the line

from the nearest active geophone group to produce an *in-line offset spread*, or a gap is left in the middle of a split spread. These arrangements are also used to provide greater offset to the farthest groups when short-offset data are not needed. Alternatively the source may be offset in the direction normal to the cable, either at one end of the active part to produce a *broadside-L* or opposite the center to give a *broadside-T spread*. Both the in-line and broadside offsets permit the recording of one to two seconds of reflection energy before the ground-roll energy arrives at the spread. *Cross spreads* consisting of two lines of geophone groups roughly at right angles to each other are used to record three-dimensional dip information.

5.3.3 *Arrays*

(*a*) *General*. The term *array* refers either to the pattern of a group of geophones which feed a single channel or to a distribution of shotholes or surface energy sources which are fired simultaneously; in the latter case it also includes the different locations of a single energy source for which the results are combined by vertical stacking. A wave approaching the surface along the direction of the vertical will affect each geophone of an array simultaneously so that the outputs of the geophones will combine constructively; on the other hand, a wave traveling horizontally will affect the various geophones at different times so that there will be a certain degree of destructive interference. Similarly, the waves traveling vertically downward from an array of shotholes or surface energy sources fired simultaneously will add constructively when they arrive at the geophones whereas the waves traveling horizontally away from the source array will arrive at a geophone with different phases and will be partially cancelled. Thus, arrays provide a means of discriminating between waves arriving at the spread from different directions.

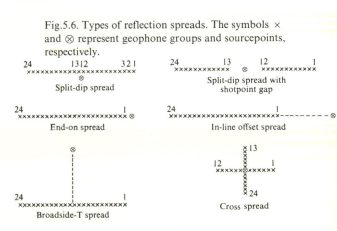

Fig. 5.6. Types of reflection spreads. The symbols $\times$ and $\otimes$ represent geophone groups and sourcepoints, respectively.

(b) *Uniform linear arrays and harmonic waves.* Arrays are classified as *linear* when the elements are spread out in line, usually along the seismic line, or as *areal* when the group is distributed over an area. The response of an array is usually illustrated by a graph of the *array response*, defined as the ratio of the amplitude of the output of the array to that of the same number of geophones concentrated at one location.

Fig. 5.7 shows an array of n identical geophones spaced at intervals Δx. We assume that a plane harmonic wave with angle of approach α arrives at the left-hand geophone at time t and that the geophone output is $A\sin\omega t$. The wave arrives at the rth geophone at time $t + r\Delta t$ where $\Delta t = (\Delta x\sin\alpha)/V$; the output of the rth geophone is $A\sin\omega(t - r\Delta t) = A\sin(\omega t - r\gamma)$ where γ is the phase difference between successive geophones, that is,

$$\gamma = \omega\Delta t = 2\pi v(\Delta x\sin\alpha)/V = (2\pi\Delta x/\lambda)\sin\alpha$$
$$= 2\pi\Delta x/\lambda_a,$$

$\lambda_a = \lambda/\sin\alpha$ being the *apparent wavelength*. The output of the array of n phones is

$$h(t) = \sum_{r=0}^{n-1} A\sin(\omega t - r\gamma)$$
$$= A\{\sin(\tfrac{1}{2}n\gamma)/\sin(\tfrac{1}{2}\gamma)\}\sin\{\omega t - \tfrac{1}{2}(n-1)\gamma\}$$

(see problem 10.11c). The array output thus lags behind that of the first geophone; for r odd, the lag is that of the central geophone, for r even, it is the mean of those of the two central geophones. The array response F depends on both n and γ:

$$F = \{\text{amplitude of } h(t)/nA\} = \left|\sin(\tfrac{1}{2}n\gamma)/\{n\sin(\tfrac{1}{2}\gamma)\}\right|$$
$$= \left|\sin\{(n\pi\Delta x\sin\alpha)/\lambda\}/[n\sin\{(\pi\Delta x\sin\alpha)/\lambda\}]\right|$$
$$= \left|\sin\{n\pi(\Delta x/\lambda)\sin\alpha\}/[n\sin\{\pi(\Delta x/\lambda)\sin\alpha\}]\right|.$$
$$(5.1)$$

Array response is often plotted using as abscissa Δx, α, λ_a, $V_a = V/\sin\alpha$ ($=$ apparent velocity; see (3.13)), *apparent dip moveout* $= \Delta t/\Delta x = (\sin\alpha)/V$, etc., other quantities remaining fixed, or using the dimensionless

Fig.5.7. Wavefront approaching linear array.

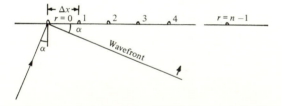

abscissa, $\Delta x/\lambda_a$ (see fig. 5.8a). The graph usually consists of a series of maxima (lobes) separated by small values. For $\Delta x = \lambda_a$, $F = 1$ (giving the first *alias lobe*) and beyond this the entire pattern repeats. The lobes between the *principal main lobe* ($\alpha = 0$) and the alias lobe are called *side lobes*. For uniform spacing the position of the first zero, or the width of the principal lobe, depends on $n\Delta x$, which is one geophone spacing greater than the distance between the end geophones, $(n - 1)\Delta x$; $n\Delta x$ is called the *effective array length*. For non-uniform arrays, the effective array length is taken as the length $n\Delta x$ of a uniform array whose principal lobe has the same width at $F = \frac{1}{2}$. The region between the points where the response is down by 3 dB, that is, where $F = 0.7$, is called the *reject region* (sometimes the reject region is defined with respect to the 6 dB points, that is, $F = 0.5$, occasionally it is defined by the nulls which separate the side lobes from the main lobe and the principal alias lobe).

Array response is often plotted in polar form as in fig. 5.9. In this case the radius vector gives the value of F as a function of the angle α.

(c) *Continuous linear source of finite length.* Consider the effect at a point P (see fig. 5.10) of a linear source of length $MN = a\lambda$ and strength σ per unit length, σ being the energy received at P per unit of length per unit of time. Assuming harmonic waves of the form $\exp[j(\kappa r - \omega t)]$ (see (2.50)), the resultant at P is

$$h(t) = \int_{x_0 - \frac{1}{2}a\lambda}^{x_0 + \frac{1}{2}a\lambda} (\sigma\,dx)\exp[j(\kappa r - \omega t)]$$
$$= \sigma\exp[-j\omega t]\int_{x_0 - \frac{1}{2}a\lambda}^{x_0 + \frac{1}{2}a\lambda} \exp[j\kappa r]\,dx.$$

For $r_0 \gg a\lambda$,

$$r \approx r_0 + (x - x_0)\sin\alpha_0 = r_0(1 - \sin^2\alpha_0) + x\sin\alpha_0$$
$$= r_0\cos^2\alpha_0 + x\sin\alpha_0.$$

Thus

$$h(t) = \sigma\exp[-j\omega t]\int_{x_0 - \frac{1}{2}a\lambda}^{x_0 + \frac{1}{2}a\lambda}$$
$$\exp[j\kappa(r_0\cos^2\alpha_0 + x\sin\alpha_0)]\,dx$$
$$= 2\sigma\exp[j(\kappa r_0 - \omega t)]\frac{\sin(\pi a\sin\alpha_0)}{\kappa\sin\alpha_0}$$
$$= \sigma a\lambda\operatorname{sinc}(\pi a\sin\alpha_0)\exp[j(\kappa r_0 - \omega t)],$$

where $\operatorname{sinc} x = (1/x)\sin x$. If the linear source were concentrated at the midpoint of MN, the effect at P would be $\sigma a\lambda\exp[j(\kappa r_0 - \omega t)]$; hence the array response F is

$$F = |\operatorname{sinc}(\pi a\sin\alpha_0)|. \qquad (5.2)$$

Fig.5.8. Response of arrays to 30 Hz signal. The effective length of the array, which controls the width of the main lobe, and the alternative scales shown in (*a*) are the same for all three arrays. The location of the secondary (alias) peak is controlled by the element spacing. Weighting increases the attenuation in the reject region. The dashed curves indicate the array response to a bell-shaped frequency spectrum peaked at 30 Hz with a width of 30 Hz. (Courtesy Chevron.) (*a*) Five in-line geophones spaced 10 m apart; (*b*) five geophones spaced 10 m apart and weighted 1, 2, 3, 2, 1; (*c*) nine geophones spaced 5.5 m apart.

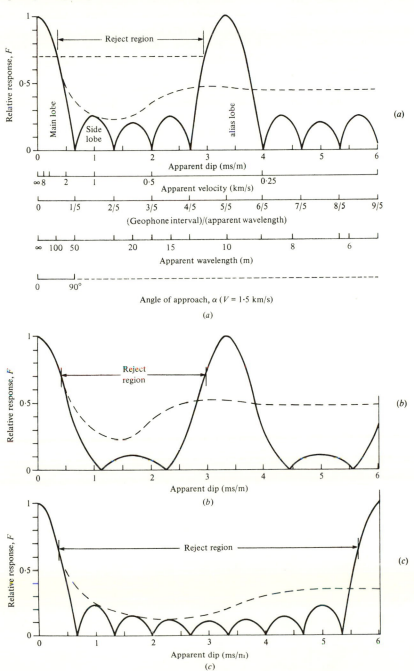

The length of the source is apt to be one wavelength ($a = 1$) or less, occasionally two wavelengths ($a = 2$); fig. 5.11 shows the directivity response for various values of a.

While fig. 5.10 implies a distributed horizontal source, the derivation also applies to a vertical source. When the dimensions are small, little directivity is achieved. Shaped charges are sometimes used to concentrate the energy traveling vertically but they are generally ineffective. Directivity is sometimes achieved by the successive detonation of parts of the charge so that energy from the newly-detonated part adds to the energy as the wave passes (see problem 5.3c); this method requires a long charge to be effective.

(*d*) *Tapered arrays*. Arrays where different numbers of elements are located at the successive positions are called *tapered arrays*. Compared with a linear array with the same overall array length, the main lobe and principal alias lobes are broadened but the response in the 'reject region' is generally smaller. The effective array length is

Fig.5.9. Directivity response of five-element linear array to 30 Hz signal for velocity of 1.5 km/s. Solid curve is for 50 m spacing (nulls at 11.5°, 24°, 37° and 53°), dashed curve for 10 m spacing (null at 90°).

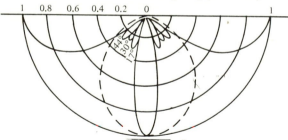

Fig.5.10. Calculating effect of continuous linear array.

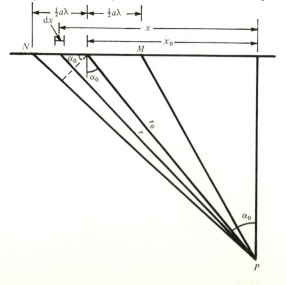

less than the actual array length. Fig. 5.8*b* shows the response of a 1, 2, 3, 2, 1 array (the numbers indicating the number of elements bunched at successive locations). Tapering can also be accomplished by varying the outputs of the individual geophones or by varying the spacing of the geophones. Tapered arrays also result from combinations of source and receiver arrays, where the effective array is the result of convolving (§8.1.2*a*) the source array with the receiver array. The Vibroseis arrangement illustrated in fig. 5.13 provides an example.

(*e*) *Areal arrays*. The principal application of linear arrays is in discriminating against coherent noise traveling more-or-less in a vertical plane through the array. Coherent noise traveling outside this plane can be attenuated by an areal array (Parr and Mayne, 1955; Burg, 1964). Some areal arrays are shown in fig. 5.12. The effective array in a given direction can be found by projecting the geophone positions onto a line in that direction; thus for the diamond array of fig. 5.12*a*, the effective array in the in-line direction is that of a tapered array 1, 2, 3, 2, 1 with element spacing $\Delta x = a/\sqrt{2}$, whereas at 45° to the line the effective array is 3, 3, 3 (or the same as a three-element uniform array) with $\Delta x = a$.

(*f*) *Effect on transients*. Actual seismic wavetrains are almost always relatively short transients rather than the harmonic waves usually assumed by array theory. An array will have different responses to the different frequency components because the apparent wavelength will change. The effective response for a bell-shaped frequency spectrum is shown by the dashed lines in fig. 5.8; effective rejection is generally poorer (except in the alias-lobe region) than the rejection of a harmonic wave.

(*g*) *Practical applications*. Response diagrams such as those in fig. 5.8 or 5.9 apply equally to arrays of geophones and arrays of sourcepoints. They also apply to the summing of traces in vertical stacking or other types of summing such as is done in data processing.

Fig.5.11. Directivity plot for continuous linear array for $a = 0.25, 0.5, 1.0, 1.5$.

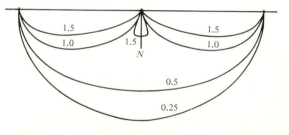

Theoretically we get the same results by using one shotpoint and sixteen geophones as by using one geophone and sixteen shotpoints spaced in the same manner and shot simultaneously with one-sixteenth the charge in each hole. However, we use multiple geophones much more than multiple shots because the cost is less, although in exceptionally difficult areas both multiple shots and multiple geophones are used at the same time. With some surface sources two or four source units may be used at the same time to provide the effect of multiple shots and a sizeable source array may be achieved in this way (fig. 5.13).

The cancelling of coherent noise by using geophone and source arrays presents a more challenging array design problem than does the cancellation of random noise. In the case of random noise the locations of the elements of the array are unimportant provided no two are so close that the noise is identical for both. For coherent noise the size, spacing, and orientation of the array must be selected on the basis of the properties of the noise to be cancelled (Schoenberger, 1970). If the noise is a long sinusoidal wavetrain, an array consisting of n elements spaced along the direction of travel of the wave at intervals of λ/n, where λ is the apparent wavelength, will provide cancellation (see problem 5.5b). However, actual noise often consists of several types arriving from different directions, each type comprising a range of wavelengths; moreover, the nature of the noise may change from point to point along the line. It is usual to resort to areal arrays in areas of severe noise problems (although the in-line distribution of elements is almost always the most important aspect). Numerous articles have been written on the subject of arrays; McKay (1954) shows examples of the improvement in record quality for different arrays.

In addition to the difficulties in defining the noise wavelengths to be attenuated, actual field layouts rarely correspond with their theoretical design (Newman and Mahoney, 1973). Measuring the location of the individual geophones is not practicable because of the time required. In heavy brush one may have to detour when laying out successive geophones and often one cannot see one geophone from another so that even the orientation of lines of geophones can be very irregular. In rough topography maintaining an array design might require that geophones

Fig.5.12. Types of areal arrays. Element locations are indicated by small circles, effective array in different directions by small triangles with effective weights.

(*a*) 3 × 3 diamond; (*b*) X-array; (*c*) rectangular array; (*d*) crow's foot array; (*e*) odd-arm star; (*f*) herringbone.

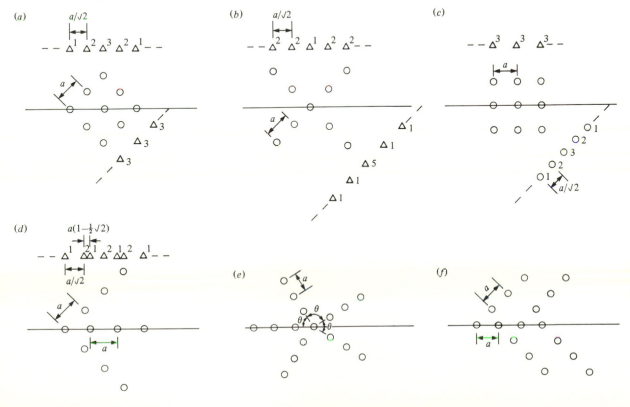

be at different elevations which may produce far worse effects than those which the array is intended to eliminate. Similar problems arise where the conditions for planting the geophones vary within a group (Lamer, 1970), perhaps as a result of loose sand, mucky soil or scattered rock outcrops. The best rules for array design are often to (1) determine the maximum size which can be permitted without discriminating against events with the maximum anticipated dip and (2) distribute as many geophones as field economy will permit more-or-less uniformly over an area a little less than the maximum size permitted, maintaining all geophone plants and elevations as nearly constant as possible even if this requires severe distortion of the layout (see also §5.3.5).

Arrays may also be of value in refraction work (Laster and Linville, 1968).

5.3.4 *Noise analysis*

Systematic investigation of coherent noise often begins with shooting a *noise profile* (also called a *microspread* or *walkaway*). This is a small-scale profile with a single geophone per trace, the geophones being spaced as

closely as 1–3 m over a total spread length of the order of 300 m or more. If the weathering or elevation is variable, corrections should be made for each trace. The corrected data, often in the form of a record section, such as shown in fig. 5.14, are studied to determine the nature of the coherent events on the records, the frequencies and apparent velocities of the coherent noise, *windows* between noise trains where reflection data would not be overridden by such noise, and so on. Once we have some indications of the types of noises present, we can design arrays or other field techniques to attenuate the noise and then field-test our techniques to see if the desired effect is achieved.

The data from nois; analyses are sometimes transformed (see §8.1.1 and §10. .1) into the frequency–wavenumber domain and displayed as v–κ plots (often called f–k plots), such as shown in fig. 5.15. Radial lines on such a plot show a constant proportionality between v and κ, that is, constant apparent velocity (by (2.48)). Such v–κ plots are helpful in seeing the characteristics of different types of events and determining the best ways of attenuating the portions regarded as noise-dominated in order to improve the signal-to-noise ratio (for example, with arrays,

Fig.5.13. An arrangement used with surface sources. Four Vibroseis units (or other sources) 33 m apart follow one another from left to right down the line, vibrating simultaneously at locations 17 m apart. The positions of the vibrators for successive sweeps are displaced vertically in the figure to avoid overlap. Six successive sweeps are summed (vertically stacked – see §5.4.9) to make an output field record. The four central groups are not used because they show too much truck noise. The recording connections are

advanced one group after sweep 6 giving the active groups shown on the center line in (*a*); these are used for sweeps 7 through 12 with the source locations shown by the solid triangles. (*a*) Position of source units and active geophone groups for successive sweeps; (*b*) effective combined array, the numbers indicating the number of sweeps contributing (this is the result of convolving (see §8.1.2) the linear group with the sum pattern for the six sweeps 13 through 18; see problem 8.13*a*).

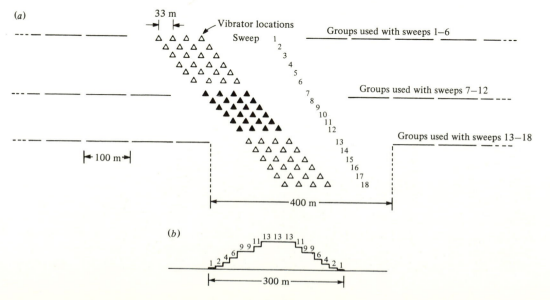

frequency filtering, apparent velocity filtering (§8.2.7), etc.).

5.3.5 *Selection of field parameters*

Field parameters should be determined in a logical way (Anstey, 1970; Sheriff, 1978), although existing equipment (number of channels and geophones and how cables and geophones are wired, etc.) usually prejudices decisions considerably.

(1) The maximum offset, the distance from source to the farthest group, should be comparable to the depth of the deepest zone of interest. This usually results in large enough normal-moveout differences to distinguish primary reflections from multiples and other coherent noise, but not offsets so large that reflection coefficients change appreciably, that conversion to shear waves becomes serious and that the approximations of the CDP method become invalid. If data quality in the deepest zone of interest is sufficiently good, the maximum offset may be increased up to the value of the basement depth.

(2) The minimum offset distance should be no greater than the depth of the shallowest section of interest. Getting sufficiently far from source-generated noise some-

times dictates a greater distance but this may cause a loss of useful shallow data.

(3) The maximum array length is determined by the minimum apparent velocity of reflections. The minimum apparent velocity usually occurs at the maximum offset. The shortest apparent wavelength (highest frequency) at this minimum apparent velocity should be just within the main lobe of the array's directivity pattern (fig. 5.8).

(4) The minimum useful in-line geophone spacing within arrays is usually determined by the ambient noise, sometimes by source-generated noise. Ambient noise characteristics can be determined experimentally by recording individual geophones spaced 0.5 to 1 m apart to determine the minimum geophone spacing for which the noise appears to be still incoherent. This minimum spacing is often 2 to 5 m, the smaller value being where noise is mostly generated locally (such as noise caused by the wind blowing grass, shrubs or trees), and the larger value where noise is mainly caused by distant sources (such as microseisms, surf noise, traffic noise, etc.). If more geophones are available, an areal distribution of phones at this minimum spacing will be more effective than crowding the phones closer together in-line. Areal arrays are rarely

Fig.5.14. Noise analysis or *walkaway*. Geophones spaced 1.5 m apart, offset to first geophone 425 m. Identification of events: 1890 m/s = refraction from base of weathering; 530 and 620 m/s = ground-roll modes; 330 m/s = airwave; 3140 m/s = refraction event. (After Sheriff, 1973.)

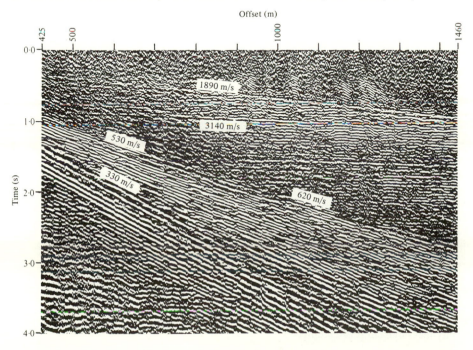

required to attenuate noise coming from the side of the line. Airwave attenuation may require closer geophone spacing.

(5) Group interval should be no more than double the desired horizontal resolution, thus providing subsurface spacing equal to the desired resolution. However, the group interval should not exceed the maximum permissible array length indicated by (3) above.

(6) The minimum number of channels required is determined by the combination of spread length and group interval decisions already reached.

(7) The minimum charge size is determined by the ambient noise late on the record. If two shots are repeated, the records should be nearly identical to a time which corresponds to a depth below the deepest section of interest. If this is not the case the charge size (or source effort) should be increased.

Special circumstances may require variations from these guidelines. For example, mapping a zone of maximum interest may require recording in an offset window where desired reflections are between wavetrains following the first arrivals and those caused by surface waves.

5.3.6 *Uphole surveys*

An *uphole survey* is one of the best methods of investigating the near-surface and finding the thickness and velocity of the low-velocity layer (LVL), D_w and V_W, and the sub-weathering velocity, V_H. An uphole survey requires a shothole deeper than the base of the LVL. Usually a complete spread of geophones plus an uphole geophone is used. Shots are fired at various depths in the hole, as shown in the lower part of fig. 5.16, beginning at the bottom and continuing until the shot is just below the surface of the ground. Arrival times are plotted against shot depth for the uphole geophone and for several distant geophones, including two or more spaced 200 m or more apart, as shown in the upper part of fig. 5.16. The plot for the uphole geophone changes abruptly where the shot enters the LVL; the slope of the portion above the base of the LVL gives V_W and the break in slope usually defines D_w clearly.

For the distant geophones the plot is almost vertical at first since the path length changes very little as long as the shot is in the high-speed layer. However, when the shot enters the LVL there is an abrupt change in slope and the traveltime increases rapidly as the path length in the LVL increases. The refraction velocity at the base of the LVL, V_H, is obtained by dividing the time interval between the vertical portions of the curves for two widely separated geophones (Δt_{17} in fig. 5.16) into the distance between the geophones. This velocity measurement is often differ-

Fig.5.15. Frequency–wavenumber sketch of data in fig. 5.14. A horizontal slice can be taken by frequency filtering, a vertical slice by means of arrays, and a pie-shaped wedge by apparent-velocity filtering.

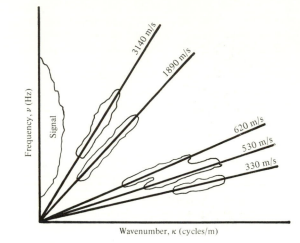

Fig.5.16. Uphole survey. At top, traveltime versus shot depth; at bottom, vertical section showing raypaths.

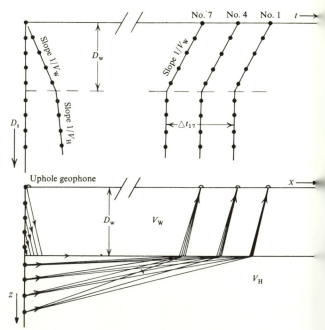

ent from that given by the slope of the deeper portion of the uphole geophone curve, partly because the latter is less accurate (since the time interval is much less than Δt_{17}), partly because the layering of beds of different velocities has little effect on Δt_{17} but may affect the uphole time, t_{uh}, substantially.

5.3.7   *Crooked line and 3-D methods*

(*a*) *Crooked-line methods.* Many interpretation criteria, such as changes in dip rate, become more difficult to use when line direction changes, so efforts are made to keep lines straight. However, it is sometimes impossible to run straight lines and access and structural complications may make it impossible to locate lines in desired directions.

Crooked-line CDP shooting is carried out in the same way as conventional CDP shooting and the depar-

tures from regularity are accommodated in subsequent processing. Source-to-geophone distance (as opposed to distance measured along the line) is calculated so that the proper amount of normal moveout can be applied, and the plotted location is at the point midway between source and geophone (fig. 5.17). A best-fit straight line (or series of straight-line segments) is drawn through the plot of such midpoints. The midpoints are then projected onto this line; various projection criteria are used, but in practice the projection is often perpendicular to the line. Then those traces whose projected midpoints fall within a group length are stacked together. The section which results has regularly-spaced traces but the stack multiplicity varies from trace to trace.

Since the actual midpoint locations are distributed over an area, they contain information about dip perpendicular to the line and in effect provide a series of

Fig.5.17. Portion of a computer-drawn midpoint plot for a crooked line. The shotpoints (indicated by squares) and geophones are laid out along a road, there being one shotpoint every third geophone group. The midpoints show as dots. The synthetic line made in

processing has cross-dashes showing the output trace spacing. The boxes show the areas of midpoints which might be combined to make a single trace by projecting (*a*) perpendicular to the line or (*b*) along the strike. (Courtesy Seiscom Delta.)

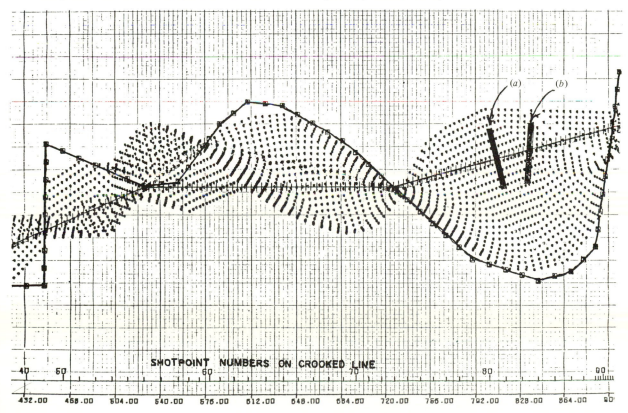

SHOTPOINT NUMBERS ON CROOKED LINE

Fig.5.18. Arrangements for 3-D surveying. (*a*) Location of ship and 48 receiver groups plotted every 25th shot; (*b*) use of paravanes to pull source units to side of ship's track so that two lines of coverage can be obtained; (*c*) 'wide-line' layout; (*d*) 'block' layout;

(*e*) 'Seisloop' layout with geophones and sourcepoints around perimeter of an area; midpoints show as dots; 40 geophone groups and 40 shotpoints might be located at 160 m intervals around the four sides of a square 1.6 km on a side.

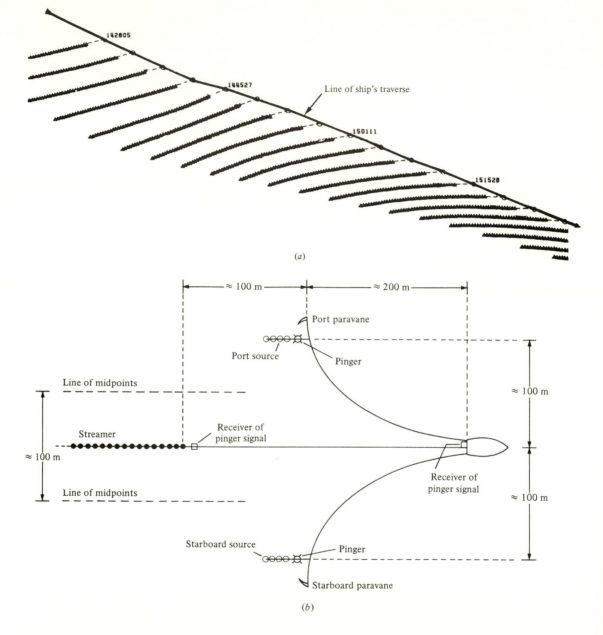

(*a*)

(*b*)

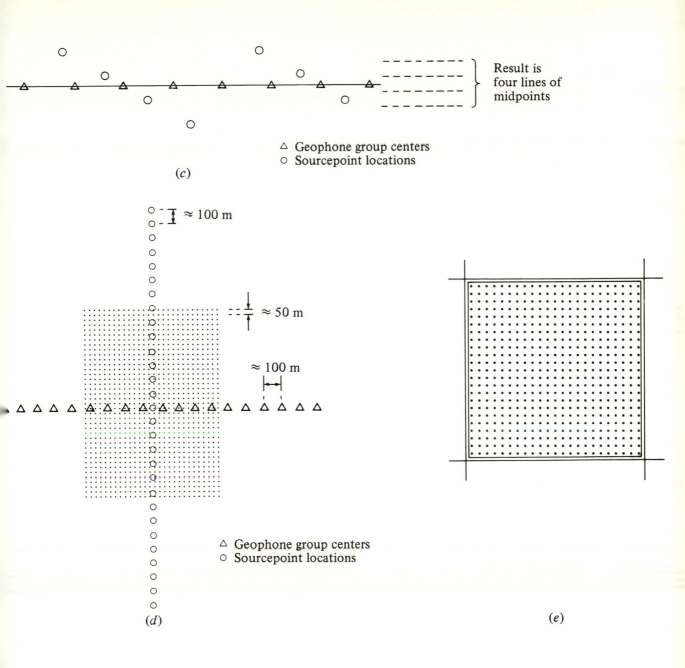

Result is
four lines of
midpoints

△ Geophone group centers
○ Sourcepoint locations

(c)

≈ 100 m

≈ 50 m

≈ 100 m

△ Geophone group centers
○ Sourcepoint locations

(d)

(e)

cross-spreads, from which the true dip can be resolved. Lines are sometimes run intentionally crooked to give cross-dip information.

(b) *Marine 3-D methods.* Ideally we would like to have data distributed arealy rather than linearly, that is, *three-dimensional* (3-D) *data.* We would further like to have (i) data distributed on a uniform grid with (ii) the same CDP multiplicity (iii) utilizing the same offset distances; the objective of these is so that data quality variations can be attributed to geological factors rather than acquisition factors. However, it is nearly impossible to achieve this and still acquire the data economically. Cross-dip information can be obtained in a one-boat operation (1) if the streamer is pulled off at an angle to the seismic line or (2) if two sources are used which are displaced to opposite sides of the line. Where natural ocean currents are present, the seismic line can be oriented perpendicular to the current so that the streamer is pulled off-line. Fig. 5.18a is a plot of streamer location for a line in such an area. The streamer location is determined by magnetic compasses built into the streamer and by sighting on the tail buoy. One has relatively little control over the amount of cross-coverage obtained, however, and in subsequent processing especial care has to be taken so that residual normal moveout does not appear as fictitious cross-dip.

Paravanes can be used to pull source units up to about 100 m off the seismic line (fig. 5.18b). With port and starboard sources used alternately, two lines of sub-surface coverage 100 m apart can be obtained in this way (Whittlesey *et al.*, 1980). The arrays are equipped with 'pingers' (transducers which send out acoustic pulses whose traveltimes to tuned receivers on the boat and in the streamer can be measured) which are used to locate the sources.

Sometimes two or more boats are used in offshore areal surveys; having a separate source boat increases the freedom in layout but also the costs. Areal coverage offshore is most commonly obtained by shooting closely-spaced (100 m) parallel lines; 100 m spacing is not sufficiently close to avoid spatial aliasing problems for even moderate dips (see §8.1.2b). The major problem is usually location uncertainty; anchored pingers (§5.5.5c) provide one of the best location systems for survey of a restricted area.

(c) *Land 3-D methods.* Many arrangements are used on land (Walton, 1972) and here we describe two general types: (i) *wide-line* or *swath methods* and (ii) block methods. With the swath methods, the coverage dimension perpendicular to the line is only a few hundred meters at most.

Usually geophones are laid out along the line in essentially conventional manner (sometimes two or more parallel lines are used, especially when many channels are available) and sources are located at different distances from the line (fig. 5.18c); often two or more source units are used, firing alternately.

With the *block methods*, coverage is obtained over a more-or-less rectangular area, and then work proceeds to the next 'block'. Sometimes perpendicular source and geophone lines are used (fig. 5.18d) to produce coverage over the block, sometimes geophones and sourcepoints are located around the perimeter of the block (fig. 5.18e) in a 'Seisloop'™ arrangement.

While these arrangements provide areal coverage, they sacrifice CDP multiplicity and result in systematic offset distributions which increase the possibility of confusing residual normal moveout (NMO) and dip effects. The loss of multiplicity can often be tolerated because the signal-to-noise threshold is smaller when data which are nearby can be consulted in deciding which events to honor. With the Seisloop arrangement, multiplicity is greater along lines of symmetry (such as along the bisectors of a square). Array directivity is also apt to differ for various points. Three-dimensional data display is discussed in §5.4.7.

All these methods are expensive both in field work and in subsequent processing, but they provide detailed information not otherwise achievable. They are used especially in detailing oil fields to aid in locating development wells.

5.3.8 *Extended resolution*

Although conventional geophones and recording systems are usually adequate for recording up to 125 Hz and normal alias filters (which cut sharply above $1/4\Delta$ where Δ is the sampling rate) permit recording up to 250 Hz for 1 ms sampling, the bandwidth of most reflection surveys is only about 10–60 Hz. Since both vertical and horizontal resolution (§4.3.2) are fixed by the high-frequency components, we must expand the bandpass upward to achieve higher resolution. Techniques for doing this are called *extended resolution*.

The high-frequency limitations are usually due to (1) limitations in the source, (2) processes within the Earth which discriminate against high frequencies, (3) conditions at or near the surface, including array effects, and (4) recording instruments.

Surface sources are often limited with respect to high frequencies because of mechanical and coupling problems as well as high near-surface attenuation (in comparison with a source in a borehole) resulting from two passes through the weathered layer. Dynamite generates waves relatively rich in high frequencies; the high-frequency content increases as the charge size decreases so that the charge size should be the minimum consistent with adequate energy.

High frequencies are attenuated in the Earth by absorption (§2.3.2) and peg-leg multiples (§4.2.2b). The practical limiting factor is the amplitude of useful high-frequency reflection energy compared to the noise level. Denham (1981) gives the empirical formula $v_{max} = 150/t$ where v_{max} is the maximum useable frequency and t is the two-way time. According to this, the upper limit should be greater than 60 Hz for events shallower than $t = 2.5$ s. The processes within the Earth are largely beyond our control and set the ultimate limit to the resolution that can be achieved.

The greatest single cause of high-frequency loss is often ground mixing owing to the use of geophone arrays (and also source arrays); random time shifts with a standard deviation of 2 ms are equivalent to a 62 Hz filter (fig. 5.19). Thus, although arrays provide one of the best ways of attenuating surface waves and ambient noise, alternative methods should be used if high resolution is desired. Low-frequency filtering and burying the phones should be considered (the best results are obtained by burying the phones below the weathered layer, but burial by only 10–50 cm often achieves dramatic improvement). Consideration must also be given to spatial aliasing (§8.1.2b) which may dictate group spacing as small as 15–20 m. Sometimes a hybrid spread is used, group intervals being longer for long-offset groups (see problem 5.10).

Fig.5.19. Filter effect of timing errors in stacking. The numbers on the curves are standard deviations of the timing differences among the traces stacked.

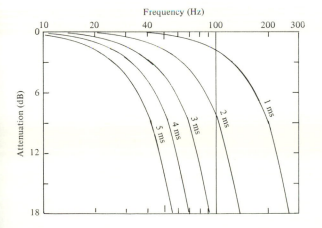

The characteristics of the recording equipment usually do not present a severe problem. If the high-frequency response is not adequate, accelerometers (which have a frequency response increasing 6 dB/octave relative to velocity geophones (§5.5.4)) can be used even though maintenance costs increase. When low-cut filtering is used to attenuate severe ground roll, a passband of at least 2 octaves should be retained in order to achieve good wavelet shape.

Significant improvement in high-frequency response may be obtainable in some areas without marked increase in acquisition costs. However, processing costs usually increase. On the other hand, extending the frequency spectrum upward usually improves both shallow and deep resolution.

5.3.9 *Special methods*

(*a*) *Vertical profiling.* Usually the seismic source and geophones are located at or very near the surface. The most common exception involves a velocity survey where the geophone is lowered into the borehole (see §7.3.1), but velocity surveys usually record only the first arrival at a limited number of depths. Vertical profiling (Kennett and Ireson, 1977) involves recording the complete waveform (sometimes the waveforms from three-component phones) at regularly- and closely-spaced depth stations. A portion of a vertical profile using a pressure phone is shown in fig. 5.20a. The slope of the first-breaks gives velocity. Reflections have a slope opposite to the first-breaks. The vertical profile provides a separation of downward-traveling waves (direct wave and multiples involving an even number of reflections) from upward-traveling waves (reflections and multiples involving an odd number of reflections). Vertical profiles may be time-shifted by the direct-wave arrival time to produce *reduced vertical profiles* which emphasize reflections (fig. 5.20b).

Vertical profiling provides a means of studying seismic phenomena such as waveform changes, attenuation, the correlation of reflections with interfaces observed in the borehole, etc. By shooting at various distances and azimuths from the borehole additional phenomena can be studied such as dip, variations of reflectivity with incident angle, converted waves, head waves, anisotropy, etc. However, this method is used but little. The cost of occupying a borehole for the time required for a study is the major deterrent. Gal'perin (1974) describes studies employing vertical profiling in the USSR.

Shooting into geophones in deep holes is also used in searching for and defining nearby features such as saltdomes (see §6.1.3).

(b) *Undershooting.* Long in-line offsets are sometimes used where one cannot shoot and record over the desired region, perhaps because of structures, river levees, canyons, cliffs, permit problems, etc.; this technique is called *undershooting.* Such techniques are also useful where raypaths are so distorted by shallow features of limited extent that sense cannot be made of deeper events, as might be the situation in mapping underneath a salt-dome, reef or local region of very irregular topography or weathering.

(c) *Expanding spread.* A much larger than usual range of shotpoint-to-geophone distances (*expanding spread*) is

sometimes used to give X^2-T^2 data (§7.3.3a) or to relate refraction and reflection events.

(d) *Use of channel waves for locating faults in coal seams.* Faults which offset coal seams seriously affect the economics of coal mining, especially with modern mining equipment which must be disassembled and moved when fault dislocation interrupts production. Prakla-Seismos have developed 'in-seam' transmission and reflection techniques to predict faults (Krey, 1963 and Millahn, 1980).

Coal measures usually have appreciably smaller velocity and density than the encasing rock, conditions which favor normal-mode propagation (§2.4.8), in this

Fig.5.20. Vertical profile. (Courtesy SSC.) (*a*) Each trace is recorded at a station in a borehole using an airgun source at the surface. (*b*) Same except each trace has been shifted by the one-way traveltime to the surface, thus aligning reflections (upcoming events) horizontally. (*c*) Portion of reflection record shot across well. (*d*) Sonic log in well.

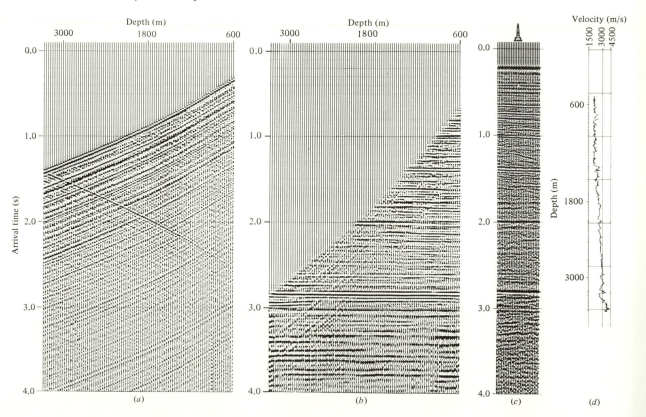

Fig.5.21. In-seam seismic methods. (From Millahn, 1980.) (*a*) Diagrammatic map (not to scale) showing location of geophones and shotpoints *B* and *C*; (*b*) transmission record from shotpoint *B*; (*c*) reflection records of common-depth-point type obtained from shots in Gallery *A*; (*d*) display of envelopes of records in (*c*); (*e*) 6-fold stack of records such as those shown in (*c*).

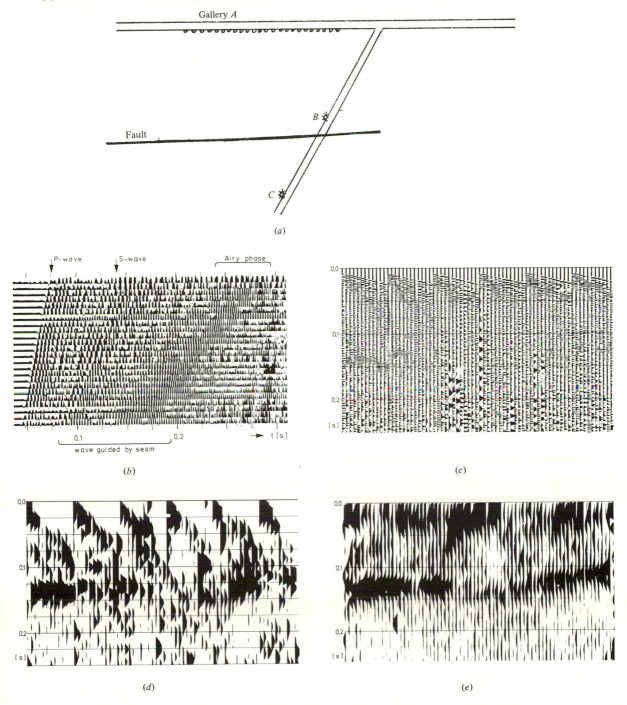

(*a*)

(*b*)

(*c*)

(*d*)

(*e*)

context usually called *channel waves*. The channel waves travel only within the low-velocity channel (coal) and have maximum amplitude at the center of the channel. They are dispersive and have a group-velocity minimum which depends on the channel thickness and produces an Airy phase.

Multichannel digital techniques are employed. Small explosives implanted in the seam provide the energy source. Two-component geophones oriented in the plane of the coal seam are planted in holes drilled in the center of the coal seam. Sampling is at 0.5 ms and frequencies up to 750 Hz are recorded.

Application of the techniques will be illustrated by an example. Assume galleries in a coal seam as illustrated in the map of fig. 5.21a with geophones planted along gallery *A*. The transmission record obtained from a shot at *B* is shown in fig. 5.21b; it shows P- and S-wave head waves which travel in the higher-velocity rock bounding the coal seam and wave-guided channel waves with a trailing amplitude build-up, the Airy phase. If a fault with throw larger than the coal seam thickness had cut the seam between the shotpoint and the receivers (as would be the case from shotpoint *C*), the channel waves would not be seen. Several reflection records obtained with shots along gallery *A* are shown in fig. 5.21c; these contain reflected channel waves but their phases are not sufficiently coherent to allow them to be picked easily. Complex-trace techniques (§8.4.2) are used to obtain the amplitude envelope (fig. 5.21d) and stacking to make the high-amplitude Airy phase stand out as a distinct arrival (fig. 5.21e). The amplitude envelope of the Airy phase is similarly determined on the transmission records, which yields the group velocity to be used in stacking the reflected records.

Borges (1969) describes measurements to locate fault zones in coal mines from the amplitude of waves traveling in the coal seams. Faults with throws of 1.5–4 m produced amplitude reductions of 50–70% for waves passing through the faults, while throws greater than 4 m resulted in 70–90% reduction in amplitude. Saul and Higson (1971) carried out an extensive evaluation of the method. Holes were drilled 2–3 m into the rock above or below the coal for both the shots and geophones. The shots were located along one roadway and the geophones along an adjacent roadway. The shots were coupled to the rock by careful stemming (tamping) while the geophones were attached to rock bolts. Their conclusions were that the method was soundly based and that faults produce attenuation; however, their results varied too widely for them to establish a relation between amplitude and the degree of disturbance.

5.4 Equipment for land surveys

5.4.1 *Drilling*

When dynamite is being used as the energy source, holes are drilled so that the explosive can be placed below the low-velocity layer. The holes are usually about 8–10 cm in diameter and 6–30 m in depth, although depths of 80 m or more are used occasionally. Normally the holes are drilled with a rotary drill, usually mounted on a truck bed but sometimes on a tractor or amphibious vehicle for working in difficult areas. Some light drills can be broken down into units small enough that they can be carried. Augers are used occasionally. In work in soft marshes, holes are sometimes jetted down with a hydraulic pump. Typical rotary-drilling equipment is shown in the photograph in fig. 5.22 and in the diagram in fig. 5.23.

Rotary drilling is accomplished with a drill bit at the bottom of a drill pipe, the top of which is turned so as to turn the bit. Fluid is pumped down through the drill pipe, passes out through the bit and returns to the surface in the annular region around the drill pipe. The functions of the drilling fluid are to bring the cuttings to the surface, to cool the bit and to plaster the drill hole to prevent the walls from caving and formation fluids from flowing into the hole. The most common drilling fluid is *mud* which consists of a fine suspension of bentonite, lime and/or barite in water. Sometimes water alone is used, sometimes air is the circulating fluid. Drag bits are used most commonly in soft formations; these tear out pieces of the earth. Hard rock is usually drilled with roller bits or cone bits which cause pieces of rock to chip off because of the pressure exerted by teeth on the bits. In areas of exceptionally hard rock, diamond drill bits are used.

5.4.2 *Explosive energy sources*

Explosives were the sole source of energy used in seismic exploration until weight dropping was introduced about 1954. While explosives are no longer dominant, they continue to be an important seismic energy source in land work.

Two types of explosives have been used principally: gelatin dynamite and ammonium nitrate. The former is a mixture of nitroglycerin and nitrocotton (which form an explosive gelatin) and an inert material which binds the mixture together and which can be used to vary the 'strength' of the explosive. Ammonium nitrate is cheaper and less dangerous since it is more difficult to detonate than gelatin dynamites. Ammonium nitrate and NCN (nitrocarbonitrite) are the dominant explosives used today (in such forms as Nitramon[TM]). Other types of explosives are also used occasionally.

Explosives are packaged in tins or in tubes of cardboard or plastic about 5 cm in diameter which usually contain 1–10 pounds (0.5–5 kg) of explosive. The tubes and tins are constructed so that they can be easily joined together end-to-end (fig. 5.24a) to obtain various quantities of explosives. Ammonium nitrate sometimes is used in bulk form, the desired quantity being mixed with fuel oil and poured directly into a dry shothole.

The velocity of detonation (that is, the velocity with which the explosion travels away from the point of initiation in an extended body of explosive) is high for the explosives used in seismic work, around 6000–7000 m/s; consequently, the seismic pulses generated have very steep fronts in comparison with other energy sources. This high concentration of energy is desirable from the point of view of seismic wave analysis but detrimental from the viewpoint of damage to nearby structures.

The US Bureau of Mines, Bulletin 656, 'Blasting vibrations and their effects on structures', states that it is the velocity of ground motion rather than displacement or acceleration which correlates best with damage.

Fig.5.22. Mayhew 1000 drilling rig. (Courtesy Gardner-Denver.)

Damage is minimal if peak velocity does not exceed 12 cm/s but a 'safe criterion' is 5 cm/s. This translates into an empirical rule of $x = k\, m^{\frac{1}{3}}$ where $k = 50$ for x in feet and m in pounds, or $k = 23$ for x in meters and m in kilograms. The International Association of Geophysical Contractors sets the following minimum distances:

Pipe lines	200 ft (60 m)
Telephone lines	40 ft (12 m)
Railroad tracks	100 ft (30 m)
Electric lines	80 ft (24 m)
Transmission lines	200 ft (60 m)
Oil wells	200 ft (60 m)
Water wells, cisterns, masonry buildings	300 ft (90 m)

Electric blasting caps are used to initiate an explosion. These consist of small metal cylinders, roughly 0.6 cm in diameter and 4 cm long (see fig. 5.24*b*). They contain a resistance wire embedded in a powder charge which deflagrates readily. By means of two wires issuing from the end of the cap, a large current is passed through the resistance wire and the heat generated thereby initiates the deflagration of the powder which causes the explosion of an adjacent explosive in the cap. The cap has previously been placed inside one of the explosive charges so that the explosion of the cap detonates the entire charge.

Primers are generally necessary in setting off the explosion in ammonium nitrate explosives. These are tins

Fig.5.23. Rotary drill. (From Sheriff, 1973.)

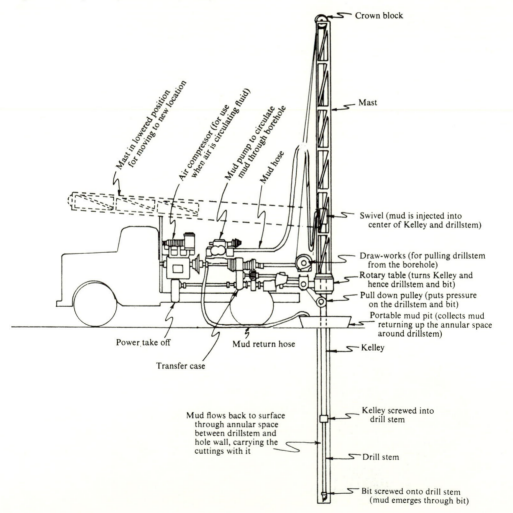

of more powerful explosive which are used as one of the elements in making up the total charge. A cap is inserted into a 'well' in the end of the tin of primer to set it off.

The current which causes the blasting cap to explode is derived from a *blaster*; this is basically a device for charging a capacitor to a high voltage by means of either batteries or a hand-operated generator and then discharg-

ing the capacitor through the cap at the desired time. Incorporated in the blaster is a device which generates an electrical pulse at the instant that the explosion begins. This time-break pulse fixes the instant of the explosion, $t = 0$. The time-break pulse is transmitted to the recording equipment by a telephone line or radio where it is recorded along with the seismic data.

Fig.5.24. Seismic explosives. (Courtesy Du Pont.)
(*a*) Cans of Nitramon joined end-to-end; (*b*) electric blasting cap.

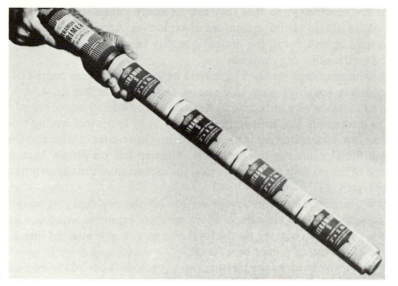

(*a*)

(*b*)

Several techniques are used at times to concentrate the energy traveling downward from an explosion. The detonating front in an explosive usually travels much faster than the seismic wave in the formation so that the seismic wave originating from the top of a long explosive charge lags behind the wave from the bottom of the charge even where the explosive is detonated at the top (which is the usual method). Explosives with low effective detonating velocity are sometimes used but these are made in long flexible tubes which are difficult to load. Delay units are sometimes used between several concentrated explosive charges to allow the wave in the formation to catch up with the explosive front; these may consist of delay caps (which introduce a fixed delay between the time the detonating shock initiates them and the time they themselves explode) or helically-wound detonating cord (so that the detonating front has to travel a longer distance) (see problem 5.3*c*). Expendable impact blasters have also been used; these detonate when they are actuated by the shock wave from another explosion.

While explosives provide the most compact high-energy source, they have many disadvantages which often preclude their use: high cost, the time and expense involved in drilling holes, potential damage to nearby buildings, wells, etc., and most important of all, restrictions about where holes can be drilled and explosives detonated.

Fig. 5.25. Plough for planting detonating cord. The cord feeds down through pipes behind blades which vibrate as they are pulled forward, planting the cord $\frac{2}{3}$ to 1 m below the surface. (Courtesy Primacord Services.)

Explosive charges near the surface are sometimes used, especially in areas remote from habitation. Explosive detonating cord such as Primacord™ is buried 0.3 to 1 m in the earth or laid in shallow water. A vibratory plough (fig. 5.25) may be used to bury the cord and up to 100 m may be used for a shot. Since the speed of detonation of the cord is about 6.5 km/s, the length of cord determines the number of caps required to detonate the entire cord within the desired time interval. Generally caps are used at both ends, sometimes in the middle.

Occasionally detonating cord or explosives are laid on the snow or mounted a meter or so high on sticks, especially in mountainous areas where operations have to be portable. Special 'flashless' cord or explosives are used so that fires do not result. An array of explosives on sticks, called *air shooting* (Poulter, 1950) causes essentially no damage in forest areas.

5.4.3   *Surface energy sources*

Many alternative energy sources have been developed for use in both land and marine work. Discussion of those which are used primarily at sea and infrequently on land will be postponed until §5.5.3.

Without exception surface energy sources are less powerful than explosives and their use on a large scale has been made feasible by stacking methods (see §5.4.9) which permit adding together the effects of a large number of weak impulses to obtain a usable result. Fig. 5.13 shows a possible field arrangement.

The earliest non-dynamite source to gain wide acceptance was the *thumper* or *weight dropper*. This method was developed largely by the McCollum Geophysical Company. A rectangular steel plate weighing about 3000 kg is dropped from a height of about 3 m. The instant of impact is determined by a sensor on the plate. Weights are often dropped every few meters so that the results of 50 or more drops are composited into a single field record. The time between release of the weight and impact on the ground is not constant enough to permit more than one source to be used simultaneously. Often two or three units are used in succession, one dropping its weight while the others lift their weights into the armed position and move ahead to the next drop point. The use of weight-dropping is now largely restricted to desert or semi-desert areas where the massive trucks can move about relatively freely.

The Dinoseis™ method (Godfrey *et al.*, 1968) developed by the Sinclair Oil and Gas Company involves the explosion of a mixture of propane and oxygen within an expandable chamber. The explosion chamber is mounted under a truck and is lowered to the ground when ready for use (see fig. 5.26). The explosion of the gas mixture by means of a spark plug creates a pressure which acts on a moveable plate forming the bottom of the chamber, thus transmitting the pressure pulse into the ground. The weight of the chamber provides the necessary reaction inertia. Several other types of gas exploders have been used in land work. As with weight-dropping, the heavy source chamber requires massive field equipment which in turn restricts usage to fairly open areas.

While the foregoing are primarily surface sources, gas guns, airguns (Brede *et al.*, 1970), and other devices are sometimes used in boreholes, especially in soft marsh where there is little risk of being unable to recover the equipment from the hole. Airguns in portable chambers of water are also used on the land surface. These airguns are modifications of the guns designed for marine use which are discussed in §5.5.3*b*.

Fig.5.26. Truck-mounted twin 36-inch Dinoseis gas exploders. (Courtesy Geo Space.)

Unlike other energy sources which try to deliver energy to the ground in the shortest time possible, the Vibroseis™ source passes energy into the ground for several seconds. A control signal causes a vibrator (usually hydraulic) to exert a variable pressure on a steel plate pressed against the ground by the weight of the vehicle (fig. 5.27). The pressure $\mathscr{P}$ generally varies according to the relation

$$\mathscr{P}(t) = A(t) \sin 2\pi t \{v_0 + (dv/dt)t\}, \qquad (5.3)$$

dv/dt being either positive or negative and constant in the usual case of a linear 'sweep'. (Nonlinear sweeps have advantages at times, but departure from linearity is equivalent to filtering and the same result can be achieved more easily in subsequent processing (Goupillaud, 1976)). The amplitude $A(t)$ is usually constant except during the initial and final 0.2 s or so when it increases from zero or decreases to zero. The sweep usually lasts for 7 to 35 s with a frequency varying from about 12 to 60 Hz (or vice versa). For a more detailed account see Waters (1978, pp. 78–99).

Since reflections occur at intervals much smaller than 7 s, the seismic record is the superposition of many wavetrains and the field records are uninterpretable even to the experienced. Subsequent data processing (see §8.1.3*d*) is necessary to resolve the data; in effect the processing (cross-correlation with the sweep) compresses each returning wavetrain into a short wavelet, thus removing much of the overlap (see fig. 8.5).

Ideally the input to the ground is a copy of the pressure applied to the steel plate but in fact crushing and compaction of the surface material (pressures are as high as 200 kg/cm^2) result in the input to the ground varying nonlinearly with the pressure exerted by the vibrator; this introduces harmonics not present in the original input signal. This effect is less serious for *downsweeps* (negative dv/dt in (5.3)) but *upsweeps* are much easier on the equipment so that there is no clear-cut preference.

Vibroseis sources produce low energy density; as a result they can be used in cities and other areas where explosives and other sources would cause extensive damage (Mossman *et al.*, 1973). Vibroseis is now used for about a third of the land seismic exploration.

An impactor such as shown in fig. 5.28 can be used as a source for shallow penetration surveys (up to 1.0 s) using the Sosie™ method (see §5.4.9). The impactor strikes the ground 5 to 10 times per second and a recording is made for about 3 minutes (therefore of 900 to 1800 impacts). The impact times can be considered random for seismic frequencies. A sensor on the baseplate records the moment of each impact for use in correlation. Random repetitive firing of other small sources, such as small Vaporchoc™ units (see §5.5.3*d*) in marine surveys, can also be used as Sosie sources.

Fig.5.27. Vibroseis equipment mounted for off-road survey. (Courtesy Conoco.)

5.4.4 *Geophones*

(*a*) *General.* Seismic energy arriving at the surface of the ground is detected by *geophones*, frequently referred to as *seismometers, detectors, phones* or *jugs*. Although many types have been used in the past, modern geophones are almost entirely of the moving-coil electromagnetic type for land work and the piezoelectric type for marsh and marine work and sometimes for measurements in boreholes. The latter will be discussed in §5.5.4 in connection with marine equipment.

The moving-coil electromagnetic geophone is shown schematically in fig. 5.29*a* while fig. 5.29*b* is a photograph of a cutaway model. The schematic diagram shows a permanent magnet in the form of a cylinder into which a circular slot has been cut, the slot separating the central South Pole from the outer annular North Pole. A coil consisting of a large number of turns of very fine wire is suspended centrally in the slot by means of light leaf

springs, *A*, *B* and *C*. The geophone is placed on the ground (in firm contact with it) in an upright position. When the ground moves vertically, the magnet moves with it but the coil, because of its inertia, tends to stay fixed. The relative motion between the coil and magnetic field generates a voltage between the terminals of the coil. The geophone output for horizontal motion is essentially zero since the coil is supported in such a way that it stays fixed relative to the magnet during horizontal motion.

(*b*) *Equations of motion.* The theory of geophones has been discussed in several places (Dennison, 1953; Washburn, 1937; Scherbatskoy and Neufeld, 1937). We let

x = displacement of the surface = displacement of geophone;

x_c = displacement of geophone coil relative to the permanent magnet;

m, r, n = mass, radius, number of turns of the coil;

Fig.5.28. Two impactors used as mini-Sosie source. Geophones on the base plates indicate the times of impacts. (Courtesy Wacker-Werke.)

i = current in the coil;

τ = mechanical damping factor, $\tau(\mathrm{d}x_\mathrm{c}/\mathrm{d}t)$ being the damping force;

S = spring constant = $f/\Delta x$, where the force f stretches the spring by Δx;

H = strength of permanent magnetic field;

$K = 2\pi rnH$;

Ki = force on coil due to current;

R, L = total resistance and inductance of the coil plus external circuit.

A geophone coil in motion is acted upon by three forces: the restoring force of the springs, the force of friction, and the force resulting from the interaction of the permanent magnetic field with the magnetic field of the current. The first two are retarding (negative) forces, while the last is positive. Newton's second law of motion gives

$$-Sx_\mathrm{c} - \tau\frac{\mathrm{d}x_\mathrm{c}}{\mathrm{d}t} + Ki = m\left(\frac{\mathrm{d}^2 x}{\mathrm{d}t^2} + \frac{\mathrm{d}^2 x_\mathrm{c}}{\mathrm{d}t^2}\right). \qquad (5.4)$$

Faraday's law of induction relates x_c to i:

$$\text{emf induced in coil} = -\frac{\mathrm{d}\phi}{\mathrm{d}t} = -\frac{\mathrm{d}\phi}{\mathrm{d}x_\mathrm{c}}\frac{\mathrm{d}x_\mathrm{c}}{\mathrm{d}t}$$

$$= -2\pi rnH\frac{\mathrm{d}x_\mathrm{c}}{\mathrm{d}t} = -K\frac{\mathrm{d}x_\mathrm{c}}{\mathrm{d}t}$$

$$= Ri + L\frac{\mathrm{d}i}{\mathrm{d}t},$$

where ϕ = flux through the coil. Solving for x_c,

$$\frac{\mathrm{d}x_\mathrm{c}}{\mathrm{d}t} = -\frac{1}{K}\left(Ri + L\frac{\mathrm{d}i}{\mathrm{d}t}\right).$$

Differentiating (5.4) and substituting for $\mathrm{d}x_\mathrm{c}/\mathrm{d}t$ gives the geophone equation of motion,

$$L\frac{\mathrm{d}^3 i}{\mathrm{d}t^3} + \left(R + \frac{L\tau}{m}\right)\frac{\mathrm{d}^2 i}{\mathrm{d}t^2} + \left(\frac{SL + \tau R + K^2}{m}\right)\frac{\mathrm{d}i}{\mathrm{d}t} + \left(\frac{SR}{m}\right)i$$

$$= K\frac{\mathrm{d}^3 x}{\mathrm{d}t^3}. \qquad (5.5)$$

Fig.5.29. Moving-coil electromagnetic geophone. (*a*) Schematic; (*b*) cutaway of digital-grade geophone (Courtesy Geo Space).

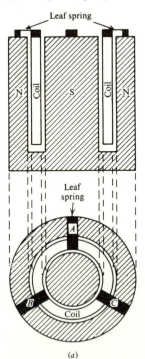

(a) (b)

For a geophone output to be independent of frequency, $L = 0$ (since inductive reactance depends on frequency). While this cannot be achieved, we assume that L is sufficiently small that we can neglect it, giving

$$\frac{d^2 i}{dt^2} + \left(\frac{\tau}{m} + \frac{K^2}{mR}\right)\frac{di}{dt} + \left(\frac{S}{m}\right)i = \left(\frac{K}{R}\right)\frac{d^3 x}{dt^3}. \quad (5.6)$$

The term involving (di/dt) represents damping, τ/m giving the mechanical damping and K^2/mR the electromagnetic damping. If the damping were zero, the system would be simple harmonic with natural frequency ν_0, where

$$\nu_0 = \frac{\omega_0}{2\pi} = \left(\frac{1}{2\pi}\right)\left(\frac{S}{m}\right)^{\frac{1}{2}}. \quad (5.7)$$

When the damping is not zero, we write

$$\left.\begin{aligned} &\frac{d^2 i}{dt^2} + 2h\omega_0\frac{di}{dt} + \omega_0^2 i = \left(\frac{K}{R}\right)\frac{d^3 x}{dt^3}, \\ &\text{where} \\ &2h\omega_0 = \frac{\tau}{m} + \frac{K^2}{mR}, \end{aligned}\right\} \quad (5.8)$$

h being the damping factor of (2.95). This is the equation for damped simple harmonic motion and the solution is given in standard texts (Wylie, 1966).

(*c*) *Transient response*. The transient solution is obtained by setting the right side of (5.8) equal to zero. Let us assume that $i = 0$, $di/dt = u_0$ at $t = 0$; then the solution has the following form depending on the value of h:

for $h > 1$ (overdamped),

$$i = [u_0/\{\omega_0(h^2 - 1)^{\frac{1}{2}}\}]e^{-h\omega_0 t}\sinh\{\omega_0 t(h^2 - 1)^{\frac{1}{2}}\}; \quad (5.9)$$

for $h = 1$ (critically damped),

$$i = u_0 t e^{-\omega_0 t}; \quad (5.10)$$

for $h < 1$ (underdamped),

$$i = [u_0/\{\omega_0(1 - h^2)^{\frac{1}{2}}\}]e^{-h\omega_0 t}\sin\{\omega_0 t(1 - h^2)^{\frac{1}{2}}\}. \quad (5.11)$$

These solutions are shown in fig. 5.30 in terms of the resonant period T_0; they are transient solutions because i eventually becomes zero owing to the exponential factor. For $h > 1$, the current starts to build up because of the sinh factor, but then decreases as the exponential factor

Fig.5.30. Free oscillation of a geophone as a function of damping factor, h.

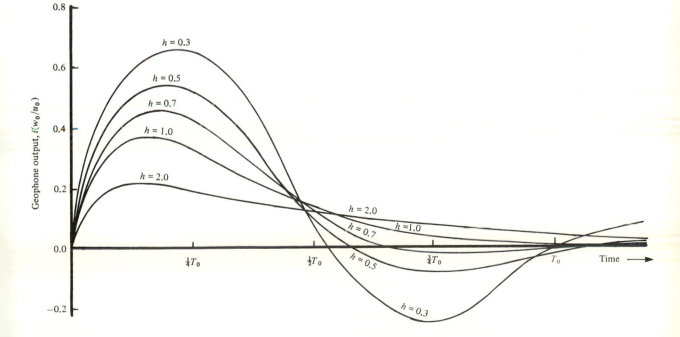

begins to dominate. When $h < 1$, the output is a damped sine wave. For $h = 1$, the *critically damped* case, the output just fails to be oscillatory. When $h < 1$, successive peaks occur at intervals

$$T_0 = 2\pi/\{\omega_0(1 - h^2)^{\frac{1}{2}}\}, \tag{5.12}$$

and the ratio of successive peaks is

$$i_n/i_{n+1} = \exp[2\pi h(1 - h^2)^{\frac{1}{2}}]. \tag{5.13}$$

The logarithmic decrement δ in nepers (see problem 2.17) is given by

$$\delta = \ln(i_n/i_{n+1}) = 2\pi h(1 - h^2)^{\frac{1}{2}}; \tag{5.14}$$

we can obtain h when $h < 1$ by measuring δ.

(*d*) *Response to a driving force.* If the geophone is subjected to a harmonic displacement such that the velocity $dx/dt = v_0 \cos \omega t$, then

$$x = \frac{v_0}{\omega}\sin \omega t; \quad \frac{dx}{dt} = v_0 \cos \omega t;$$

$$\frac{d^2x}{dt^2} = -\omega v_0 \sin \omega t; \quad \frac{d^3x}{dt^3} = -\omega^2 v_0 \cos \omega t,$$

and (5.8) becomes

$$\frac{d^2i}{dt^2} + 2h\omega_0\frac{di}{dt} + \omega_0^2 i = -\frac{\omega^2 K v_0}{R}\cos \omega t. \tag{5.15}$$

The solution of this equation is made up of two parts, a transient solution given by (5.9) to (5.11) plus a solution representing the forced motion of the geophone resulting from the motion of the ground. The latter is

$$i = (v_0/Z)\cos(\omega t + \gamma), \tag{5.16}$$

where

$$Z = (R\omega_0^2/K\omega^2)[\{1 - (\omega/\omega_0)^2\}^2 + (2h\omega/\omega_0)^2]^{\frac{1}{2}},$$

$$\tan \gamma = (2h\omega/\omega_0)\{(\omega/\omega_0)^2 - 1\}. \tag{5.17}$$

Thus the amplitude of i for a given geophone depends upon v_0, ω/ω_0, R, K, and h. When $\omega \to \infty$, $Z \to R/K$ and the amplitude of i becomes $i_\infty = v_0 K/R$.

One of the most important factors of merit of a geophone is the output voltage per unit velocity of the case. We can define the geophone sensitivity, Γ (also called the *geophone transduction constant*), by the relation

$$\Gamma = \frac{\text{amplitude of output voltage}}{\text{amplitude of geophone velocity}}. \tag{5.18}$$

Assuming the geophone is connected to an amplifier with essentially infinite input impedance (the usual case),

the output voltage is the voltage across R_s, the shunt resistance. Using (5.16) and (5.17) we get

$$\Gamma = R_s(v_0/Z)/v_0 = R_s/Z$$

$$= K(R_s/R)f(\omega/\omega_0), \tag{5.19}$$

where

$$f(\omega/\omega_0) = 0, \quad \text{when } \omega = 0,$$

$$= 1, \quad \text{when } \omega = \infty,$$

$$= \tfrac{1}{2}h, \quad \text{when } \omega = \omega_0.$$

For practical purposes the geophone sensitivity is determined largely by K and h, that is, by the radius and number of turns of the coil, by the magnetic field strength and by the damping. Modern geophones have sensitivities of about 0.7 V/cm/s.

Curves of Γ are shown in fig. 5.31*a* for various values of h. For $h = 0$, the output becomes infinite at the natural frequency; obviously this is merely a theoretical result since zero damping can never be achieved. As h increases, the output peak decreases in magnitude and moves towards higher frequencies. When $h \approx 0.7$ the peak disappears and the range of flat response has its maximum extent. As h increases beyond this value, the low-frequency response falls off. The generally accepted choice of 70% of critical damping for geophones thus results in more-or-less optimum operating conditions with respect to amplitude distortion in the geophone output. Obviously the damping of a geophone is a key factor in determining its performance. The damping factor (expressed by h in (5.8)) can be increased by winding the coil on a metal 'former' so that eddy currents induced in the former by motion of the coil will oppose the motion; h can be increased to about 0.3 by this means. The damping is usually further increased by a resistance in parallel with the coil (inside the case).

The output of a geophone is shifted in phase with respect to the input as shown in fig. 5.31*b*. The phase shift γ will change the waveshape, that is, produce phase distortion, since the seismic signal comprises a range of frequencies.

Fig. 5.31*a* shows that for $h = 0.7$ the distortionless signal band extends from about $1.2\omega_0$ upward, hence the lower the natural frequency, the wider the distortionless band. The natural frequency of geophones employed in petroleum exploration (v_0) is usually 7 to 28 Hz for reflection work, 4.5 Hz for refraction. The decrease of sensitivity below the natural frequency (fig. 5.31*a*) often provides the lower limit to the passband to be recorded.

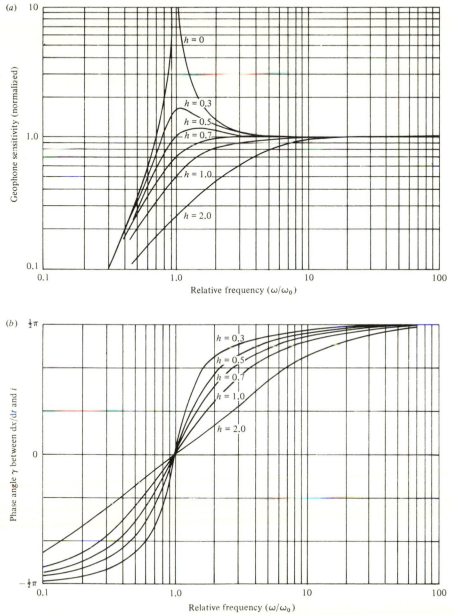

Fig.5.31. The dependence of geophone response on damping factor h. (After Dennison, 1953.) (a) Sensitivity; (b) phase response.

(a)

Geophone sensitivity (normalized)

$h = 0$

$h = 0.3$

$h = 0.5$

$h = 0.7$

$h = 1.0$

$h = 2.0$

Relative frequency (ω/ω_0)

(b)

Phase angle γ between dx/dt and i

$h = 0.3$

$h = 0.5$

$h = 0.7$

$h = 1.0$

$h = 2.0$

Relative frequency (ω/ω_0)

(e) *Other aspects.* Geophone coils are often divided into two parts which are wound in opposite directions and so wired that signals due to motion add whereas those which result from stray electrical pickup in the coils cancel; this feature is called *hum-bucking.* Geophones also have spurious resonances because of modes of motion other than that intended but these usually occur at frequencies above the seismic passband.

Usually several closely-spaced geophones are connected in a series–parallel arrangement to produce a single composite output. The entire geophone group is considered to be equivalent to a single geophone located at the center of the group. However, the damping of each geophone will be affected by the presence of the other geophones because of the change in resistance of the circuit. An exception is an arrangement of n parallel branches, each containing n identical geophones in series, which has the same resistance as a single geophone and hence the same damping.

We have assumed that the geophone follows exactly the motion of the surface of the ground but the geophone is not rigidly fastened to the ground and the coupling of a geophone to the ground also affects the response. Coupling often has a natural frequency in the 100–200 Hz range, being as low as 30–40 Hz in swampy ground (O'Brien, 1965). The coupling can be improved by pushing a spike fastened to the geophone into the ground or by increasing the base area of the geophone.

5.4.5 *Amplifiers*

Except for very strong signals arriving soon after the shot is fired, the output of the geophone is too weak to be recorded without amplification. Also, the useful range of amplitudes of the geophone output extends from a few tenths of a volt at the beginning of the recording to about 1 μV near the end of the recording several seconds after the shot (signals weaker than about 1 μV are lost in the system noise), a relative change or *dynamic range* of about 10^5 (100 dB). Therefore, besides amplifying weak signals, the amplifier usually is called upon to compress the range of signals as well. In addition, amplifiers are used to filter the geophone output to enhance the signal relative to the noise. A good discussion of seismic amplifiers is given by Evenden and Stone (1971).

Seismic amplifiers generally employ solid-state circuitry which allows them to be very compact. While they are usually mounted in a recording truck or other vehicle, they can also be carried where necessary (fig. 5.32a).

A block diagram of an analog amplifier is shown in fig. 5.32b; the arrangement of circuit elements and the number of amplification stages vary from manufacturer to manufacturer. The cable from the geophones may be connected to a balance circuit which permits adjusting the impedance to ground so as to minimize the coupling with nearby power lines, thus reducing pickup of noise at the power-line frequency (*high-line pickup*). The next circuit element usually is a filter to attenuate the low frequencies which arise from strong ground roll and which otherwise might overdrive the first amplification stage and introduce distortion.

Seismic amplifiers are multi-stage and have very high maximum gain, usually of the order of 10^5 (100 dB), sometimes as much as 10^7 (140 dB); 100 dB means that an input of 5 μV amplitude appears in the output with an amplitude of 0.5 V. Lower amplification can be obtained by means of a multi-position master gain switch which reduces the gain.

The amplifier gain is varied during the recording interval starting with low amplification during the arrival of strong signals at the early part of the record and ending up with the high gain value fixed by the master gain setting. This variation of gain with time (signal compression) can be accomplished with *automatic gain* (*volume*) *control*, usually abbreviated AGC or AVC. This is accomplished by a negative feedback loop, a circuit which measures the average output signal level over a short interval and adjusts the gain to keep the output more-or-less constant regardless of the input level. If the time between a change of amplitude and the consequent change of gain is too short, the output amplitude will be nearly constant and reflection events will not stand out; if the time is too long, subsequent reflections will not stand out. In either case, amplitude information will be lost. The use of AGC was standard prior to the 1960s, and AGC is still used, especially in making displays.

It is important in making corrections for near-surface effects that we be able to observe clearly the *first-breaks*, the first arrivals of energy at the different geophones. (For a geophone near the shotpoint, the first arrival travels approximately along the straight line from the shot to the geophone; for a distant geophone the first arrival is a head wave refracted at the base of the low-velocity layer – see the discussion of weathering corrections, §5.6.2.)

If we allow the AGC to determine the gain prior to the first arrivals, the low input level (which is entirely noise) will result in very high gain; the output will then be noise amplified to the point where it becomes difficult to observe the exact instant of arrival of the first-breaks. This problem is solved by using *initial suppression* or *presuppression.* A high-frequency oscillator signal (about 3 kHz) is fed into the AGC circuit which reacts by reducing

Fig.5.32. Analog seismic instruments. (*a*) Portable
instruments (Courtesy TI). (*b*) Block diagram of analog
seismic amplifier.

(*a*)

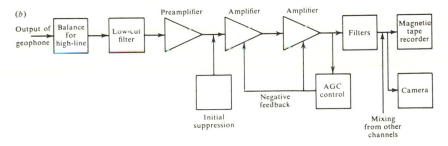

(*b*)

the gain so that the noise is barely perceptible; the high-frequency signal is subsequently removed by filtering so that it does not appear in the output. With the reduced gain the relatively strong first-breaks stand out clearly. As soon as the first-breaks have all been recorded, the oscillator signal is removed, usually by a relay triggered by one of the first-breaks. Thereafter, the AGC adjusts the gain in accordance with the seismic signal level.

Seismic amplifiers are intended to reproduce the input with a minimum of distortion and hence the gain (without filters) should be constant for the entire frequency spectrum of interest. For reflection work this range is about 10–100 Hz while for refraction work the range is about 1–50 Hz; most amplifiers have flat response for frequencies from about 1 Hz to 200 Hz or more.

Frequency filtering refers to the discrimination against certain frequencies relative to others. Seismic amplifiers have a number of filter circuits which permit us to reduce the range of frequencies which the amplifier passes. While details vary, most permit the selection of the upper and the lower limits of the passband. Often it is possible to select also the sharpness of the *cutoff* (the rate at which the gain decreases as we leave the passband). Fig. 5.33 shows typical filter response curves. The curves are specified by their *cutoff frequencies*, that is, the frequency values at which the gain has dropped by 3 dB (30% of amplitude, 50% of power); the curve marked 'Out' is the response curve of the amplifier without filters.

Seismic amplifiers may include circuitry for *mixing* or *compositing*, that is, combining two or more signals to give a single output. Mixing in effect increases the size of the geophone group and is sometimes used to attenuate certain types of surface waves. The commonest form, called '50% mixing', is the addition equally of the signals from adjacent geophone groups. Magnetic tape recording now has virtually eliminated the need to mix during recording since we can always mix in playback.

The time-break signal often is superimposed on one of the amplifier outputs where it appears as a sharp pulse which marks the point $t = 0$ for the record. When explosives are being used, the output of an *uphole geophone* (a geophone placed near the top of the shothole) is also superimposed on one of the outputs; the interval between the time-break and the uphole geophone signal is called the *uphole time* (t_{uh}); it measures the vertical traveltime from the shot to the surface and is important in correcting for near-surface effects.

High-resolution or HR amplifiers are used in engineering and mining problems to map the top 200 m or so. To get resolution of a few meters, we must use short wavelengths; accordingly these amplifiers have essentially uniform response up to 300 Hz, sometimes to 500 Hz, and the AGC time constants are correspondingly short. To permit recording very shallow reflections, small offsets are used and the initial suppression permits recording events within 0.050 s or so after the first break.

5.4.6 *Analog data recording*

For the first thirty years or so of seismic exploration, the outputs of the amplifiers were recorded directly on

Fig.5.33. Response of seismic filters.

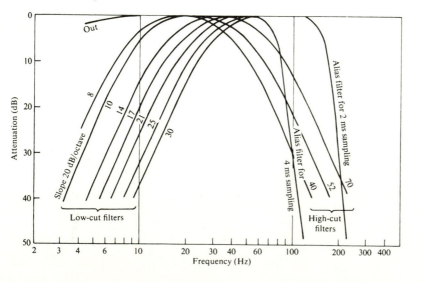

photographic paper by means of a camera. However, about 1952 recording on magnetic tape began and today it is nearly universal (see fig. 1.20). The feature which originally led to widespread use of magnetic recording was the ability to record in the field with a minimum of filtering, automatic gain control, mixing, etc., and then introduce the optimum amounts of these on playback. Later a more important advantage turned out to be the ability to produce record sections (see §5.6.3) which proved to be powerful aids in interpretation. However, magnetic-tape recording did not develop its full potential until the introduction of digital techniques during the early 1960s.

Analog magnetic-tape recorders usually have heads for recording 26 to 50 channels in parallel. In the early years direct recording was used; the output from the amplifier went directly to the recording head, the intensity of magnetization of the tape being proportional to the current in the recording head and hence proportional also to the signal strength. Later, direct recording was displaced by frequency modulation and pulse-width modulation techniques since these are more noise-free and can accept a wider range of signal strengths.

5.4.7   *Data display*

The data recorded on magnetic tape must be presented in visual form for monitoring and for interpretation. This is done most commonly by a camera. The main elements of a camera are (1) a series of galvanometers, one for each geophone group, which transform the electrical signals into intense spots of light moving in accordance with the signals, (2) a device for recording accurate time marks, and (3) a means for recording the positions of the light spots on a moving piece of paper. In the past the latter was accomplished mainly by photographic methods and this is still used somewhat today. On some crews dry-write (where a latent image on paper gradually develops upon exposure to daylight) has replaced wet-process photography (using liquid developer and fixer). More widely used, however, are electrostatic cameras wherein the light produces an electrical charge image and printing powder adheres to the paper where it is charged. Electrostatic cameras use ordinary paper which is cheaper than the special papers required in photographic or dry-write processes. Plotters used in fixed installations (and sometimes by field crews) are often of the *raster* type wherein a matrix of very fine dots is used to create the image; usually an intense, very fine beam of light (often from a laser) is swept across the paper, the beam being turned on and off very rapidly to produce the dots. With a raster plotter the information from the different channels is

formatted in a minicomputer and individual galvanometers are no longer used.

Each individual graph representing the motion of a geophone (or the average of a group of geophones) is called a *trace*. A simple graph of amplitude against arrival time is called a *wiggly-trace* mode of display (fig. 5.34c). Where part of the area under a wiggly-trace curve is blacked in, the display is called *variable-area* (fig. 5.34b); sometimes a half-cycle is blacked in, sometimes the portion between the trace and a reference value called the *bias* is blacked in. Sometimes the light intensity is varied instead of the light-spot position to produce *variable-*

Fig.5.34. Modes of displaying seismic data. (Courtesy Geo Space.) (*a*) Wiggle superimposed on variable area; (*b*) variable area; (*c*) wiggle; (*d*) variable density; (*e*) wiggle superimposed on variable density.

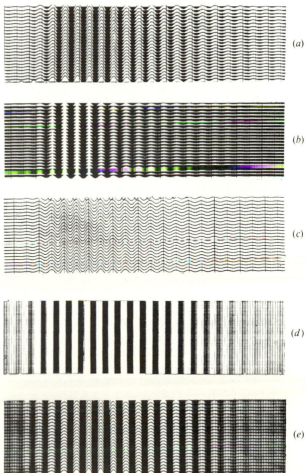

density mode (fig. 5.34*d*). Superposition of modes are also used (figs. 5.34*a*, *e*).

The mode and parameter choice for display greatly affect what an interpreter sees in the data. Among the display parameters are horizontal and vertical display scales, width, amplitude and clip level (maximum amplitude which will be plotted) of wiggly-traces, degree of blackness, bias and clip level of variable-area traces, etc. Usually the effective vertical scale is greater than the horizontal scale, that is, sections are horizontally compressed (the vertical scale is of course variable with depth when time is plotted linearly, as is usually the case). However, scale ratios of approximately 1 : 1 are most helpful when making structural interpretation. Color is sometimes superimposed on sections to display additional information.

Three-dimensional (3-D) data (amplitude as a function of north and east coordinates and arrival time) provide special display problems because of the large amount of data involved; the data occupy a volume (fig. 5.35*a*). Usually the data are displayed as a series of seismic sections, including sections in arbitrary directions through the data set. The data at one particular arrival time for the entire area are also displayed as a time-slice or Seiscrop™ map (figs. 5.35*b* and 5.36). Lineups on time-slice maps constitute time contours on the respective events.

5.4.8 *Digital recording*

Digital recording was first introduced into seismic work early in the 1960s and by 1975 was almost universal (see fig. 1.20). Whereas analog devices represent the signal by a voltage (or other quantity) which varies continuously with time, *digital recording* represents the signal by a series of numbers which denote values of the output of the geophone measured at regular intervals, usually 2 or 4 ms. Digital recording is capable of higher fidelity than analog recording and permits numerical processing of the data without adding appreciably to the distortion. Digital processing has proven to be so effective in improving seismic data that it has gained widespread acceptance and analog recording may be completely superseded eventually. However the beginning (geophone response) and end (display) of the recording process continue to be analog.

Before describing digital recording, we shall discuss digital representations. While we could build equipment to handle data using the scale of ten which forms the basis of our ordinary arithmetic, it is more practical to operate on the *binary scale* of 2. The binary scale uses only two digits, 0 and 1; hence only two different conditions are required to represent binary numbers, for

example, a switch opened or closed. Binary arithmetic operations are much like decimal ones. The decimal number 20873 is a shorthand way of saying that the quantity is equal to 3 units plus 7×10 plus 8×10^2 plus 0×10^3 plus 2×10^4. Similarly, the binary number 1011011 is equal to 1 unit plus 1×2 plus 0×2^2 plus 1×2^3 plus 1×2^4 plus 0×2^5 plus 1×2^6, which is the same as the decimal number 91. We can use positive and negative square pulses to represent 1 and 0 or represent them in other ways. Each pulse representing 1 or 0 is called a *bit* and the series of bits which give the value of a quantity is called a *word*.

Figure 5.37 shows the interior of a recording cab equipped with a floating-point amplifier system. The observer communicates with the system via the typewriter console and video-tube display. A schematic dia-

Fig.5.35. Three-dimensional data obtained from a set of closely-spaced N–S lines. (*a*) Isometric diagram of the volume these data occupy, the easternmost section is shown along with an E–W section made from the southernmost traces from each line; (*b*) isometric diagram of the data set with top portion removed; the top constitutes a time-slice (Seiscrop) map.

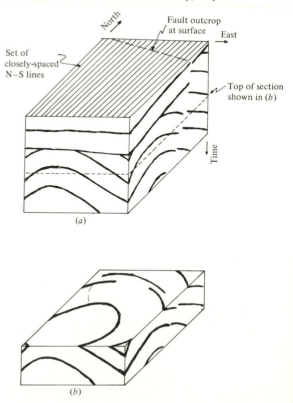

(*a*)

(*b*)

Fig. 5.36. Series of time-slice maps. The area is 3.6 × 8.0 km. (Courtesy GSI.) (a) through (g): Maps at t = 1.580 to 1.604 s at 4 ms intervals. (h) Time-contour map made by tracing one contour from each slice (a) through (g), starting with the outside of the central area on the shallowest map (a).

(a)

(b)

(c)

(d)

(e)

(f)

(g)

(h)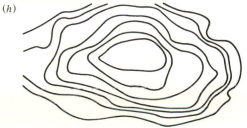

gram of a digital recording system is shown in fig. 5.38. Systems as of 1981 recorded up to 120 channels at 1, 2 or 4 ms sample rate. Specifications are given in appendix F.

The tape recorder usually writes nine bits at a time (a *byte*) onto magnetic tape a half inch wide; two bytes are used to record each sample of data (one *word*). One of the bits in each byte is usually reserved for a *parity* bit; a 0 or 1 is placed here so that the total number of 1s in the byte is an odd number if the parity convention is for odd parity, or even if for the opposite convention. Subsequent counting of the number of bits in each byte is used as a check against loss of information. The formatter distributes the various bits among the heads in a fixed pattern known as the *format*. Tape speeds in digital recording range from 10 to 150 in/s, depending upon the sampling interval and the format arrangement (North-

wood *et al.*, 1967; Meiners *et al.*, 1972; Barry *et al.*, 1975). The tape speed is adjusted so that the density of data along a single track is constant (800 or 1600 bits/inch, sometimes 6250 bits/inch).

Digital recording (and processing) involves a long series of operations which take place sequentially on a time scale measured in microseconds. The entire sequence is controlled by an electronic 'clock', a crystal-controlled oscillator operating in the megacycle range which furnishes a continuous series of pulses whose shape and spacing are accurately maintained. Time is measured by counting these pulses and the operating cycles of the component units (such as the multiplexer and the formatter) are controlled by circuits which count the clock pulses and operate electronic switches when the count reaches predetermined values.

Fig.5.37. Digital field equipment mounted in a truck; the recording system includes some data processing capabilities. (Courtesy TI.)

5.4.9   *Field processing*

The records from a series of weak seismic sources are often added together to give the effect of a single strong source. Because the successive source locations are usually within a fairly small area with dimensions no more than double the geophone group spacing (fig. 5.13 shows about the maximum used), successive records change in only minor ways and can be simply added together without applying any corrections, except for time shifts to align the source instants (vertically stacked). While vertical stacking is sometimes done in a data-processing center, it is often done in the field. Stacking requires a memory device to hold the partial sum while additional records are being obtained. Magnetic-tape loops, discs, drums and solid-state memories are used to hold the information in digital form until the sum is completed and can be written on the output magnetic tape. Such memories may have capacities up to several megabits.

With the Sosie method (Barbier and Viallix, 1973), a geophone group's output begins adding into the summation registers again each time a new impulse is applied to the Earth. Since a number of impulses occur within the recording length, the signal is being added at several places simultaneously. Source-generated energy thus adds

Fig.5.38. Block diagram of digital recording system. Each channel has its own components prior to the multiplex switch. The *line filter* reduces radiofrequency static picked up by the geophone cables. The *pre-amplifier* increases the signal level by a constant amount while providing impedance matching. The *low-cut filter* supplements geophone filtering in removing very low frequencies in areas of excessive ground roll. The *high-cut filter* prevents aliasing (see §8.1.2b); the slope is typically 72 dB/octave. The *notch filter* reduces 50 or 60 Hz power-line pickup (or $16\frac{2}{3}$ Hz electric railroad pickup). The *multiplexer* connects each geophone sequentially to the *quaternary-gain amplifier* which automatically adjusts its gain in 4 : 1 steps until the signal amplitude falls within a prescribed range, after which a 3-bit word specifying the gain is sent to the formatter. The *A/D converter* measures the signal amplitude, the output being one bit for polarity and fourteen bits for the magnitude. The *formatter* arranges the data for writing on magnetic tape by the *tape transport*. Separate *read heads* read the magnetic tape immediately after the data have been written. The output is amplified in the *digital AGC unit*, converted to analog form in the *D/A converter*, after which the *camera* gives a monitor paper record.

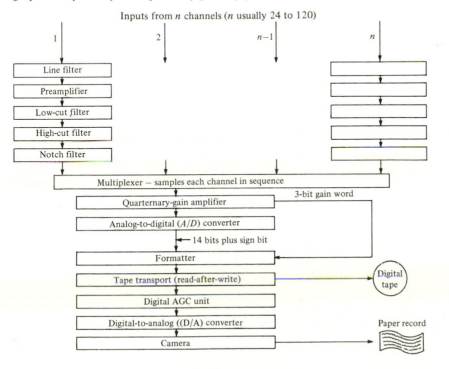

in-phase with respect to the proper arrival time but is random with respect to the other times at which it is added into the registers (see problem 5.21).

As field operations get more complex and continuous, the likelihood of human error and the incentive for automation increase. Small minicomputers in the recording truck carry out most routine operations. A computerized field system is shown in fig. 5.39.

When both memory and computer are available, Vibroseis correlation (§8.1.3*d*) can be done in the field so that the field output may be correlated tapes. This also permits vertical stacking after correlation. It may also be possible to read the memory in a different sequence so that the field output tapes may be demultiplexed, that is, the data may be recorded in a channel-by-channel (*trace-sequential*) sequence rather than in a time-after-source (*time-sequential*) sequence; this lessens subsequent processing time. Additional field processing is usually impractical on a large scale, although sometimes done on an experimental basis.

5.4.10 *At-the-geophone digitization*

Many geophone channels are necessary to record (i) redundant shallow data and long-offset deep data simultaneously, (ii) stations distributed over an area (rather than along a line) for three-dimensional analysis, (iii) individual geophones rather than arrays to improve high-frequency response (minor time shifts between differ-

ent geophones within a group result in attenuation of higher-frequency components; see (§5.3.8), (iv) closer geophone spacing along the line to improve horizontal resolution.

For more than a hundred channels or so, cables to carry the electrical signals to the recording instruments become cumbersome. Distortion also occurs in transmission through long cables, especially as they age. Digitizing near the geophone in remote data units (RDU) and then transmitting the digitized data in multiplexed form over one or a few pairs of wires overcomes many of these difficulties. The tape transport presently limits data rates to about 1.5 megabits/s with 1600 bits/inch recording and to 6 megabits/s with 6250 bits/inch recording (see problem 5.20).

One such system uses individually-addressable battery-powered field units which can be turned on or off remotely. Each digitizes the signal from four geophones (or groups) and transmits a 76-bit burst at 640 kilobit/s rate in a time interval reserved for it. Another system uses digitizer–repeater units connected in series, with power supplied from the recording truck via an extra pair of wires. Each unit relays information at a 4 megabit/s rate, adding the digitized output of the geophones connected to it at the tail end of the information stream. Another system uses sign–bit recording as described in §8.1.3*e*. Still another system (fig. 5.40) transmits the data from the RDUs to the recording unit by radio, and yet another

Fig.5.39. Computerized field system. Data can be transferred to and from various units. The computer's main function is to control data flow, much of the processing being done in other units of the system.

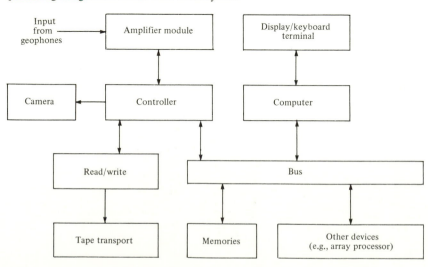

system stores the information on magnetic-tape cassettes within each RDU, to be retrieved subsequently.

5.5   Marine equipment and methods

5.5.1   *Marine operations*

Marine seismic operations usually imply water more than 10 m deep and sufficiently clear to allow freedom of movement for ships 30 to 70 m in length. A typical seismic ship is shown in fig. 5.41. The ship carries enough fuel, water, and other supplies to operate at sea for 30 days. The ship is manned by about 25 men. These include a ship's crew of about nine consisting of the Captain, Mate, Chief Engineer, Second Engineer, two deck hands, a cook, a mess man, and a cabin attendant. The seismic crew is made up of the party Manager, Chief Observer, Observer, three or four Junior Observers, the Instrument Technician, one or two Navigation Engineers, the Chief Mechanic, and three or more mechanics.

Marine seismic operations differ from those on land and in shallow water primarily because of the speed with which they take place. Normal production shooting proceeds at about 6 knots (11 km/hr) and can proceed on a 24 hr/day basis. Hence with a 2400 m 48-channel streamer, 24-fold CDP coverage of 250 km/day (6000 shots/day) would be possible if all the time were spent shooting. This much production is never achieved because much time is spent traveling to the line or from the end of one line to the start of the next, waiting for good weather, or because of other factors. Nonetheless, production rates are of the order of 4 profiles/minute and each profile may consist of 2 component *subshots* (separate records which are then vertically stacked); shots are thus spaced about every 25 m at normal operating speeds. When a ship is ten or more kilometers from the work location, the *streamer* (§5.5.4), a tube several kilometers long containing the detectors and connecting cables is unreeled and positioned in the water, followed by the seismic source units. The navigation computer commands the steering of the ship, and, at the proper times, issues 'on location' commands to tell the seismic recording system to begin recording and to fire the sources. The subsequent recording of the various sensors is largely automatic, the chief functions of the observers, navigators, and others being to see that everything is functioning properly.

Fig.5.40. The Opseis™ system. The system is capable of sequentially recording up to 4 lines with 2 spreads/line and 1016 channels/spread. (Courtesy L. Denham.) (*a*) Remote telemetry unit serving 4 geophone groups. The seismic traces are stored in the unit's memory until instructed by the central unit to transmit them. This allows many remote units to operate on one radio channel and also circumvents tape-transport data-rate limitations. Systems provide automatic identification.

The shooter can plug into any remote unit. An alarm is broadcast if a unit is moved. (*b*) Program unit at the central recording station. The keyboard allows the operator to enter data or instructions, and panels of light-emitting diodes display information to the observer. Communication between central and remote units is by horizontally-polarized radiofrequency waves.

(*a*)

(*b*)

The streamer and other gear which are towed behind the ship can be kept on location only while the ship is moving. Thus the ship cannot stop to make an adjustment or effect a repair without having to reshoot part of the seismic line. To repeat a profile the ship must circle and return to the line about 10 km before the required position is reached so that the streamer will be straightened out; this procedure results in the loss of about two hours. Operations are normally continued despite minor malfunctioning, such as the loss of one or two airguns (§5.5.3*b*), failure of one group of hydrophones, etc., and malfunctions are remedied as soon as operational constraints permit. Airguns might be brought aboard for repair, for example, while operations continue with the remaining guns. A single defective group of hydrophones would probably not be repaired until the streamer is reeled in at the conclusion of the unit of work.

The monthly cost of a marine crew is large but the high production cuts the unit cost of marine seismic data to about 10% of that of land data (see fig. 1.23). The high production rate requires special emphasis on efficiency in operations. Source and receiver units are towed into place and forward travel does not stop during a recording. While detailed monitoring of data quality is not possible at the pace of operations, the relatively constant water environment surrounding the sources and receivers and

the general absence of the low-velocity weathering layer which is usually present on land lessen variations in data quality.

5.5.2 *Bubble effect*

An underwater explosion produces a bubble of gases at high pressure. As long as the gas pressure exceeds the hydrostatic pressure of the surrounding water, the net force will accelerate the water outward away from the shot. The net force decreases as the bubble expands and becomes zero when the bubble expansion reduces the gas pressure to the value of the hydrostatic pressure. However, at this point the water has acquired its maximum outward velocity and so continues to move outwards while decelerating because the net force is now directed inward. Eventually the water comes to rest and the net inward force now causes a collapse of the bubble with a consequent sharp increase in gas pressure, and the process repeats itself. Thus the bubble will oscillate and seismic waves will be generated on each oscillation. The waveshape generated by the explosion of a small charge is shown in fig. 5.45*e*.

As the bubble loses energy and rises toward the surface its oscillation period decreases (Kramer *et al.*, 1968). For 'conventional' dynamite charges of $16\frac{2}{3}$ lb (7.5 kg) the bubbles produced by the explosion in effect

Fig.5.41. A vessel equipped for marine exploration. A large reel on the rear deck holds the 3–4 km of streamer; the remainder of this deck is occupied by compressors to provide air for the airguns. The enclosed part of this deck level houses the instrument rooms, workshops and galley; the upper level lounges and cabins for the crew. A heliport is above the cable reel. Note the various antennae for satellite and radio navigation, communications with shore and sonobuoys, etc. The ship can stay at sea for about a month. (Courtesy Seiscom Delta.)

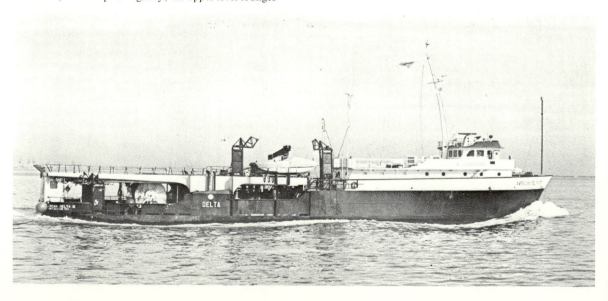

generate additional seismic records every 0.2 to 0.4 s. These records are superimposed on each other so that one cannot tell from which oscillation a reflection event comes. The practice when using conventional explosives as the source, therefore, was to shoot within 2 meters or so of the surface so that the bubble would vent to the surface. This produced spectacular plumes of water but was inefficient in generating useful seismic energy.

The bubble effect is important in determining the source waveshapes of almost all seismic sources, even for those designed expressly to minimize the effect.

5.5.3 *Marine energy sources*

(*a*) *General.* Marine seismic reflection work consists mostly of two types, common-depth-point and profiler work; these differ considerably in cost, size of source, effective penetration, and various other aspects. We shall discuss here the larger energy sources commonly used in common-depth-point recording: the smaller energy sources used in profiling will be described in §5.5.7.

(*b*) *Airguns.* The most widely used large energy source is the *airgun*, a device which discharges air under very high pressure into the water (Giles, 1968; Schulze-Gattermann, 1972). Pressures up to 10 000 psi (70 MPa) are used although 2000 psi (14 MPa) is most common. The airgun shown in fig. 5.42*a* is in the armed position, ready for firing. Chambers *A* and *B* are filled with high-pressure air which entered *A* at the top left and passed into *B* through an axial opening in the 'shuttle'. The latter is held in the closed position by the air pressure (since flange *C* is larger than flange *D*, resulting in a net downward force). To fire the gun, the solenoid at the top opens a valve which allows high-pressure air to reach the underside of flange *C*. This produces an upward force which is large enough to overcome the force holding the shuttle in the closed position and consequently the shuttle opens rapidly. This allows the high-pressure air in the lower chamber to rush out through four ports into the water. The bubble of high-pressure air then oscillates in the same manner as a bubble of waste gases resulting from an explosion. However, since the energy is smaller, the oscillating frequency is in the seismic range and therefore has the effect of lengthening the original pulse (rather than generating new pulses as with dynamite).

The upward motion of the shuttle is arrested before it strikes the top of chamber *A* because the upward force falls off rapidly as the air enters the water and the downward force of the air in the upper chamber increases. The shuttle then returns to the armed position and the lower chamber again fills with air. The explosive release of the air occurs in 1–4 ms while the entire discharge cycle

requires 25–40 ms. The lower chamber is often divided into two parts connected by a small orifice (Mayne and Quay, 1971), which results in a delayed discharge of the air in the lowermost chamber. The flow of this air into the bubble continues for some time after the initial discharge, retarding the violent collapse of the bubble and diminishing the subsequent bubble effect. The waveshape generated by a single airgun is shown in fig. 5.45*a*.

Usually several airguns are used in parallel. Because the dominant frequency of the pulse depends on the energy (that is, on the product of the pressure and volume of air discharged), mixtures of gun sizes (the gun size is the volume in cubic inches of the lower chamber) from 10 to 2000 in^3 (0.16 to 33 liters) are often used to give a broader frequency spectrum. An array of 14 guns is shown in fig. 5.43. The firing of the different guns is synchronized so as to align the first pressure peak for a downgoing wave; this produces some cancellation of the secondary effects. The waveshape from an array is shown in fig. 5.45*b*.

(*c*) *Explosive sources.* The *sleeve exploder* (also called *Aquapulse*^{TM} and *Deltapulse*^{TM}) utilizes the explosion of a mixture of propane and oxygen in a closed flexible chamber. A heavy rubber sleeve (the *boot*) is fastened around a steel frame which is filled with the explosive mixture. The mixture is fired by a spark plug and the products of the explosion expand the rubber boot. A valve opens following the explosion so that the contraction of the boot vents the gases to the surface; this attenuates the bubble effect. A sleeve exploder waveform is shown in fig. 5.45*c*.

Several other arrangements utilizing explosive gas mixtures have been used. In a marine version of the Dinoseis, the explosion of gas in a metal chamber drove a piston outward against the confining water pressure. One method used a long neoprene tube in which an explosive gas mixture was fired. Spark plugs in each 6 m module permitted the entire gas mixture to be exploded at once, thus achieving a linear seismic source. Several types of gas exploders used firing chambers open to the water. The chambers were filled with an explosive gas mixture and fired by a spark plug at the top of the chamber, the waste gases being vented into the water. If placed shallow in the water, such open-chamber exploders were inefficient; if placed deep, they generated severe bubble oscillation.

The *Flexotir*^{TM} method utilizes the explosion of a small (2 oz or 60 g) dynamite charge within a *cage*, a thick-walled cast-iron spherical shell about 60 cm in diameter with many perforations spaced around the shell. The charge is placed at the center of the cage by pumping it down a hose leading from the ship; it is then fired electrically. Water is forced out through the perforations in the shell by the expanding gases. The water flow out

Fig. 5.42. Airgun. (Courtesy Bolt Associates.)
(*a*) Charged and ready for firing; (*b*) firing; (*c*) photograph of airgun.

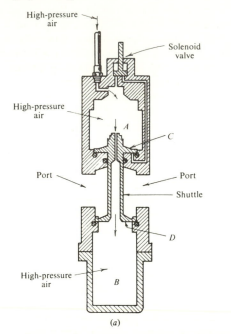

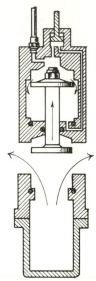

(*a*)

(*b*)

(*c*)

of and back into the shell as the bubble of gases oscillates dissipates the energy and dampens subsequent oscillations while having little effect on the initial expansion (Knudsen, 1961; Lavergne, 1970). Flexotir cannot be used in shallow water because, without a head of hydrostatic pressure, the bubble diameter becomes larger than the spherical shell, resulting in its destruction. Cages are sometimes used with large airguns to serve the same function of damping bubble oscillation.

The *Maxipulse*™ method records the bubble oscillation and uses this information to 'deconvolve' the bubble effect (see §8.1.2d) in subsequent processing. A cartridge containing a percussion delay detonator and about 200 g of explosive is pumped down a hose hydraulically. At the lower end of the hose the percussion cap strikes a firing wheel and after a one-second delay detonates the charge. After the one-second interval the hose is about 5 m away from the charge so that the explosion does not destroy the hose and accessory equipment. A pressure detector capable of withstanding high pressures is mounted on the hose to record the instant of detonation and the subsequent bubble waveform. The source pressure signature (waveform) is recorded on an auxiliary channel at fixed gain after alias filtering; a waveform is shown in fig. 5.45e. The first bubble collapse often creates more seismic energy than the initial explosion and the second bubble collapse often 50% as much. The explosion time is not sufficiently

predictable to permit the simultaneous use of multiple sources.

The *Aquaseis*™ system uses up to 100 m of detonating cord which trails behind the ship. When fired by blasting caps (several may be placed along the charge at different places to shorten the detonation time). This produces a linear charge involving about 0.5 kg of explosive per 30 m. Secondary bubble pulses are small because of the low explosive content per unit of length.

(*d*) *Imploders*. Several types of imploders are sometimes used. *Imploders* operate by creating a region of very low pressure; the collapse of water into the region generates a seismic shock wave. With the Flexichoc™ an adjustable-volume chamber is evacuated while the walls of the chamber are kept fixed by a mechanical restraint; upon removal of the restraint the hydrostatic pressure collapses the chamber and so generates a seismic pulse (fig. 5.45f) relatively free of spurious bubbles. Air is then pumped into the chamber to expand it again whereupon the mechanical restraint holds it open while the air is evacuated, ready for the next collapse. With the Hydrosein™, two plates are driven apart suddenly by a pneumatic piston, creating between them a very-low-pressure region into which the water rushes. With the Boomer™, two plates are forced apart suddenly by a heavy surge of electrical current through a coil on one of the plates which generates eddy

Fig.5.43. Layout of a 14-airgun array. The guns are towed two to a line behind the ship's stern. The sizes of the guns are indicated. The guns are spaced to prevent

interaction of the bubbles of the individual guns. (Courtesy Seiscom Delta.)

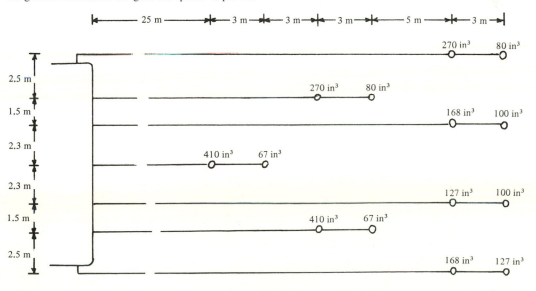

currents in the other plate, resulting in its being suddenly repelled. The Boomer produces less energy than Flexichoc or Hydrosein.

The *watergun* is a type of imploder. Compressed air is used to drive a shuttle which propels water from the gun. Voids are formed behind the high-velocity jets of water and the implosion into these voids generates the seismic pulse (fig. 5.45*g*). Waterguns are used in arrays similar to airguns.

The *Vaporchoc*™ or *steamgun* injects superheated steam into the water. As shown in fig. 5.44 the superheated steam under high pressure passes through an insulated pipe to a submerged tank. When the valve is opened the steam emerges into the water where it forms a bubble. The bubble collapses and disappears because the steam condenses; thus there is no subsequent bubble oscillation. The steam injection time is usually between 10 and 50 ms and shooting rates of 5 to 10 shots per minute are used. The time from valve opening to bubble collapse is not constant, however, so that multiple units are not used simultaneously. The steam can be released into the water through several vents to produce several bubbles and give a composite pulse of broader frequency content. The steam injection generates a forerunner seismic pulse when the valve is opened but the main seismic pulse is generated by the bubble collapse. The forerunner precedes the main pulse by about 50 ms and may have 20% of the amplitude of the main pulse. The seismic tape is referenced to the time of valve opening; the time difference between this and the generation of the main pulse is removed in data processing. The waveform for Vaporchoc is shown in fig. 5.45*d*.

(e) Other sources. A marine version of the Vibroseis has been used, often employing several source units simultaneously. The main problem with the marine Vibroseis was that the sweep length had to be kept short (since the

Fig.5.44. Schematic diagram of Vaporchoc equipment. (Courtesy CGG.)

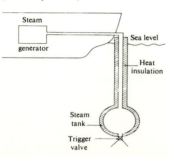

Fig.5.45. Far-field waveshapes generated by marine seismic sources. (*a*) Single 120 in³ airgun; (*b*) array of airguns of different sizes selected to attenuate bubble effects by destructive interference; (*c*) sleeve exploder; (*d*) Vaporchoc; (*e*) Maxipulse; (*f*) Flexichoc; (*g*) watergun; (*h*) 5 kJ sparker. Curves are intended to show features of waveshape, not amplitude relationships. *B* indicates bubble effects; the interval between successive bubbles becomes smaller with time; *I* indicates implosion. (*a*) and (*g*) are from McQuillin *et al.* (1979); (*b*), (*c*) and (*e*) from Wood *et al.* (1978); (*d*) from Farriol *et al.* (1970); (*f*) from manufacturer's literature; (*h*) from Kramer *et al.* (1968).

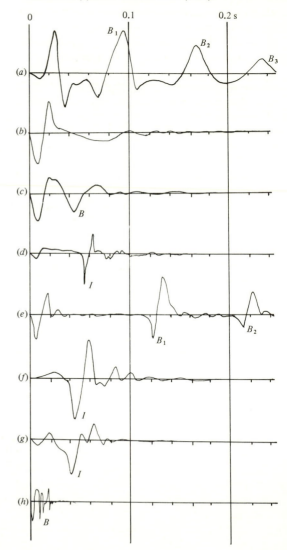

ship is continually moving, long sweeps result in large source 'points' and undesirable 'smearing') so that not enough energy could be injected.

In 1981 most seismic ships used airguns as energy sources (see table 1.4). A few used Vaporchoc, sleeve exploders, Maxipulse or other sources, and new sources were still being introduced occasionally.

(*f*) *Source considerations.* Several marine source units are often used in arrays, the firing of the different elements being synchronized by a controller system which monitors the firing of each source unit and uses feedback criteria to 'tune' the array. One such controller (Roark, 1976) monitors the waveform in the water near each source unit with hydrophones and aligns the peaks so as to maximize the peak pressure. A composite waveform synthesized by this controller is recorded and used in subsequent processing to remove minor time variations (see §8.2.1*e*).

The marine source characteristics most sought after are high peak pressure and low secondary oscillations. These can be determined empirically by firing the source in deep water (so that water-bottom reflections will not confuse the results) and observing the waveform with a calibrated hydrophone 75–100 m below the source. While the experiment sounds easy, its implementation is difficult. Keeping the hydrophone at the desired distance below the source is very difficult when the ship is moving, and static test conditions are apt to be very different from operational conditions. Standard calibrated hydrophones, whose outputs depend on passband, impedence matching, damping, etc., are available but the significance of results may be doubtful since the source waveform depends on source depth, spacing of array elements, cycle time, etc. A source can be made to look better by measuring peak energy with the source deep and a very wide passband, by measuring bandwidth with the source shallow (so that the reinforcement of the surface ghost occurs at a high frequency), and by selecting the best-case example of minimum secondary wave as the firing time of elements is varied. The source waveforms (signatures) for several energy sources are shown in fig. 5.45.

Rayleigh (1917) while studying the sounds emitted by oscillating steam bubbles related bubble frequency to bubble radius, pressure, and fluid density, and Willis (1941) while studying underwater explosions expressed the relationship in terms of source energy (the Rayleigh–Willis formula),

$$T = 36\rho^{\frac{1}{2}}\mathscr{P}_0^{-\frac{5}{6}}E^{\frac{1}{3}}, \qquad (5.20)$$

where T is the period of bubble oscillation in seconds, ρ is fluid density in g/cm^3, $\mathscr{P}_0$ is the absolute hydrostatic pres-

sure in pascals (N/m^2) and E is the energy in joules. If we assume a density of 1.024 g/cm^3 for sea water and replace $\mathscr{P}_0$ with $(h + 10)$ where h is the depth in meters (10 m is one atmosphere), the formula becomes

$$T = 0.017E^{\frac{1}{3}}(h + 10)^{-\frac{5}{6}}. \qquad (5.21)$$

Fig. 5.46 shows the energies of various sources versus dominant frequency. In general, large energy involves low frequency and vice versa.

5.5.4   *Marine detectors*

Hydrophones or marine pressure geophones are usually of the piezoelectric type (Whitfill, 1970). Synthetic piezoelectric materials, such as barium zirconate, barium titanate or lead metaniobate are generally used. A sheet of piezoelectric material develops a voltage difference between opposite faces when subjected to mechanical bending. Thin electroplating on these surfaces allows electrical connection to be made so that this voltage can be measured. Disc hydrophones (fig. 5.47*a*) are essentially two circular plates of piezoelectric ceramic mounted on the ends of a hollow brass cylinder. Electrical connections are made so that if both bend inward, as they would in response to an increase of pressure outside the unit, the induced voltages add, whereas if the plates bend in the same direction, as they would in response to acceleration, they cancel (fig. 5.47*b*). This feature is called *acceleration cancelling.* Cylindrical hydrophones (fig. 5.47*c*) are essentially thin hollow piezoelectric ceramic cylinders closed at the ends by brass caps. A change in pressure outside the cylinder induces stresses in the ceramic and hence a voltage difference between the inside and outside of the cylinder.

The sensitivity of each hydrophone element is small so that 3 to 50 elements are usually combined in series to make up a hydrophone group; these are distributed over 3–50 m. Piezoelectric hydrophones have high impedance, so an impedance-matching transformer is usually included with each group. Sometimes charge amplifiers are used instead of transformers.

Conventional geophones respond to velocity (see (5.16)) whereas hydrophones respond to pressure changes, that is, acceleration (see (2.85)). Thus a hydrophone's response differs from a geophone's by the factor $j\omega$ (see (2.86)), resulting in a 90° phase difference (due to j) and a rise of 6 dB/octave (due to ω). Since pressure is non-directional, an individual hydrophone's output is independent of wave direction whereas reversing the direction of travel inverts a geophone's output. The amplitude of the pressure change is maximum at a depth $\frac{1}{4}\lambda$ (see problem 2.30).

The hydrophones are mounted in a long streamer

towed behind the seismic ship at a depth often between 10 and 20 m. A streamer is shown diagrammatically in fig. 5.48 and a photograph of a portion is shown in fig. 5.49. The hydrophones, connecting wires and a stress member (to take the strain of towing) are placed inside a neoprene tube which is then filled with sufficient lighter-than-water liquid to make the streamer neutrally buoyant, that is, so that the average density of the tube and contents equals that of the sea water. A lead-in section 100 m or more in length is left between the stern of the ship and the first group of hydrophones. Dead sections are sometimes included between different hydrophone groups to give the spread length desired. The last group is often followed by a tail section to which is attached a buoy which floats on the surface; visual or radar sighting on this buoy is used to determine the amount of drift of the streamer away from the track of the seismic ship (caused by water currents). This buoy also helps retrieve the streamer if it should be broken accidentally. The total length of streamer in the water is 1000–2400 m, occasionally more than 2400 m. Depth controllers (one is shown in fig. 5.49) are fastened to the streamer at 5 to 12 places. These sense the hydro-

static pressure and tilt vanes so that the flow of water over them raises or lowers the streamer to the proper depth; they are ineffective when the streamer is not in motion. The depth which the controllers seek to maintain can be controlled by a signal sent down the streamer so that the streamer depth can be changed to accommodate changes in water depth or tó allow a ship to pass over the streamer.

When not in use, the streamer is stored on a large motor-driven reel on the stern of the ship. Depth detectors may be included at several places within the streamer to verify that the depth is correct. Water-break detectors are also included at several places along the streamer; these are high-frequency (500–5000 Hz) hydrophones which detect energy from the shot traveling through the water. Knowing the velocity of sound in the water permits converting the water-break traveltime into the offset distance. Remote-reading magnetic compasses may also be included in the streamer to indicate the streamer's orientation. A current perpendicular to the direction of a seismic line can separate what are intended to be common-depth points and allow dip perpendicular to the seismic line to masquerade as velocity variation (see problem 5.25).

Fig. 5.46. Energy–frequency relationships for marine sources at 9 m depth. (After Kramer *et al.*, 1968.)

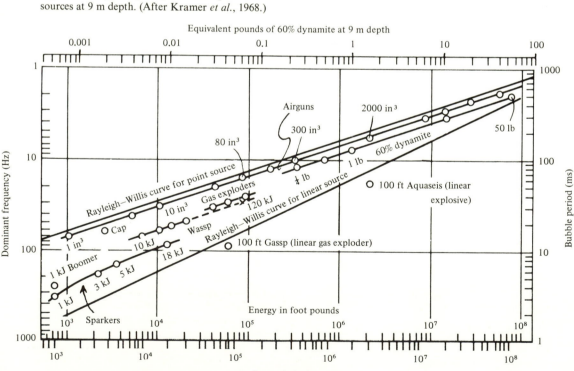

A marine detection system picks up noise of several kinds (Bedenbender *et al.*, 1970): (1) ambient noise due to wave action, shipping, marine life, etc.; (2) locally-caused water-borne noise such as that caused by the turbulence generated by motion of the lead-in cable, depressor paravane, depth controllers and tail buoy through the water, and energy radiated from the ship because of propellers, motors, and other machinery; and (3) mechanically-induced noise traveling in the streamer such as results from cable strumming, tail-buoy jerking, etc. Usually (3) is dominant except in rough weather when (1) dominates. Towing noise is reduced by (*a*) making the streamer system as smooth as possible and keeping depth controllers and other deviations from a smooth streamer at least 3 m

Fig.5.47. Hydrophones. (*a*) Disc hydrophone; (*b*) acceleration-cancelling feature of disc hydrophone; (*c*) cylindrical hydrophone.

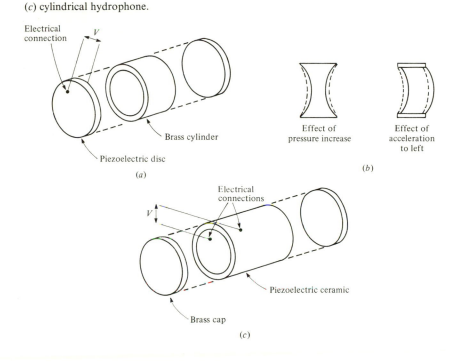

Effect of
pressure increase

Effect of
acceleration
to left

(*b*)

Electrical
connection

V

Brass cylinder

Piezoelectric disc

(*a*)

Electrical
connections

V

Piezoelectric ceramic

Brass cap

(*c*)

Fig.5.48. Disposition of streamer during operations.
(After Sheriff, 1973.)

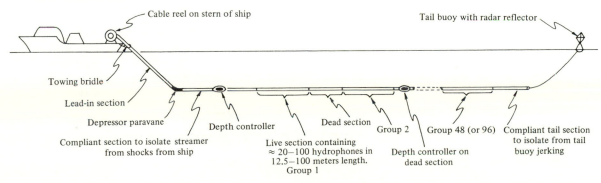

Cable reel on stern of ship

Tail buoy with radar reflector

Towing bridle

Lead-in section

Depressor paravane

Compliant section to isolate streamer
from shocks from ship

Depth controller

Live section containing
≈ 20−100 hydrophones in
12.5−100 meters length.
Group 1

Dead section

Group 2

Depth controller on
dead section

Group 48 (or 96)

Compliant tail section
to isolate from tail
buoy jerking

from the nearest hydrophone, (*b*) using a lead-in section to increase the distance between the ship and the nearest hydrophone group, and (*c*) using compliant and stretch sections with nylon rather than steel tensile members to reduce energy transmitted along the streamer. A separate, small, short streamer is sometimes used to record short-offset traces since there is usually an appreciable distance from the ship to the nearest group in the main streamer.

As of 1981, most streamers in use were 48 to 96 channel, some with up to 500 channels. The increase in the number of channels permits groups to be smaller and still encompass a large range of offsets. With more than 96 channels, the data are usually digitized in the streamer and the digital data transmitted through a single-channel coaxial cable. This reduces distortion produced by leakage and transmission-line variability.

5.5.5 *Marine positioning*

(*a*) *General requirements.* Marine seismic navigation involves two aspects: (*a*) placing the ships at a desired position, and (*b*) determining the actual location afterwards so that the data can be mapped properly. Sometimes (as with reconnaissance surveys) it is not too important that the data be obtained exactly at predetermined loca-

tions provided that one can subsequently determine accurately the actual locations which were occupied. In assessing the accuracy of a navigation method, we must distinguish between absolute and relative accuracy. Absolute accuracy is important in tying marine surveys to land surveys and in returning to a certain point later, for example, to locate an offshore well. Relative accuracy is important primarily to ensure the proper location of one seismic profile relative to the next. Relative accuracies of ± 15 m are desirable whereas absolute accuracies of ± 100 m are usually sufficient. Three-dimensional surveys (where cross-dip is to be determined accurately by comparing closely-spaced lines) may require ± 25 m or better accuracy and site surveys (where bottom sediments are being studied for engineering purposes) ± 40 m. The actual accuracies obtained in a survey (which are usually very difficult to assess) depend upon the system and equipment used, the configuration of shore stations, the position of the mobile station with respect to the shore stations, variations in the propagation of radio waves, instrument malfunctioning, operator error and so on. Systems capable of giving adequate accuracy under good conditions may not realize such accuracy in geophysical surveys unless considerable care is exercised at all times (Sheriff, 1974).

Fig.5.49. Diagram and photo of seismic streamer. Plastic spacers, *a*, are connected by three tensile cables, *b*; a bundle of electrical conductors, *c*, passes through holes in the spacers. The hydrophone is at *d*. The streamer covering is a soft plastic filled with a liquid to make the streamer neutrally buoyant. The depth controller, *e*, is clamped over the streamer. (Courtesy Seismic Engineering.)

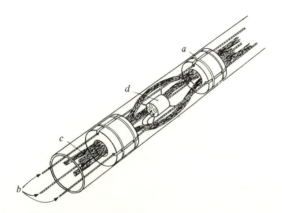

Navigation systems generally measure one of the following: (1) the time between transmission and receipt of a signal, which gives a distance; (2) the difference between the time of receipt of two signals, which gives a difference in distances; (3) a difference in frequency because of Doppler shift, which gives a velocity; (4) acceleration by means of oriented accelerometers; (5) direction with respect to north, usually determined with a gyrocompass. With velocity and acceleration measurements, position is determined by integrating. Navigation systems can also be classified according to the way locations are determined: (1) *piloting* wherein location is determined with respect to known locations; (2) *dead reckoning* wherein location is determined with respect to a known starting point and known course; and (3) *celestial* measurements, based on measurements of the altitude of the Sun or stars at a known time or measurements with respect to a navigation satellite.

Many types of navigation systems are available, including radiopositioning, sonic, inertial, satellite observations, etc. Each has advantages and disadvantages and combinations of systems are often used so that an advantage of one system may compensate for a disadvantage of another.

(b) Radiopositioning. Radiopositioning, which depends upon the use of radio waves, is used to locate many marine surveys relative to fixed shore stations. Radiopositioning methods can be divided into two basic types depending upon the type of measurement: (a) systems which measure the time required for a radiofrequency pulse to travel between a mobile station and a shore station (examples are radar, shoran, the rho-rho mode of Loran-C), and (b) systems which measure the difference in traveltime (or phase) of signals from two or more shore stations (these include Raydist™, Lorac™, Decca Navigator™ (Mainchain™), Pulse-8™, Hi-fix™, Toran™, ANA™, Argo™, the phase mode of Loran, and Omega). Angular measurements are not ordinarily used in radiopositioning because direction cannot be determined with enough precision with antennae of reasonable size.

Radar and shoran are similar in principle. *Radar* depends upon the reflection of pulses by a target, the distance to the target being equal to one-half the product of the two-way traveltime of the reflected pulse and the velocity of the radio waves. *Shoran* differs from radar in that the target is a shore station which receives the pulse and rebroadcasts it with increased power so that the return pulse is strong. Two or more shore stations are used and the position of the mobile station is found by swinging arcs as in fig. 5.50.

Radar and shoran are high-frequency systems, radar frequencies being in the range 3000–10 000 MHz, shoran in the range 225–400 MHz. Since such high frequencies are refracted only very slightly by the atmosphere these methods are basically line-of-sight devices. With normal antennae heights of about 30 m, the range for shoran is roughly 80 km. If the shore stations can be located on hills adjacent to the sea greater ranges can be obtained. By using very sensitive equipment (directional antennae and preamplifiers), ranges of 250 km can be obtained; this variation is called *extended-range* or *XR shoran*. The extension of range beyond the line-of-sight appears to be due to refraction, diffraction and scattering from the troposphere. In some tropical or subtropical regions strong temperature gradients in the atmosphere refract the radio waves so that ranges of 300 km or more can be obtained.

The distance between the ship station and each shore station is normally measured within ± 25 m (± 0.2 μs), sometimes within ± 5 m. The error in location depends mainly upon the angle between lines joining the shore stations to the mobile station, as shown in fig. 5.50; angles between 30° and 150° are usually considered acceptable.

Several devices utilize the same principles as shoran but use the higher radar frequencies; they 'interrogate' a small *transponder*, a device which emits a signal immediately upon receipt of the interrogating signal. These include RPS™, Miniranger™, and Trisponders™, which use frequencies around 9500 MHz, and Autotape™ and Hydrodist™, which use frequencies around 3000 MHz. The effective range of these devices is strictly line-of-sight but they are extremely portable. Their accuracy is often excellent, of the order of 5 m.

Fig.5.50. Effect of station angle on errors in shoran position. θ = station angle, A = mobile station, B and C = shore stations. Point A can be anywhere inside the 'parallelogram' formed by the four arcs. (Note: range errors are not to scale.)

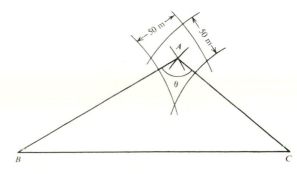

Loran-C involves the broadcast of a coded sequence of pulses of frequency 100 kHz, the broadcast times being controlled very accurately by atomic clocks. The stability of relatively cheap atomic clocks makes it feasible to carry such a standard on a seismic ship so that the instant of signal transmission can be determined and hence the range to the transmitter. Such range determination is called the *rho* mode, or the *rho-rho* or *rho-rho-rho* modes if ranges are determined to one, two or three transmitters. Despite the 3 km wavelength, ranges can be determined to 20–100 m. Long travel paths may be involved, however, so that very minor variations in the speed of the radio waves because of variations in the conductivity of the ground or moisture in the atmosphere can introduce sizeable *propagation errors* (see problem 5.27). To minimize such errors, the system should be calibrated in the local area. The shipboard atomic clock may drift slowly so that the drift has to be checked every few days.

If two shore stations simultaneously broadcast a radio pulse or coded sequence of pulses, a mobile station can measure the difference in arrival times and so find the difference in distances to the two shore stations. The locus of points with constant difference in distance from two shore stations (*A* and *B* in fig. 5.51, for example) is a hyperbola with foci at the two stations; thus a single measurement determines a hyperbola *PQ* passing through the location of the mobile station, *R*. If the difference in arrival times for a second pair of stations (*B* and *C*) is measured, the mobile station is located also on the hyperbola *VW* and hence at the intersection of the two hyperbolae.

This principle forms the basis of the phase-comparison mode of Loran and Omega, long-range radio-navigation systems maintained by the US Government. Omega is a world-wide system but its long wavelength (23–30 km) and seasonal and diurnal variations in the ionosphere preclude achieving accuracy greater than about 1 km. Loran-C is available over much of the northern hemisphere, especially in American and European waters. With care, the accuracy of Loran-C phase comparisons may be nearly that of its rho-rho mode. Decca Mainchain

Fig.5.51. Hyperbolic coordinates for radionavigation system.

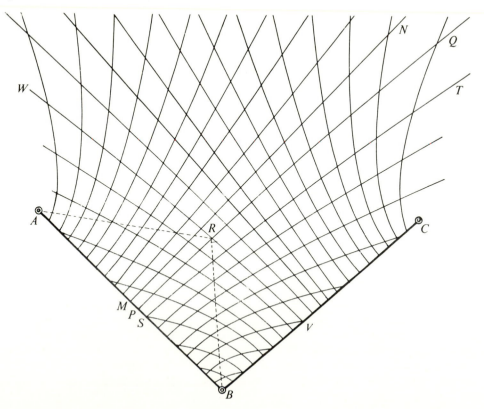

(or Navigator) is another system generally comparable to Loran-C, used mainly in Western Europe.

Many medium-frequency radiopositioning systems utilize the broadcasting of continuous waves (CW) from several stations, locations being determined by comparison of phase. *Phase-comparison systems* used in seismic exploration generally operate in the frequency band 1.5–4.0 MHz and have ranges up to 650 km.

Referring again to fig. 5.51, shore stations A and B transmit steady continuous sinusoidal signals which are exactly in phase at M, the midpoint of the baseline AB. A mobile station with a phase-comparison meter will show zero phase difference at M and at all points on the perpendicular bisector MN. If $MP = \frac{1}{2}\lambda$ the phase-comparison meter indicates zero phase difference at P also; if the mobile station moves away from P in such a direction that the phase difference remains constant, it traces out the hyperbola PQ. In general, a point R moving in such a way that

$$RA - RB = n\lambda, \quad n = 0, \pm 1, \pm 2, \pm 3, \dots$$

traces out the family of hyperbolae shown in the diagram.

The zone between two adjacent zero-phase-difference hyperbolae is called a *lane*. If we start from a known point and maintain a continuous record of the phase difference, we know in which lane the mobile station is located at any given time. By using a second pair of stations (one of which can be located at the same point as one of the first pair of stations) transmitting a different frequency we obtain a second family of hyperbolae, hence another hyperbolic coordinate of the mobile station. The accuracy of location decreases with increasing lane width as we go farther from the base stations, also as the angle of intersection of the hyperbolae decreases. Location accuracy is of the order of 30–100 m. If the continuous count of lanes is lost, however, one could be considerably off location as the phase-difference meters give only the position within a lane and do not indicate in which lane. The major factor governing accuracy in actual system usage therefore is maintaining the lane count accurately. Interference with signals reflected from the ionosphere becomes variable around sunrise and sunset when the ionospheric layering changes because of sunlight-induced ionization, and it is sometimes difficult to maintain accurate lane count during such periods. Improvements in circuitry (such as the use of phase-lock loops) have appreciably reduced loss-of-lane-count problems. In addition, the transmissions may be coded in various ways to aid in lane identification. The frequency can be changed periodically with consequent changes in the phase at a given position, which can be used to identify the lane.

Atomic clocks are also used with some medium-frequency systems to allow their use as range measuring devices. Toran-O and ANA are systems utilizing atomic clocks.

Translocation can be used to improve accuracy by removing the effect of propagation variations; this involves using variations in observations made simultaneously at a fixed station to correct determinations at a mobile station. Differential Omega, such a method, can improve Omega accuracy by a factor of five.

(c) *Acoustic positioning.* Acoustic or sonar positioning methods include the use of sonar range and frequency-shift measurements. For surveys of restricted areas several acoustic transponders, also called *pingers*, are anchored in the area. The ship to be located transmits a sonar pulse and the transponders emit coded responses when they sense the interrogating pulse. With most systems, the two-way traveltime is measured, though sometimes the phase difference at separated sensors on the ship is used to determine direction (much as moveout gives the apparent direction of a seismic ray). Four or more transponders might be set about 1–6 km apart where water depths are 20–500 m. The range is improved if the transponders are 5–10 m above the sea floor. Recoverable transponders having lifetimes of 5 years are available.

Once transponders have been set their locations have to be verified, not only because of uncertainties in transponder locations but also for definition of local velocity and propagation variations. Verification is usually done by criss-crossing over the area while using some other navigation system. The transponder locations should also be verified periodically since anchored transponders sometimes move, especially during storms. Acoustic transponders permit relative positioning of ± 5 m, while the absolute accuracy depends mainly on the method used to position the transponders.

Doppler-sonar is a *dead-reckoning* system, that is, it determines position with respect to a starting point by measuring and integrating the ship's velocity. The ship's velocity is measured by projecting sonar beams against the ocean floor in four directions from the ship (fig. 5.52). These beams are reflected back to the ship but their frequencies undergo a Doppler-shift because of motion of the ship with respect to the ocean floor. The frequency shift in each beam thus gives the component of the ship's velocity in that direction. The Doppler effect relates to the compression of wavefronts ahead of a moving source or as seen by a moving observer. If V is the velocity in the medium and V_s the component of a ship's velocity in the direction of the acoustic beam, the wavelength transmitted will be $(V - V_s)/v_s$ but a stationary observer would see it as V/v_1; hence $v_1 = v_s V/(V - V_s)$. If an observer with component of velocity V_s is moving toward a stationary source, he would observe $v_2 = v_s(V + V_s)/V$. For a moving ship, both source and observer are moving, hence

$$v_2 = v_s(V + V_s)/(V - V_s). \tag{5.22}$$

The fore and aft measurements are averaged to minimize the effects of pitching motion of the ship, and starboard and port measurements to minimize rolling motion. The four beams often actually look in 45° directions to the ship's course, which gives improved sensitivity, rather than in-line and perpendicular to the ship's course. These measurements can be resolved to give the ship's

actual velocity (in conjunction with direction information from a gyrocompass) and the velocity can be integrated to give the ship's position. Small errors in velocity measurement accumulate in the integration resulting in position uncertainty of the order of 100 m/hr. The requisite accuracy has to be maintained by periodic *updates*, i.e., periodic determinations of location by independent measurements. In deep water, scatter of the sonar beams by inhomogeneities in the water dominates and the Doppler-shifts give a measure of the velocity with respect to the water rather than the ocean floor, resulting in considerable loss of accuracy. A 300 kHz Doppler-sonar system can usually 'see' bottom shallower than about 200 m while a 150 kHz system can see to depths of 400–500 m.

Fig.5.52. Doppler-sonar navigation. (From Sheriff, 1973.)

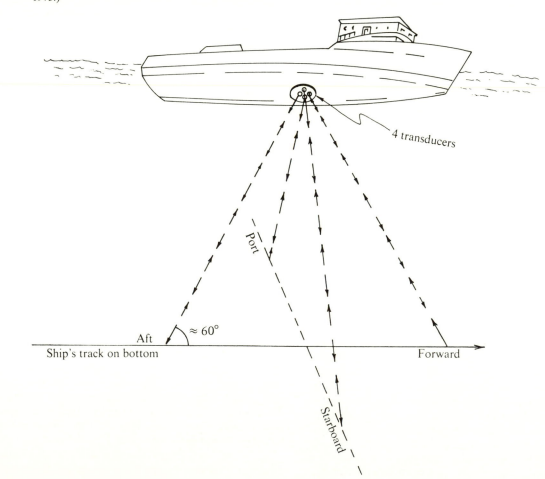

(*d*) *Inertial positioning. Inertial navigation* can be accomplished by measuring acceleration in orthogonal directions, integrating once to get velocity and a second time to get location relative to a known starting point. The accelerometers are usually located on a stable platform which is kept horizontal by a leveling-feedback system and whose direction in space is maintained by a gyro-feedback system. Periodic fixes from an independent navigation system minimize the accumulation of systematic error. The uncertainty with inertial systems in geophysical use increases at a rate of about 200 m/hr.

(*e*) *Satellite positioning.* Many seismic ships are equipped to determine their location from observations of navigation satellites. The US Navy has a number of *Transit*

satellites (five as of 1981) in polar orbits 1075 km above the Earth. Each satellite takes about 107 minutes to circle the Earth, being in sight of a point under its orbit for about 18 minutes (horizon to horizon). Each satellite transmits continuous waves of frequencies 150 and 400 MHz. The frequencies measured by a receiver on the ship are Doppler-shifted because of the relative motion of the satellite with respect to the ship. The differences between the ship's longitude and latitude and the satellite's longitude and latitude at closest approach (see fig. 5.53) are calculated from the Doppler-shifts. The satellite transmits information which gives the satellite's location every two minutes. A small computer on the ship combines this information with the Doppler-shift measurements and the speed and course of the ship to give the ship's location.

Fig.5.53. Position-fix from navigation satellite. (After Sheriff, 1973.)

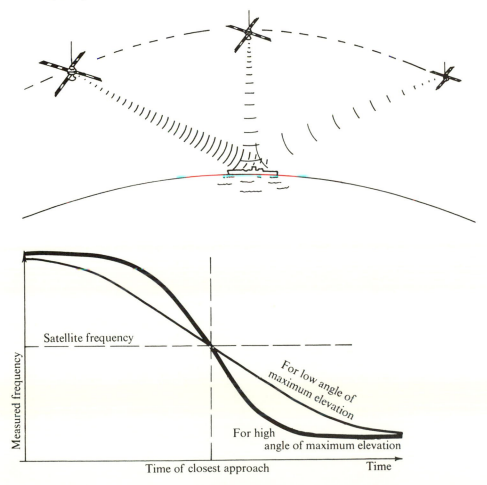

Each satellite can be observed on four or more orbits each day, hence 20 or more determinations of position are possible each day. However, the satellites are not uniformly spaced and do not have precisely the same orbital period so that sometimes more than one satellite is visible while at other times several hours may intervene without any satellite being visible. One may expect about two-thirds of the 'passes' to result in satisfactory *fixes* or determinations of position. Satellite fixes may be accurate within ± 50 m, provided the ship's velocity is accurately known (Spradley, 1976). The principal disadvantage of satellite navigation for seismic purposes is that it gives no information about position during the interval between fixes.

Commonly Doppler-sonar, gyrocompass and satellite navigation are combined (Kronberger and Frye, 1971), or else radio and satellite navigation. The satellite gives the periodic updating information needed to maintain the Doppler-sonar accuracy or to remove ambiguities or propagation error effects from radionavigation, while the Doppler-sonar along with the gyrocompass and/or radio systems give the velocity information needed for an accurate satellite fix.

While locations may be determined with accuracy sufficient for the immediate purposes with a minimum system, redundancy of location systems is highly desirable. Over-determination of position provides a needed check on malfunctioning and an assessment of the amount of accuracy actually being achieved, both of which may not be ascertainable otherwise. Redundancy also provides back-up navigation in event of failure. The history of marine geophysical surveying includes many instances of work lost in whole or in part because of survey failures.

Location uncertainty can be reduced significantly by analysis of the data afterwards. Data acquired later in a survey may be used to reduce the uncertainty of data acquired earlier. Much positioning uncertainty is systematic and analysis of the entire body of data may clarify the nature of errors and permit correcting for them. Post-plot accuracy as of 1981 is about 15 m.

The Navstar satellite navigation system (also called Global Positioning System, GPS) is scheduled to be operational for military use in 1988, but when it will be available for civilian use has not been determined, nor the accuracy for such use. It is to comprise eighteen satellites at 20000 km altitude, at least three of which will always be visible so the system will give 3-coordinate continuous location determination. Each satellite will include an atomic clock and be in synchronous orbit.

5.5.6 *Data flow in marine surveys*

A marine geophysical vessel is equipped with many sensors which are integrated together as a complete system, rather than being merely a collection of subsystems. Location, seismic data, variations in the magnetic and gravity fields, high resolution sub-bottom data, water depth, and other information are recorded. The flow of data and commands between the elements of the system is shown schematically in fig. 5.54. Often three types of magnetic tape result from a survey: (i) navigation data, (ii) seismic data, and (iii) auxiliary data, including magnetometer, gravimeter, fathometer, plus peripheral data such as readings of streamer depth and direction sensors, latitude and longitude, *Julian Day* (number of day within calendar year, referred to Greenwich), time, and seismic recorder data such as seismic line and shot point numbers, file numbers, and instrument settings.

In addition to data on magnetic tapes, data usually include monitor records of the seismic data, a variable-density section from a single seismic channel, magnetometer, gravimeter, water depth, and high-resolution sub-bottom records. Additional plots and records are made of the ship's course and logs are kept by the helmsman, navigator, observer and others.

5.5.7 *Profiling*

Marine profiling differs from conventional CDP marine shooting in that its objective is to map only the shallow portion of the sedimentary section, sometimes only the unconsolidated sediments and the upper surface of competent bedrock (the usual engineering objective), sometimes detailed analysis of the upper 1000 m. Profiling is also widely used in oceanographic work to survey large areas cheaply.

Profiling employs smaller ships and weaker energy sources and hence is much cheaper than conventional marine work. These smaller energy sources are usually rich in high-frequency content so that resolution is usually much higher than for conventional seismic work. Often only a single hydrophone group is employed, and shooting and recording take place at such short intervals that an essentially continuous record is obtained. The technique is similar to that of the continuous recording of water depth using a fathometer (echo sounder). Most profilers, being single hydrophone devices, cannot discriminate between events on the basis of normal moveout and so the useful data window is often limited by the depth of water, which determines the arrival time of water-bottom multiples.

The energy sources most commonly used for profiler

Fig.5.54. Schematic of the flow of data in marine operations. Individual ships will not utilize all of the sensors shown nor communicate in the precise manner illustrated. (From Sheriff and Lauhoff, 1977.)

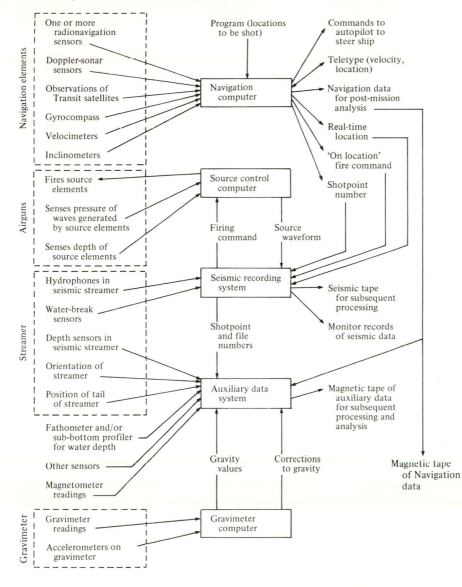

work are high-powered fathometers, electric arcs, airguns and imploders. The high-powered fathometers are usually piezoelectric devices employing barium titanate or lead zirconate. Such materials not only generate electric fields when they are compressed (as when they are used as hydrophones) but they also change dimensions when subjected to an electric field. They are thus *transducers* because of their ability to transform electrical energy to acoustic energy and vice versa. Fathometers used for profiling operate at lower frequencies and higher power levels than fathometers used for water depth measurements; frequencies in the range 2–10 kHz and power levels of roughly 100 W are commonly used. A repetition rate of once every 2 s or so is used and penetration of 20–100 m is generally achieved. Sub-bottom reflections usually permit the mapping of the various layers of mud and silt overlying bedrock, as shown in fig. 5.55. Reflection character can sometimes be interpreted to indicate the nature of the sediments, for example, to find sand layers which can support structures erected on pilings. Surveys for other

purposes, such as ones designed to locate pipe-lines buried in the mud, are also made at times.

Electric arcs used as sources for profiler work, usually called *sparkers*, utilize the discharge of a large capacitor to create a spark between two electrodes located in the water. The heat generated by the discharge vaporizes the water creating an effect equivalent to a small explosion. Several sparker units are often used in parallel to give increased penetration. The penetration obtained by earlier models was small but modern sparker arrays deliver as much as 200 kJ at 50–2000 Hz and achieve penetrations of 1000 m or so (although a 5 kJ source with a penetration of less than 300 m is more common; fig. 5.45*h* shows the waveshape from a 5 kJ sparker). A variation of the sparker called *Wassp*™ involves connecting the electrodes by a thin wire which is vaporized by the energy discharge. This increases the duration of the bubble and consequently its low-frequency content. It is used in fresh water where conduction is insufficient to initiate a discharge between unconnected electrodes on a consistent basis.

Fig.5.55. Profiler record showing sub-bottom deposits. (After King, 1973.)

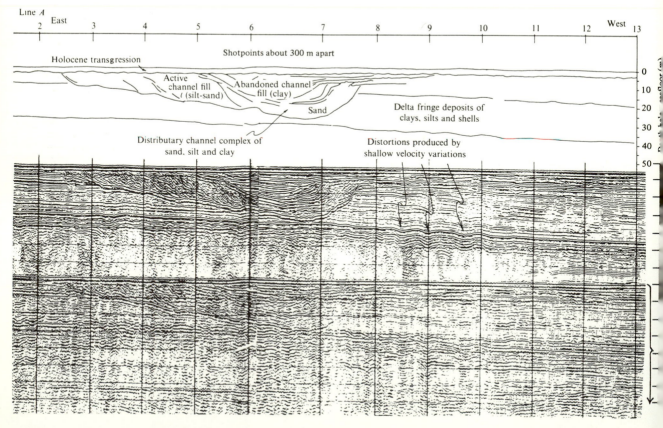

The airguns used in profiling are similar to those used in CDP marine work except that they are smaller, involving as little as one cubic inch of air at 1000 psi and a dominant frequency around 250 Hz. The imploders used in profiling include the Boomer already described and deliver about 200 J of energy from 50 Hz to a few kilohertz.

Profiler data frequently are recorded on electro-sensitive paper using a strip recorder. A wire is wrapped once around a cylinder in helical fashion. The cylinder rotates so that the point of contact of the wire with a metal plate tangent to the cylinder changes linearly as the cylinder rotates. An electrical current passing from the wire through sensitive paper on the plate between the cylinder and the plate produces a recording on the paper. The paper moves a small distance perpendicular to the cylinder's axis during the time of a recording so that subsequent recordings are made adjacent to each other. The source is fired when the wire is touching the paper at the top, and the current is proportional to the seismic energy picked up by the hydrophone. The interval between source impulses is an integral multiple of the rotation time of the cylinder (often 2 s) and a recording may not be made on every rotation of the cylinder. Sometimes reflections with arrival time greater than the cylinder rotation time are written on top of the shallower recording, an effect sometimes called 'paging' (see fig. 5.56). Usually it is clear which data belong with which page (or set of arrival times) although occasionally paging creates ambiguity.

Multiples of the seafloor often are very strong and make data arriving after them unusable (Allen, 1972). In engineering work where interest is concentrated in the relatively thin layer of unconsolidated sediments overlying bedrock such as shown in fig. 5.55 this usually does not create a problem; where the unconsolidated sediments are thick the seabottom is apt to be soft (not involving a large acoustic impedance contrast) and hence the sea-bottom multiple relatively weak. In deep-water oceanographic work a long period of time (a wide *window*) elapses before the ocean bottom multiple arrives (as in fig. 5.56) so that appreciable data are recorded without ambiguity.

Profiling has received considerable stimulus as the result of US legal requirements for archaeological surveys of sites before permits for the construction of facilities (including well locations) can be obtained. Typically, a grid of profiler lines spaced 150–500 m apart with cross-lines at 500–1000 m is run over the site area. While the archaeological value of such surveys is doubtful, the engineering information obtained has proven sufficiently valuable that such surveys have become common over the regions surrounding prospective well sites in many parts of the world. The high resolution achieved by such surveys

is useful in delineating foundation problems, shallow faults, unstable bottom conditions, regions of gas seeps and shallow pockets of high-pressure gas (by their amplitude standout; see §9.8) which constitute drilling hazards.

Profilers have grown more complicated as more detailed information has been sought with them. The source power has been increased to achieve greater penetration, but usually at a loss in resolution. The number of hydrophone groups has been increased to provide common-depth-point data and data are recorded on magnetic tape to allow subsequent data processing such as is used with conventional seismic work. In consequence, the distinction between profiler and conventional work is sometimes obscure.

5.6   Data reduction
5.6.1   *Preliminary processing*

The initial step in processing paper records is to 'write up' the records. The information labeled on the seismic record or record section includes (*a*) identifying information such as the name of the company and prospect, (*b*) line and profile numbers, (*c*) survey data (location of the shotpoint and geophone groups), (*d*) description of the spread (number and spacing of geophones per group, in-line and perpendicular offsets of the shotpoint), (*e*) source information (size and depth of explosive charge or type of source and source pattern), (*f*) amplifier information (filter and gain settings), (*g*) time of shooting the profile, (*h*) history of any processing which may have been done, etc. Such information may be already printed on computer-generated record sections.

Weathering and elevation corrections are calculated as described in the following section. A time reference, $t = 0$, is marked on the records; this is the shot instant shifted in order to take into account corrections for weathering and elevation so that arrival times measured with respect to this time reference give depths measured from the datum. Timing lines are labeled at 0.1 s intervals starting from the time reference.

5.6.2   *Elevation and weathering corrections*

Variations in the elevation of the surface affect traveltimes and it is necessary to correct for such variations as well as for changes in the low-velocity-layer (LVL). Usually a *reference datum* is selected and corrections are calculated so that in effect the shotpoints and geophones are located on the datum surface, it being assumed that conditions are uniform and that there is no LVL material below the datum level.

Many methods exist for correcting for near-surface effects. These schemes are usually based on (1) uphole times, (2) refractions from the base of the LVL, or (3) the smoothing of reflections. We shall describe several of these methods which are simple to apply and adequate to cover most situations. Automatic statics-correction schemes which usually involve statistical methods of smoothing reflections will be discussed in §8.2.2. We shall assume that V_W and V_H, the velocities in the LVL and in the layer just below it, are known; these can be found from an uphole survey or the refraction first-breaks, as will be discussed later in the section. In what follows we also assume that the shot is placed below the base of the LVL; if this is not true, modifications have to be made in the following equations in this section (see problem 5.33).

Fig. 5.57 illustrates a method of obtaining the correction for t_0, the shotpoint arrival time. E_d is the elevation of the datum, E_s the elevation of the surface at the shot-

Fig.5.56. Profiler record, Offshore Japan. The water-bottom reflection, *A*, with traveltime 1.0 to 2.0 s indicates water depths of 750–1500 m. The ship traveled 8.5 km between the 30 minute marks at the top of the record. Most primary reflections are obscured by the first and second water-bottom multiples, *B* and *H*. More than a kilometer of sediments are indicated by the reflections near *C*; multiples of these appear paged-back at *D*. *E* indicates a fault scarp on the ocean floor. *F* are diffractions, probably from seafloor relief slightly offset from the line. Note the onlap and thinning above *G*. (Courtesy Teledyne.)

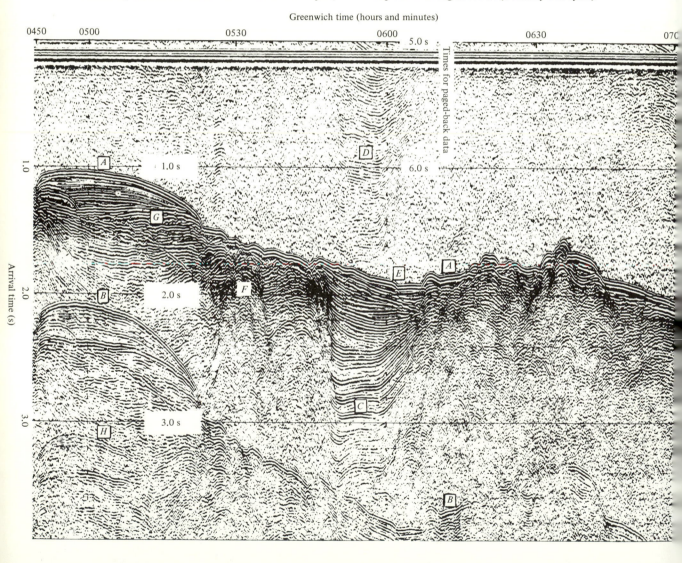

point, D_s the depth of the shot below the surface, and t_{uh} is the uphole time, the time required for energy to travel vertically upward from the shot to a geophone on the surface at the shotpoint (in practice the uphole geophone is within 3 m or so of the shothole). The deviation of reflection paths from the vertical is usually small enough that we can regard the paths as vertical; therefore the time required for the wave to travel from the shot down to the datum is Δt_s, where

$$\Delta t_s = (E_s - D_s - E_d)/V_H. \tag{5.23}$$

Similarly, the time for the wave to travel up from the datum to a geophone on the surface at B is Δt_g where

$$\Delta t_g = \Delta t_s + t_{uh}. \tag{5.24}$$

The correction, Δt_0, for the traveltime at the shotpoint is then

$$\Delta t_0 = \Delta t_s + \Delta t_g = 2\Delta t_s + t_{uh}$$
$$= 2\{(E_s - D_s - E_d)/V_H\} + t_{uh}. \tag{5.25}$$

Subtraction of Δt_0 from the arrival time t_0 is equivalent to placing the shot and the shotpoint geophone group on the datum plane, thereby eliminating the effect of the low-velocity-layer if the shot is beneath the LVL. At times the shot may be so far below the datum plane that Δt_0 will be negative.

When (3.11) is used to calculate dip, the dip moveout must be corrected for elevation and weathering. The values needed in (3.11) are the differences in arrival time at A' and C'. The correction to be applied, Δt_c, called the *differential weathering correction*, is the difference in arrival times at opposite ends of a split-dip spread for a reflection from a horizontal bed. Referring to fig. 5.57, the raypaths from the shot B down to a horizontal bed and back to geophones at A and C have identical traveltimes except for the portions $A'A$ and $C'C$ from the datum to the surface. Assuming as before that $A'A$ and $C'C$ are vertical, we get for Δt_c the expression

$$\Delta t_c = (\Delta t_g)_C - (\Delta t_g)_A$$
$$= (\Delta t_s + t_{uh})_C - (\Delta t_s + t_{uh})_A, \tag{5.26}$$

where we assume that A and C are shotpoints so that the quantities in brackets are known. If we take the positive direction of dip to be down from A towards C, then Δt_c must be subtracted algebraically from the observed difference in arrival times at A and C to obtain that for A' and C'.

The following calculation illustrates the effect of the correction. We take as datum a horizontal plane 200 m above sea level; V_H is 2075 m/s. Data for three successive shotpoints, A, B and C, such as those in fig. 5.57 are as follows:

		Shotpoint C	Shotpoint B	Shotpoint A	
Surface elevation,	E_s (m)	248	244	257	⎫
Depth of shot,	D_s (m)	15	13	20	⎬ measured
Uphole time,	t_{uh} (ms)	48	44	53	⎭
Shot-to-datum time,	Δt_s (ms)	16	15	18	⎫
Datum-to-geophone,	Δt_g (ms)	64	59	71	⎬ calculated
Correction time,	Δt_0 (ms)	80	74	89	⎭
Differential corr.,	Δt_c (ms)		−9		

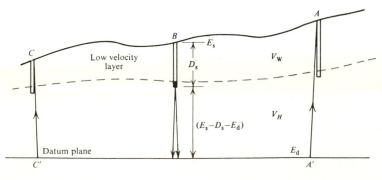

Fig.5.57. Calculation of weathering corrections.

Let us suppose that a reflection on a split profile from shotpoint B gives the following data: $t_0 = 2.421$ s, $t_A = 2.419$ s, $t_C = 2.431$ s. Then, the corrected value of t_0 is $2.421 - 0.074 = 2.347$ s and the corrected Δt_d is

$$\Delta t_d = 2.431 - 2.419 - (-0.009) = 0.012 + 0.009$$
$$= 21 \text{ ms.}$$

Suppose that (for another reflection) $t_0 = 1.392$ s, $t_A = 1.401$ s, $t_C = 1.395$ s; then the corrected value of t_0 is 1.318 s and the corrected Δt_d is

$$\Delta t_d = 1.395 - 1.401 - (-0.009)$$
$$= -0.006 + 0.009 = +3 \text{ ms.}$$

Thus the correction can change the apparent direction of dip as well as the dip magnitude. Accurate corrections are essential.

Corrections are often required for geophones in between shotpoints where uphole data are not available. The first-breaks can be used for this purpose. In fig. 5.58, G is a geophone intermediate between adjacent shotpoints A and B for which we have first-break traveltimes. Let t_{AG} and t_{BG} be the first-break times for the paths $A'C'G$ and $B'C''G$. Almost always GC' and GC'' are within $20°$ of the vertical and $C'C''$ is therefore small. Thus, we can write the approximate relation

$$t_{AG} + t_{BG} \approx (A'B'/V_H) + 2t_W \approx (AB/V_H) + 2t_W,$$

t_W being the traveltime through the weathered layer at G. Thus,

$$t_W \approx \tfrac{1}{2}\{t_{AG} + t_{BG} - (AB/V_H)\}. \tag{5.27}$$

Subtracting t_W from the arrival times in effect places the geophone at the base of the LVL; to correct to datum we must subtract the additional amount, $(E_g - E_d - D_w)/V_H$, where E_g is the mean elevation of the geophone group, D_w being found by multiplying t_W by V_W.

Occasionally special refraction profiles are shot to obtain data for making corrections for intermediate geophones. These profiles may be of the standard type using small charges placed near the surface or a non-dynamite source on the surface; these are interpreted using standard methods such as Wyrobek's (see §6.2.2d) to find the depth and traveltime to the base of the LVL. Alternatively, a shot may be placed just below the LVL as shown in fig. 5.59; in this event we must modify (3.30) since the shot is at the base rather than the top of the upper layer. Thus,

$$t = \frac{x - D_w \tan\theta}{V_H} + \frac{D_w}{V_W \cos\theta} = \frac{x}{V_H} + \frac{D_w \cos\theta}{V_W}. \tag{5.28}$$

Most near-surface correction methods require a knowledge of V_H and sometimes of V_W as well. The former can be determined by: (1) an uphole survey, (2) a special refraction survey as described above, or (3) analysis of the first-breaks for distant geophone groups (since these are equivalent to a refraction profile such as that shown in fig. 5.59). The weathering velocity V_W can be found by (1) measuring the slope of a plot of the first-breaks for geophones near the shotpoint (correcting distances for obliquity), (2) dividing D_s by t_{uh} for a shot placed near the base of the LVL, (3) an uphole survey, or (4) firing a cap at the surface and measuring the velocity of the direct wave.

5.6.3 *Picking reflections and preparing cross-sections*

While the following procedures were used prior to the era of computer-processed record sections, they are also used occasionally today, and the concepts which they involve continue to be central to today's techniques. When the record quality is poor almost any alignment may be mistakenly identified as a reflection. The best criterion in such cases is often the geological picture which results. If this picture does not make sense, we should re-examine

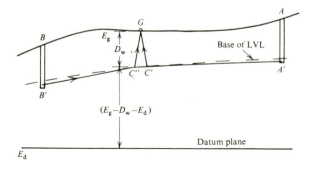

Fig.5.58. Datum correction for geophone in between shotpoints.

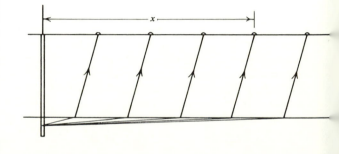

Fig.5.59. Refraction weathering profile.

the geophysical data with more skepticism. Naturally, 'making sense' does not mean that the result must fit our preconceived ideas but rather that it must be geologically plausible.

When the interpreter decides that an event is a legitimate reflection, he usually marks it and 'times' it. When working with individual split-dip records, the arrival times at the center of the record and at the two outside traces (or the difference between these outside times, Δt_d), usually corrected for weathering and elevation, are often written directly on the record.

Besides timing reflections, the interpreter may assign a *grade* to each, for example, VG, G, P,? (for very good, good, poor, questionable). These grades refer to the certainty that the event is a primary reflection and the accuracy of measurement of the arrival times. Sometimes a two-letter grading system is used to separate the grading of certainty from the grading of timing accuracy.

The process of identifying events on a seismic record and selecting and timing the reflections is referred to as *picking* the records. Fig. 5.60 shows a picked record.

The next stage after the picking of individual records is to prepare a composite representation of all the data for a given line. This can be done by plotting a seismic cross-section on a sheet of graph paper. The shotpoint locations are marked at the top of the sheet on a horizontal line indicating the datum plane. The reflection events for each record are plotted below the corresponding shotpoint according to their arrival times. Plots are often made also of the surface elevation, the depth of shot, the depth of the base of the LVL and the first-break arrival times.

Cross-sections are called *time sections* if the vertical scale is linear with time or *depth sections* if linear with depth. Occasionally the horizontal scale is in units of time (obtained by dividing horizontal distances by V_H, the sub-weathering velocity). A cross-section is *unmigrated* when reflections are plotted vertically below the shotpoint. A *migrated section* (such as fig. 5.61) is one for which we assume that the seismic line is normal to strike so that the dip moveout indicates true dip and we then attempt to plot in their actual locations the segments of reflecting horizons which produced the recorded events. The scale on hand-migrated sections is usually linear with depth.

Unmigrated sections give a distorted picture of the subsurface, the distortion increasing with the amount of dip; they are used for interpretation only in areas of gentle dip or when an accurate structural picture is not required, as for example in a reconnaissance survey. When the dips exceed 5° or so, the distortion inherent in an unmigrated section makes it difficult to determine the subsurface geometry correctly, and even with very small dips subtle

features such as pinchout evidences are apt to be seriously distorted. Fig. 5.62 shows the effect of lack of migration on simple structures. The anticline at the left would appear on an unmigrated section as the dotted line $R'ST'$ while the syncline would appear as the dotted line $T'UV'$. Failure to migrate decreases the curvature of an anticline and increases that of a syncline. Failure to migrate would also position a fault incorrectly.

Several methods of migration are used. The simplest is to assume a constant velocity down to the reflector and swing an arc whose radius is half the arrival time t_0 multiplied by the average velocity. The radius O_1R in fig. 5.62 makes an angle with the vertical equal to the reflector dip (calculated using (3.11)) and a straight-line segment equal to half the spread length is drawn at R perpendicular to the radius (tangent to the arc) to represent the reflecting segment (see also fig. 1.3b). This method of migrating positions the reflector segments incorrectly and gives them the wrong dip, compared with correctly allowing for a velocity gradient (see problem 5.35).

Migration is sometimes carried out with simple plotting machines. Some of these assume a functional form for the velocity, such as one described by Rockwell (1967) which uses (3.29) assuming a linear increase of velocity.

Probably the commonest method of migrating reflections uses a wavefront chart, a graph showing wavefronts and raypaths for an assumed vertical distribution of velocity. Fig. 5.63 is a simplified version of such a chart. It shows the location of a wavefront at different times after the shot instant, the successive positions being labeled with the two-way traveltime. Raypaths are also shown; these are found by applying Snell's law at each change in velocity. Raypaths are labeled in terms of dip moveout, $\Delta t_d/\Delta x$. Dip moveout can be measured from corrected record sections such as stacked sections, as the difference in arrival time, Δt_d, between points a distance Δx apart. To plot a cross-section, the wavefront chart is placed under transparent graph paper with the chart origin at the appropriate shotpoint. A reflection with certain values of t_0 and $\Delta t_d/\Delta x$ is plotted by interpolating between the wavefronts and rays (actual wavefront charts have more closely-spaced wavefronts and rays than those shown in fig. 5.63, so that interpolation is more accurate) and drawing a straight line of length equal to half the spread length tangent to the wavefront at the point (t_0, $\Delta t_d/\Delta x$). The reflection denoted by the symbol –o– in fig. 5.63 corresponds to $t_0 = 2.350$ s $\Delta t_d/\Delta x = 110$ ms/km.

With an asymmetric spread, we correct the measured moveout for the difference in normal moveout between the two ends of the spread to find the dip moveout,

Fig.5.60. Seismic record. (Courtesy Chevron.)

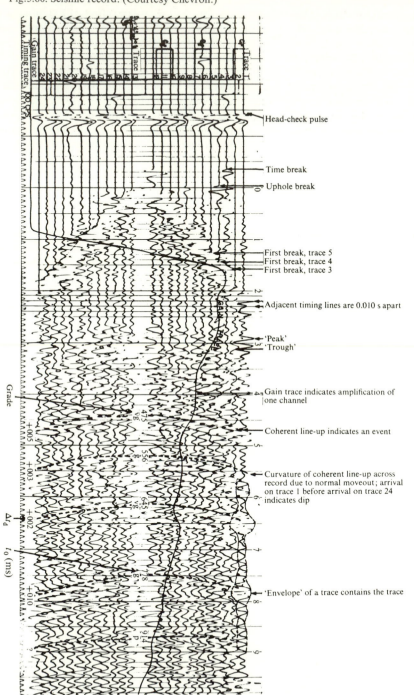

Head-check pulse

Time break

Uphole break

First break, trace 5
First break, trace 4
First break, trace 3

Adjacent timing lines are 0.010 s apart

'Peak'
'Trough'

Gain trace indicates amplification of one channel

Coherent line-up indicates an event

Curvature of coherent line-up across record due to normal moveout; arrival on trace 1 before arrival on trace 24 indicates dip

'Envelope' of a trace contains the trace

$\Delta t_d/\Delta x$. With in-line offsets we find $\Delta t_d/\Delta x$ by applying normal-moveout corrections (and correcting the arrival time to find t_0 when the offsets are large). On record sections (see below), we may measure the dip over any distance over which the horizon is dipping uniformly.

Identifying a reflection seen on one record as representing the same interface as a reflection on another record is called *correlation*. It is based partly on similarity of character and partly upon the dip and the agreement in arrival times (*time-tie*). The travel paths for reflections on the outside traces of adjacent profiles which provide continuous subsurface coverage are the same except that they are traversed in opposite directions; hence the travel-times should be the same provided that adequate weathering and elevation corrections have been applied. For example, in fig. 5.4, trace 1 on the profile shot from O_2 (with path $O_2 BO_3$) should have the same corrected travel-time as trace 24 on the profile shot from O_3.

If the record quality is too poor to provide continuous reflections or if the continuous reflections which are present are not in the part of the section where information is required, we draw *phantoms*, lines on the section so that they are parallel to adjacent dip symbols which the interpreter considers valid. Where data conflict or are absent, the phantom is drawn in the manner which seems most reasonable to the interpreter on the basis of whatever fragmentary evidence may be available.

Record sections can be regarded as a form of time section obtained by placing successive records side by side. Record sections are usually corrected for elevation, weathering and normal moveout. They display a large amount of data in compact form. A potential disadvantage with record sections is that they are so graphic that people who lack understanding of the significance of the data tend to attach incorrect meanings to them, as though the section were a 'photograph' of the rock formations in the ground.

5.7   Exploration with S-waves

Nearly all seismic exploration is carried out with P-waves, S-waves merely contributing to the noise. The laws of reflection, refraction and other phenomena apply equally to S-waves and P-waves. The velocity of an S-wave depends only on μ (and ρ) while the P-wave velocity

Fig.5.61. Migrated cross-section. (Courtesy Chevron.)

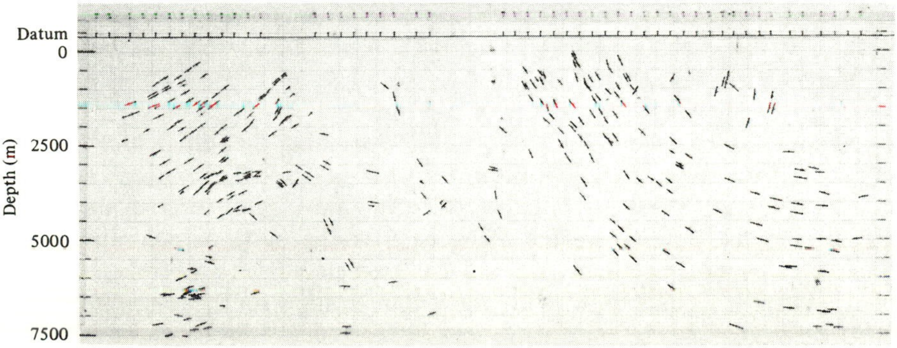

Fig.5.62. Effect of lack of migration.

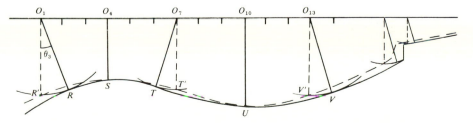

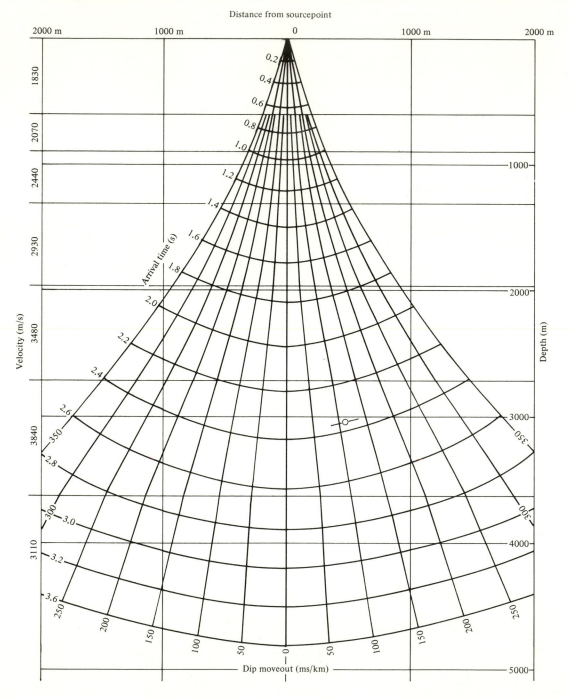

Fig.5.63. Simplified wavefront chart, for horizontal velocity layering.

Fig.5.64. Horizontal vibrator for generating S-waves. (Courtesy Conoco.) (*a*) Truck-mounted vibrator; the weight of the truck is used to keep the vibrator in firm contact with the ground; (*b*) detail of vibrator pad showing the 'teeth' which are triangular in cross-section to sustain ground coupling as the horizontal movement of the pad compacts the soil during a sweep.

(*a*)

(*b*)

depends also on Lamé's constant λ. The velocity ratio β/α can provide information about lithology (see fig. 7.3), especially of shale versus other rock types, and the nature of interstitial fluids (fig. 7.14). The shear modulus, which can be calculated from β, is an important factor in designing the foundation of structures and so S-wave studies are important in engineering geophysics. The shear modulus along fault zones seems to change in anticipation of earthquakes. Thus appreciable effort is being devoted to developing S-wave techniques but a methodology has not yet evolved.

The direct generation of S-waves invariably involves the generation of P-waves also. To minimize confusion as to wave identity, one usually attempts to generate SH-waves (so that converted waves are not involved) and then look only for the SH-component with horizontal geophones. For short distances as used in engineering studies, a hammer blow on the side of a block which is held in firm contact with the Earth by the weight of the vehicle is often used. A large truck-mounted hammer (Marthor^{TM}) which delivers a blow of 60 kJ is sometimes used. A horizontal vibrator whose pad is coupled to the Earth by triangular teeth (fig. 5.64) is used in a Vibroseis method. The Syslap^{TM} method involves the use of three adjacent holes (fig. 5.65). The center hole is shot first to give a conventional P-wave record. The subsequent firing of the side holes is affected by the disturbed zone of the center shot and the asymmetry causes SH- and SV-waves to be generated as well as P-waves. However, the shot asymmetry is reversed for the two side holes so that the SH-waves from them are opposite in polarity; subtracting the records from the two side shots thus adds the SH-wave effects and cancels SV- and P-wave effects.

The generation of SV-waves by mode conversion is efficient at liquid–solid interfaces (Tatham and Stoffa, 1976). Fig. 5.66 shows how the SV-wave amplitude depends on P-wave velocities in sediments below a water layer for various angles of incidence. The SV-waves reconvert to P-waves on reemergence at the solid–liquid interface in an equally efficient conversion and so can be detected with conventional pressure detectors. However, the long group length of conventional marine streamers discriminates against waves which approach the spread at the emergent angles and the streamer for SV-wave detection has to be specially designed.

S-wave records are usually displayed in the same manner as P-wave records except that the time scale is compressed by two, so that the S-wave and the P-wave records may be compared more easily (fig. 5.67).

Problems

5.1. (a) Replot fig. 5.5b as a subsurface stacking chart, showing lines of traces with the same geophone, same offset, same midpoint and same shotpoint. (b) Assume a reflector 2.0 km beneath the midpoint and a dip of $20°$ with constant overburden velocity; how much does the reflecting point move between members of the common-midpoint set for offsets of 0, 0.5, 1.0, 1.5 and 2.0 km?

5.2. For an airwave with a velocity of 330 m/s and two geophones separated by 5 m, at what frequency is maximum attenuation achieved?

5.3. (a) An explosion initiated at the top of a column of explosives of length $a\lambda_r$ travels down the column with velocity V_e. By comparison with the same amount of explosive concentrated at the center of the column and exploded instantaneously at the same time as the column, show that the array response F is

$$F = -\text{sinc}\,\{\pi a(\sin\alpha_0 - V_r/V_e)\},$$

V_r being the velocity in the rock and α_0 the same as in fig. 5.10. Under what circumstances does this result reduce to that of (5.2)? (b) Calculate F for a column 10 m long given that $\lambda_r = 40$ m, $V_e = 5.5$ km/s, $V_r = 2.1$ km/s, $\alpha_0 = 0°, 30°, 60°, 90°$. (c) If the column in (b) is replaced by six charges each 60 cm long equally spaced to give a total length of 10 m, the charges being connected by spirals of detonating cord with velocity of detonation 6.2 km/s, what length of detonating cord must be used between adjacent charges to achieve maximum directivity downward? (d) What are the relative amplitudes (approximately) of the waves generated by the explosives in (c) at angles $\alpha_0 = 0°, 30°, 60°, 90°$ when $\lambda_r = 40$ m. (e) Discuss the application of (5.1) to energy in a non-vertical plane passing through the line of geophones.

5.4. Reflections in the zone of interest have apparent velocities around 6.5 km/s whereas the velocity just below the uniform LVL is 2.1 km/s. If we wish to avoid cancellation below 80 Hz when using an array, what is the maximum in-line array length?

5.5. (a) Under what conditions is the response of a linear array of n evenly-spaced geophones zero for a wave traveling horizontally (such as ground roll)? (b) If the n geophones are distributed uniformly over one full wavelength, show that the response is $F = 1/n$. (c) What is the response of the array in (a) when the waves arrive perpendicular to the line of geophones? (d) What is the response of the array in (a) when the waves arrive at $45°$ to the line and $n = 8$? (e) Repeat (d) for $n = 16$.

5.6. Show that (5.1) and (5.2) are consistent.

5.7. Array tapering is sometimes achieved by (i) doubling elements at some locations, (ii) weighting equally-

Fig.5.65. The Syslap method. (Courtesy CGG.)
(*a*) Explosion in center hole generates mainly P-waves;
(*b*) because of the asymmetry produced by the explosion in the center hole, the explosion in the right-hand
hole generates SH-waves as well as P-waves; (*c*) the
left-hand hole produces P-waves plus SH-waves of
opposite polarity to those in (*b*).

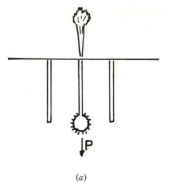

(*a*)

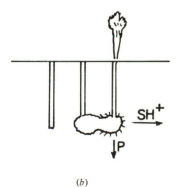

(*b*)

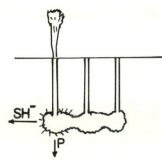

(*c*)

spaced elements either (*a*) within the element or (*b*) in a
mixing box in the field, or (iii) using unequal spacing of
elements. What arguments would you use for or against
these approaches?

5.8.   Fig. 5.68 shows the directivity effect of a group
length typical of end-on marine shooting. How will the
curves change (*a*) with arrival time, (*b*) as offset increases,
(*c*) for greater stacking velocity? Is it better to have
shooting proceed in the updip or downdip direction?

5.9.   Assume that you wish to map objectives 3 to 5 km
deep in an area with topography ranging from flat to
gentle hills (surface gradients usually less than 3 m/100 m).
Dips at objective depth may be up to 30°. The velocity at
the base of the low-velocity layer is 2 km/s, that at objective
depth is 4 km/s, and at basement (8 km) probably about
6 km/s. Five surface-source units and 48-channel recording

Fig.5.66. Conversion efficiency at water–solid interface. (After Tatham and Stoffa, 1976.) (*a*) Conversion
coefficient versus angle of incidence in the water; the
curves are labeled with seafloor velocity in km/s;
(*b*) maximum amplitude and width of main lobe versus
seafloor velocity.

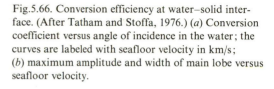

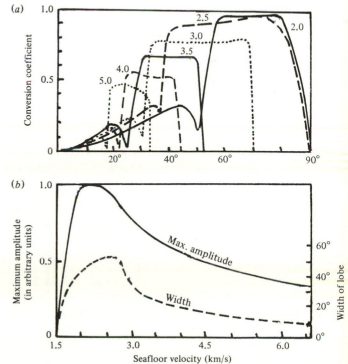

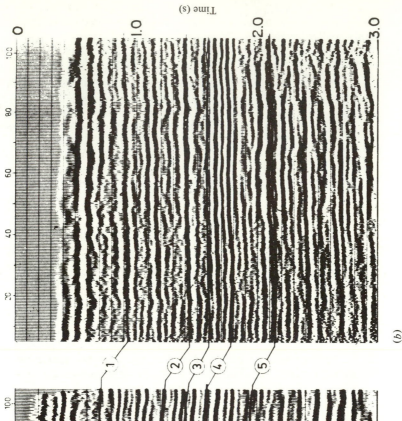

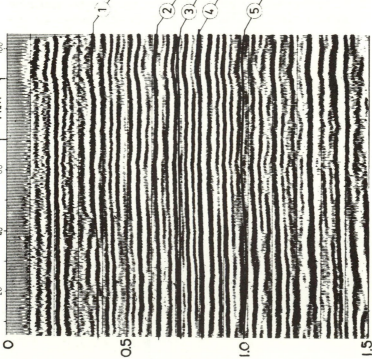

Fig. 5.67. Comparison of P- and S-wave records. (Courtesy CGG.) (a) P-wave record; (b) S-wave record; the S-wave record is displayed at double the timing speed to make comparison of events easier.

equipment are available. Both ground roll ($V_R \approx 800$ m/s) and airwaves may be problems, but the low-velocity layer (about 10 m thick with a velocity about 600 m/s) is probably fairly uniform. The area is fairly noisy and moderate effort will probably be required to achieve adequate data quality. Propose field methods and explain the bases for your proposals.

5.10. Seismic field work is usually carried out in a uniform manner with group intervals everywhere the same and the spread either all on one side or symmetric about the sourcepoint. The layout dimensions tend to be determined by what equipment is at hand or by habit rather than the nature of the problem to be solved; for example, the length of geophone flyers (several geophones for a single group being permanently wired together) may dictate the geophone interval and the equipment available the number of channels and thus the effective spread length. Sometimes hybrid spread arrangements are used to make fuller use of equipment. Assume that you have more channels available than the number given by following the guidelines of §5.3.5; what circumstances might lead you to use the extra channels to (*a*) extend the spread length beyond that given by guideline (1); (*b*) fill in the space between the shotpoint and the minimum offset given by (2); (*c*) interleave additional groups somewhere in the middle of your spread; (*d*) layout a partial spread on the other side of the shotpoint where an end-on arrangement is being used; (*e*) layout a short cross arm? If you have almost but not quite enough channels to use a split arrangement compared to an end-on, what are the advantages and disadvantages of using the split with: (*f*) longer group intervals than given by guideline (5); (*g*) shortening the maximum offset; or (*h*) increasing the minimum offset?

Fig.5.68. Response to apparent dip of a tapered array with effective length of 50 m. Reflection arrival time = 1.0 s, stacking velocity = 1.5 km/s, offset = 300 m. (After Savit and Siems, 1977.)

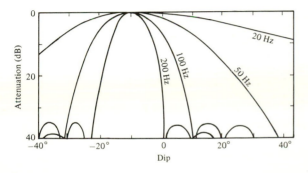

5.11. Uphole surveys in five different (unrelated) areas give the following uphole-time versus depth information:

Depth	Area *A*	Area *B*	Area *C*	Area *D*	Area *E*
5 m	0.012 s	0.011 s		0.012 s	0.008 s
8			0.020 s	0.016	
10	0.025	0.023	0.024	0.018	
12		0.024	0.027		0.020
15	0.030		0.031	0.022	
18		0.028	0.034	0.030	0.030
21	0.034		0.036	0.033	0.031
25	0.036	0.032		0.035	0.032
30	0.039	0.035	0.039		
35		0.037		0.039	0.036
40	0.046		0.044	0.042	
50	0.051	0.044	0.048	0.047	0.043

Explain the possible velocity layering for each case. How reliably are velocities and depth of weathering defined?

5.12. In one 3-D shooting technique, shotpoints ($\times$) and geophones ($\circ$) are laid out as shown in fig. 5.69; all the geophones are recorded for each shotpoint. This arrangement employs 48 geophone stations, spaced 50 m apart, and 48 shotpoints, spaced 100 m apart. (*a*) Locate all the 'depth points' and determine their respective multiplicity. [Hint: utilize symmetry to save work in the calculations.] (*b*) Note that some of the depth points fall outside the square. If this layout is repeated with common geophone lines, these points will fit in adjacent squares. What effect will this have on multiplicity?

5.13. Denham's high-frequency limit (§5.3.8) is related both to the loss of high frequencies and to the dynamic range of the recording system. Reconcile this limit with a loss by absorption of 0.15 dB/λ (§2.3.2), spreading, and high-frequency loss because of peg-leg multiples as illustrated in fig. 4.10. Take 84 dB as the dynamic range of the recording system.

5.14. A well encounters a horizon at a depth of 3 km with a dip of 7°. Shots are fired from a point 200 m updip from the well into a geophone at depths of 1.0 to 2.6 km at intervals of 400 m. Plot the raypaths and traveltime curves for the primary reflection from the 3 km horizon and its first multiple at the surface. Assume $V = 3.0$ km/s.

5.15. Fig. 5.70 illustrates filter characteristics. Evaluate the importance of (*a*) low-frequency cut, (*b*) high-frequency cut, (*c*) bandwidth and (*d*) filter slope on: (i) time delay to a point which could be timed reliably; (ii) apparent polarity; and (iii) ringing. The conclusions can be generalized for filters of other design types.

5.16. Fig. 5.71 shows waveshape changes produced by

Fig.5.69. A layout for 3-D coverage. Shotpoints are indicated by ×, geophone group centers by ○.

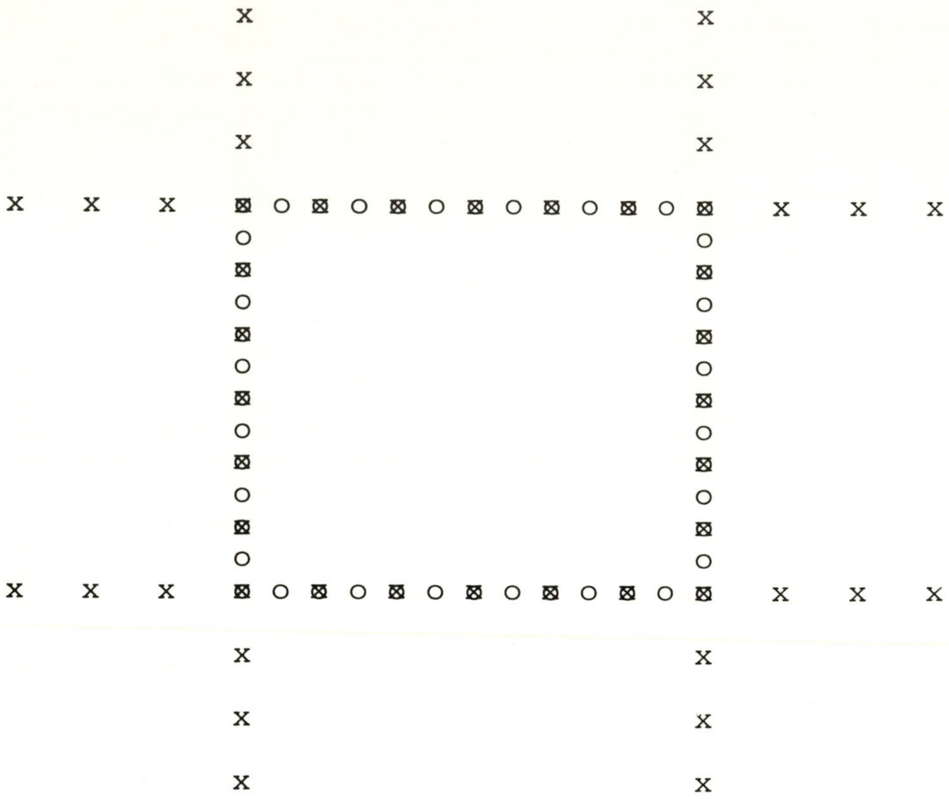

Fig.5.70. Impulse responses of minimum-phase filters. The respective rows differ in filter slopes, the columns in passbands (specified by 3 dB points). (Courtesy Seiscom Delta.) (*a*) Effect of low-cut filtering; (*b*) effect of high-cut filtering.

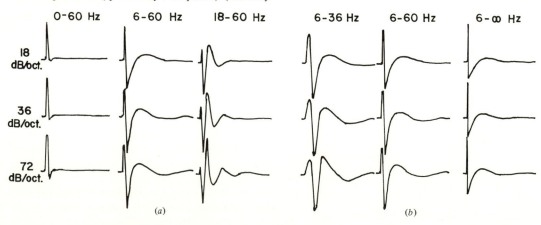

the analog filtering in modern digital instruments. What can you conclude about the effects on picking?

5.17. (a) Take the wavelet which has values at successive 4 ms intervals of $0, \ldots, 0, 8, 7, -8, -6, 0, 4, 2, 0, \ldots, 0$ (with 10 zeroes at each end) and add random noise (a random number table is given in appendix C) in the range from $+10$ to -10, that is, with signal/noise ≈ 1. Do this five times for different noise values and plot the results, shading positive values as in a variable-area display. (b) Repeat for noise ranging from $+20$ to -20 (signal/noise $\approx \frac{1}{2}$). What can you conclude about the extraction of signal buried in noise? (c) Sum the five waveforms in (a) and also in (b) to show how stacking enhances coherent signal versus random noise. (d) Replace the elements in the wavelets in (a) and (b) with $+1$ or -1 as the values

Fig.5.71. Far-field airgun signatures through various instrument filters. (a) No extra filtering; (b) out-124 Hz, 72 dB/octave; (c) out-62 Hz, 72 dB/octave; (d) out-62 Hz, 18 dB/octave; (e) 8–124 Hz with slopes of 18 and 72 dB/octave on low- and high-frequency sides respectively; (f) 18–124 Hz with 18 and 72 dB/octave slopes; (g) 8–62 Hz with 36 and 72 dB/octave slopes. Timing marks are 10 ms apart.

Time (s)

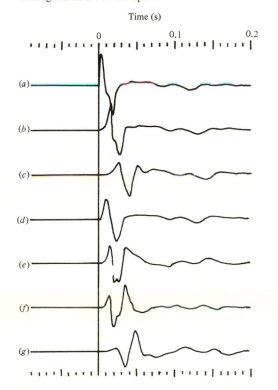

are positive or negative (sign-bit expression) and repeat part (c).

5.18. Using the data in the legend of fig. 5.38, show that both the quaternary-gain amplifier and the A/D converter have dynamic ranges of 84 dB.

5.19. Express the numbers 19 and 10 as binary numbers. (a) Add the binary numbers together and convert the sum to a decimal number. (b) Multiply the two binary numbers and convert to decimal. (Note that mathematical operations are carried out in binary arithmetic in the same way as in decimal arithmetic.)

5.20. Assume a 96-channel seismic system recording with 2 ms sampling and 25 s Vibroseis records. What is the data rate (samples/second) and the number of bits/record? How does the data rate compare with the capacity of 9-track magnetic tape moving at a 6250 bytes/inch rate, using 4 bytes/sample? How many bits of memory are required to store one channel of data? What is the effect of the header and ancillary information, and parity bits?

5.21. Imagine an impulsive source striking the ground at times $n\Delta$ apart, where n is a random number between 10 and 20 and Δ is the sampling interval. Given reflections with amplitude 5 at 0, 2 at 5Δ, -1 at 13Δ, $+3$ at 29Δ, $+1$ at 33Δ, -2 at 42Δ, add the reflection sequence as would be done with Sosie recording (§5.4.9) for 10, 20 and 30 impulses to see how the signal builds up as the multiplicity increases.

5.22. How much energy is released (approximately) per shot by the airgun array in fig. 5.43 when the initial pressure is 2000 psi (14 MPa). [Energy released = work done by the expanding gas $= \int \mathscr{P} d\mathscr{V}$. Assume that the change is adiabatic, that is, $\mathscr{P} \mathscr{V}^{1.4} = $ constant, and that the final pressure is 2 atmospheres.]

5.23. The dominant period of a marine seismic waveshape is often determined by the depth of the source, that is, by reinforcement of the second half-cycle by the ghost reflected from the surface. Assuming that this is true for the source signatures shown in fig. 5.45, determine their depths.

5.24. Land cables are built in sections which are identical. The connections to the plugs at each end of the section effectively rotate the system by the number of groups to be connected to that section. Thus if pins 1, 2, 3 at one end of the section are connected to takeouts for 3 groups in a section, at the other end of the section they will connect with pins 4, 5, 6 of the next section. There are more sets of wires than channels being used at any one time, e.g., perhaps 96 independent pairs of wires for use with 48 channels. However, occasionally so many sections and geophones are laid out that a distant group of phones is connected to the same channel as a nearer group; this mistake is called *rollover*. Sketch a possible arrangement for the connections in one section with takeouts for 3 channels and explain how rollover could appear on the seismic records.

5.25. A 48-channel streamer with 50 m groups has the hydrophones spaced uniformly throughout its length. The lead-in and compliant sections together are 200 m in length and the tail section and buoy connection are 150 m. Assume a ship's speed of 5.8 knots (3.0 m/s) and a current perpendicular to the direction of traverse with a speed of 1.9 knots. (*a*) What are the perpendicular and in-line components of the distance to the farthest active group with respect to the traverse direction? (*b*) If the velocity to a reflector 2.00 km below the ship is 3.00 km/s and if the reflector dips 20° perpendicular to the traverse direction: (i) by now much will the arrival time be changed for the far trace? (ii) if this should be attributed to a change in velocity rather than cross-dip, what velocity would it imply? (*c*) Assume that the amount of *streamer feathering* (drift of the streamer to one side) is ascertained by radar-sighting on the tail buoy with an accuracy of only ±3°: (i) how much uncertainty will this produce in locating the far group? (ii) how much change in arrival time will be associated with this uncertainty? (*d*) Over what distance will the midpoint traces which are to be stacked when making a CDP stack be distributed?

5.26. Use figs. 5.31 and 5.33 to determine the filter equivalent to a geophone with $v_0 = 10$ Hz and $h = 0.7$ feeding into an amplifier with a 10–70 Hz bandpass filter and a 4 ms alias filter.

5.27. The velocity of radio waves has the following values (km/s) over various terrains: normal sea water, 299 670; fresh water, 299 250; normal farmland, 299 400; dry sand, 299 900; mountainous terrain, 298 800. If range calculations are based on travel over normal sea water, what are the errors in range per kilometer of path over the various terrains?

5.28. If the error in Shoran time measurements is ±0.1 μs, what is the size of the parallelogram of error in fig. 5.50 when: (*a*) $\theta = 30°$; (*b*) $\theta = 150°$? Take the velocity of radiowaves as 3×10^5 km/s.

5.29. Assume that a ship traveling at 5 knots is determining its position by a Doppler-sonar system transmitting at 300 kHz where the beams make a 60° angle with the seafloor (see fig. 5.52). (*a*) What are the Doppler-shifted frequencies? (*b*) If the ship is pitching ±15° in a one-minute cycle in a rough sea in water 100 m deep, what is the range of frequencies? Since the pitching effects on aft and forward measurements are 180° out of phase (the 'Janus effect'), how can this be used to improve velocity determination?

5.30. Sieck and Self (1977) summarize 'acoustic systems' as shown in table 5.1. For each of these calculate (*a*) the wavelengths and (*b*) the penetration given by Denham's rule (§5.3.8) and reconcile with the stated purposes. (*c*) Trade literature claims 30 cm resolution with imploders and 2–5 m resolution with sparkers; how do these figures compare with the resolvable limit (§4.3.2*a*)? [Note that absorption in water is very small so that effectively absorption does not begin until the seafloor is reached.]

5.31. A satellite is in a stable orbit around the Earth when the gravitational force (*mg*) pulling it earthward equals the centrifugal force mV^2/R, where g is the acceleration of gravity, m and V the satellite's mass and velocity, and R the radius of its orbit about the center of the Earth. (*a*) Determine the acceleration of gravity at the orbit of a Transit satellite 1070 km above the Earth, knowing that g at the surface of the Earth is 9.81 m/s² and that the gravitational force varies inversely as the square of the distance between the centers of gravity of the masses. (*b*) What is the satellite's velocity if its orbit is stable? (*c*) How long does it take for one orbit? (*d*) How far away is the satellite when it first emerges over the horizon? (*e*) What is the maximum time of visibility on a single satellite pass? (Assume the radius of the Earth is 6370 km.)

5.32. Given that the base of the weathering is flat in fig. 5.72 and that the weathering velocity is 500 m/s, find: (*a*) the sub-weathering velocity; (*b*) the thickness of the

Table 5.1. *Acoustic systems.*

System	Frequency	Purpose
Fathometers	12–80 kHz	To map water bottom
Water-column bubble detectors	3–12	To locate bubble clusters, fish, etc.
Side-scan sonar	38–250	To map bottom irregularities
Tuned transducers	3.5–7.0	To penetrate 30 m
Imploders	0.8–5.0	To penetrate 120 m and find gas-charged zones
Sparker	0.04–0.15	To map to 1000 m

Fig.5.72. Split-spread Vibroseis record with 30 m group spacing. To the left of sourcepoint 620 the land is flat but a valley lies between 610 and 620, and the land is slightly rolling to the right of 610. (Courtesy SSC.)

Location along line

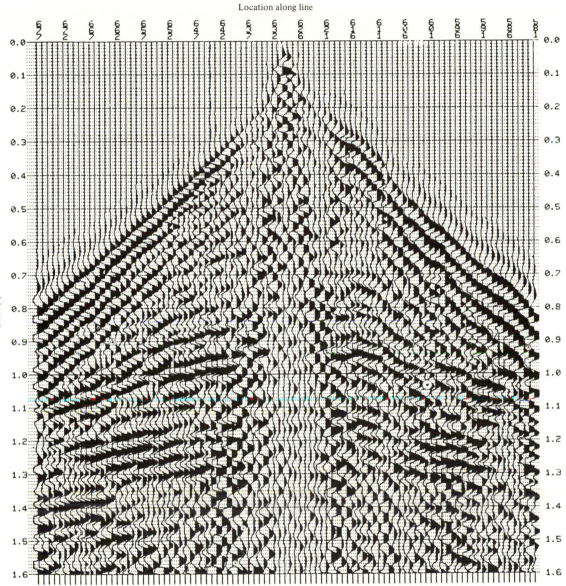

weathering; (*c*) the ground-roll velocity; (*d*) the stacking velocity and approximate depth associated with flat reflections at about 0.62, 0.93, and 1.12 s; (*e*) the depth of the valley at about location 615.

5.33. The correction methods discussed in §5.6.2 assume that the shot is below the base of the LVL. What changes are required in the equations of this section if this is not the case?

5.34. Fig. 5.73 shows the first arrivals at geophone stations 100 m apart from shots 25 m deep at each end of the spread. (There are actually 11 geophone stations with the shotpoints being at the 1st and 11th stations; however, the geophone group at each shotpoint is not recorded because of hole noise.) The uphole geophone is recorded on the third trace from the right. The weathering velocity is 500 m/s. (*a*) Estimate the sub-weathering velocity V_H by averaging the slopes of lines approximating the first-breaks. The valley midway between the shotpoints produces a change in the first-break slopes, as if two refractors are involved, which is not the case. How can we be sure of

the latter? [Hint: Plot an elevation profile as an aid in determining the best-fit line through the first breaks to give V_H.] (*b*) Determine the weathering thicknesses at the two shotpoints from the uphole times. (*c*) What corrections Δt_0 should be applied to reflection times at the two shotpoints for a datum of 1100 m? (*d*) Calculate the weathering thickness and the time correction for each geophone station. (*e*) Plot corrected reflection arrival times in an X^2–T^2 plot and determine the depth, dip and average velocity to the reflector giving the reflection at 0.30 and 0.21 s.

5.35. The arrival time of a reflection at the shotpoint is 1.200 s and the difference in arrival times at geophones 1000 m on opposite sides of the shotpoint is 0.150 s, near-surface corrections having been applied. Assume the line is perpendicular to the strike. Determine the reflector depth, dip and horizontal location with respect to the shotpoint assuming: (*a*) no dip moveout and the average velocity associated with a vertical traveltime is 2630 m/s; (*b*) the observed dip moveout and $\bar{V}$; (*c*) straight-ray travel at the angle of approach and $\bar{V} = V_H = 1830$ m/s; (*d*) straight-

Fig.5.73. Reflection first-breaks.

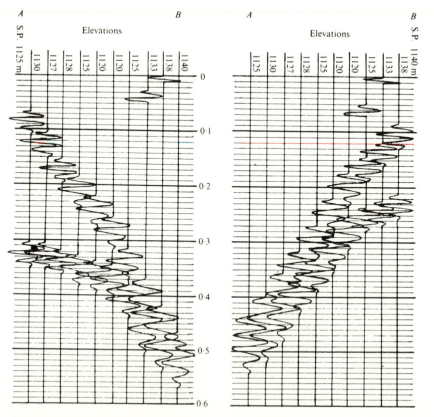

ray travel at the local velocity above the reflector, 3840 m/s; (*e*) the migrated position determined from the wavefront chart of fig. 5.63.

5.36. In a marine survey the water depth is 1 km and a reflector is 3 km below the seafloor. Use fig. 5.66 to determine the optimum range of offset for S-wave generation. The P-wave velocity just below the seafloor is 2.8 km/s and the water velocity is 1.5 km/s.

6
Refraction methods

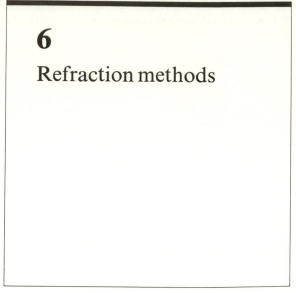

Overview

Refraction and reflection work are similar in many aspects, much different in others. The similarities are sufficient that reflection field crews sometimes do refraction profiling, though often not with the efficiency of a crew specifically designed for refraction. The differences between reflection and refraction field work mostly result from the long shot-to-geophone distances employed in refraction. The energy input to the ground must be larger for refraction shooting and explosives continue to be the dominant energy source although other seismic sources are also used. The longer travel paths result in the higher frequencies being mostly absorbed so that refraction data are generally of low frequency compared with reflection data. Consequently refraction geophones have lower natural frequencies than reflection geophones, although the response of the latter is often adequate for satisfactory refraction recording. Most digital seismic equipment can be used for refraction but some of the older analog equipment does not have adequate low-frequency response.

Most refraction techniques involve head waves and the mapping of members whose velocities are significantly higher than those of any overlying rocks. Such units are not always present and maps of high-velocity units which are present may not be related to petroleum objectives, so refraction methods are not applicable in many situations. Even where applicable, refraction shooting is usually slower than reflection shooting because the large offset distances involve more moving time and create problems of communications and logistics. However, refraction profiles are often not as closely spaced as reflection lines and hence the cost of mapping an area is not necessarily greater.

Most refraction work involves in-line profiling (§6.1.1), especially the use of reversed profiles. Broadside and fan-shooting and placing the geophone in a deep borehole are methods used for certain objectives. Small-scale refraction is used in studies for the foundations of structures and other engineering problems. Marine refraction involves special operational problems.

The computation of refraction data is discussed in §6.2; data have to be corrected for elevation and near-surface variations as with reflection data. The essential in refraction interpretation (§6.3) is correlating events which involve the same refracting layer; ambiguities can often be removed if more data are available. Once refraction events have been correlated, refractor depths and dips can be found using the formulas given in chapter 3, but in addition a variety of methods is available for more complicated situations and for routine interpretation of large amounts of data.

These methods are divided into delay-time and wavefront methods. Interpretation of engineering refraction data (§6.4) is generally much simpler and more straightforward than that of large-scale surveys because the near-surface layers are relatively few and little detail is usually required.

6.1   Refraction profiling
6.1.1   *In-line refraction profiling*

The basic refraction field method involves shooting reversed refraction profiles, a long linear spread of many geophone groups shot from each end, the distance being great enough that the dominant portion of the travel path is as a head wave in the refractor or refractors being mapped. Usually it is not practical to record simultaneously so many geophone groups spread over such a long distance, and hence refraction profiles are usually shot in segments. Referring to fig. 6.1a which shows a single refractor, the spread of geophone groups might be laid out between C and D and shots at C and G fired to give two records; the spread would then be moved between D and E and shots fired at C and G as before, and so on to develop the complete reversed profile $CDEFG$. The charge size is often varied for the different segments because larger charges are required when the offset becomes greater. Usually one or two groups will be repeated for successive segments to increase the reliability of the time-tie between segments.

The shothole at C can also be used to record a profile to the left of C and the shothole at G a profile to the right of G. Note that the *reciprocal time* t_r is the same for the reversed profiles and that the intercept times for profiles shot in different directions from the same shotpoint are equal. These equalities are exceedingly valuable in identifying segments of complex time–distance curves where several refractors are present. In simple situations

Fig.6.1. Reversed refraction profiles. (*a*) Time-distance plot for continuous reversed profiling; (*b*) section showing single refractor.

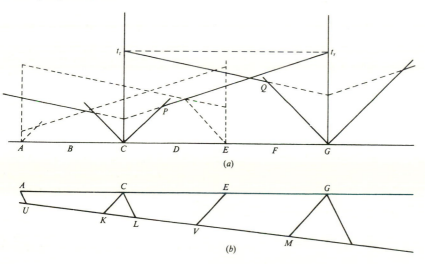

(a)

(b)

the reversed profile can be constructed without having to actually shoot it by using the reciprocal time and intercept time information. However, usually situations of interest are sufficiently complicated that this procedure cannot be carried out reliably.

The *reversed profiles* shot from C and G allow the mapping of the refractor from L to M. The reversed profile to the left of C permits mapping as far as K but no coverage is obtained for the portion KL. Hence continuous coverage on the refractor requires an overlap of the reversed profiles; a reversed profile between A and E (shown dotted in fig. 6.1a) would provide coverage between U and V, thus including the gap KL as well as duplicating the coverage UK and LV. With perfect data the duplicate coverage does not yield new information but in actual profiling it provides valuable checks which increase the reliability of the interpretation.

In cases where reversed profiles are not essential, a split refraction spread (ACE in fig. 6.1 with shotpoint at C) can be used with a saving in the number of shotpoints required. However, since the updip and downdip apparent velocities are not obtained from the same part of the refractor, faulting, curvature of the refractor, lateral velocity variations, etc. can render the method useless.

If we have the two-refractor situation in fig. 6.2, first-break coverage on the shallow refractor is obtained from L to K and from M to N when the shots are at C and G; the corresponding coverage on the deeper refractor is from Q to S and from R to P.

If we are able to resolve the refraction events which arrive later than the first-breaks, called *second arrivals* or *secondary refractions*, we can increase the coverage obtained with a single profile. However, it is difficult, and sometimes impossible, to adjust the gain to optimize both the first-breaks and the second arrivals at the same time; if the gain is too low the first-breaks may be weak and ambiguities in timing may result whereas if the gain is too high the secondary refractions may be unpickable. Because of this difficulty, prior to magnetic-tape recording, refraction mapping was generally based on first-breaks only. With magnetic tape, recording playbacks can be made at several gains so that each event can be displayed under optimum conditions.

In order to economize on field work, the portions of the time–distance curves which do not add information necessary to map the refractor of interest often are not shot where they can be predicted reasonably accurately. Thus, the portions CP and GQ of the reversed profile in fig. 6.1a are often omitted.

Where a single refractor is being followed, a series of short refraction profiles are often shot rather than a long profile. In fig. 6.3 geophones from C to E are used with shotpoint C, from D to F with shotpoint D, etc. The portions of the time–distance curves attributable to the refractor being mapped are then translated parallel to themselves until they connect together to make a composite time–distance curve such as that shown by the dashed line. The composite curve may differ from the curve which would actually have been obtained for a long profile from shotpoint C because of refraction events from other horizons.

6.1.2 *Broadside refraction and fan-shooting*

In *broadside refraction* shooting, shotpoints and spreads are located along two parallel lines (see fig. 6.4) selected so that the desired refraction event can be mapped with a minimum of interference from other events. Where the refraction event can be clearly distinguished from other arrivals, it provides a very economical method of profiling because all the data yield information about the refractor. However, usually the criteria for identifying the refraction event are based on in-line measurements (such as the apparent velocity or the relationship to other events) and these criteria are not available on broadside records where the offset distance is essentially constant. Thus if the refractor should unexpectedly change its depth or if another refraction arrival should appear, one might end up mapping the wrong horizon. Consequently broadside refraction shooting is often combined with occasional in-line profiles in order to check on the identity of the horizon being mapped.

The first extensive use of refraction was in searching for saltdomes by the fan-shooting technique (see §1.2.2). A saltdome inserts a high-velocity mass into an otherwise

Fig.6.2. Reversed refraction profiles for two-refractor case. (a) Time–distance plot; (b) section showing the two refractors.

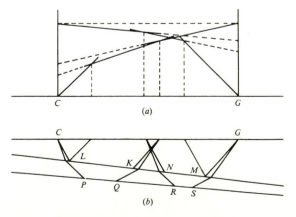

low-velocity section so that horizontally traveling energy arrives earlier than if the saltdome were not present; the difference in arrival time between that actually observed and that expected with no saltdome present is called a *lead*. In *fan-shooting* (fig. 6.5) geophones are located in different directions from the shotpoint at roughly the same offset distances, the desire to maintain constant offset distance usually being sacrificed in favor of locations which are more readily accessible. The leads shown by overlapping fans then roughly locate the high-velocity mass. This method is not used for precise shape definition.

6.1.3 *Gardner's method of defining saltdomes*

Much of the petroleum associated with production from the flanks of saltdomes lies close to the salt flanks, so that the accurate mapping of the flanks is of considerable economic interest. The method of Gardner (1949) involves locating geophones in a deep borehole drilled into the salt for this purpose and shooting from various locations on the surface. The travelpath for each shot is partially through relatively low-velocity sediments and partially through high-velocity salt. For given traveltime and geophone location, the locus of possible points of entry of the travel paths into the salt is a surface (*aplanatic surface*) which is roughly a paraboloid (fig. 6.6a). A surface tangent to the paraboloids for all measurements for various combinations of source–detector locations defines the saltdome.

Variations of Gardner's method employ shooting from surface locations into geophones located in a deep well near to but not within the salt. These methods are also used occasionally to define bodies other than salt.

6.1.4 *Engineering surveys on land*

Engineering projects generally do not allow much money for geophysical studies and so the equipment and methods are usually simple. The energy source might be a hammer striking a steel plate on the ground, the instant of impact being determined by an inertial switch on the ham-

mer, a hand-operated 'tamper', a weight dropped on the ground, a small explosion, or an impactor using the Sosie method (fig. 5.28). (Such sources are also used for reflection engineering surveys; see Meidav, 1969.) The energy is usually detected by moving-coil geophones similar to those already described.

Often only a few channels are used, one to six usually. The amplifiers and the camera generally weigh only a few kilograms and often are contained in a small metal suitcase. In some systems the recorded data are displayed on a small oscilloscope tube and are photographed with a Polaroid camera to give a permanent record. In other systems time counters are started at the instant the energy is delivered to the ground and stopped when the first-break energy arrives at the geophones, thus giving direct readings of traveltimes. More modern recording systems add together (vertically stack) the records from several successive source impulses, which effectively doubles the range (the useful range of a hammer blow can be increased from 50 to 100 m in this way).

Fig.6.4. Broadside refraction profiling.

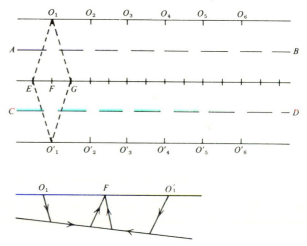

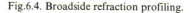

Fig.6.3. Unreversed refraction profiles for a single refractor.

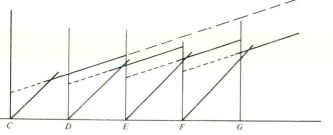

6.1.5 *Marine refraction*

Because refraction recording requires that there be appreciable distance between the source and the recording locations, two ships have usually been required for marine refraction recording. To shoot a reversed refraction profile in one traverse requires three ships – a shooting ship at each end while the recording ship travels between them. For the shooting ships to travel the considerable distances between shotpoints takes appreciable time because of the relatively low maximum speed of ships and hence the high production rates which make marine reflection work economical are not realized in refraction shooting. Consequently marine refraction work is relatively expensive.

The *sonobuoy* (fig. 6.7) permits recording a refraction profile with only one ship. The sonobuoy is an expendable listening station which radios the information it receives back to the shooting ship. The sonobuoy is merely thrown overboard; the salt water activates batteries in the sonobuoy as well as other devices which cause a radio antenna to be extended upward and one or two hydrophones to be suspended beneath the buoy. As the ship travels away from the buoy, shots are fired and the signals

received by the hydrophones are radioed back to the ship where they are recorded. The arrival time of the wave which travels directly through the water from the shot to the hydrophone is used to give the offset distance. After a given length of time the buoy sinks itself and is not recovered. Sonobuoys make it practical to record unreversed refraction profiles while carrying out reflection profiling, the only additional cost being that of the sonobuoys.

6.2 Computation of refraction data

6.2.1 *Refraction data reduction*

Refraction data have to be corrected for elevation and weathering variations, as with reflection data. The correction methods are essentially the same except that often geophones are too far from the shotpoint to record the refraction at the base of the LVL and thus there may be no weathering data along much of the line. Additional shots may be taken for special refraction weathering information.

Where complete refraction profiles from zero offset to large offsets are available, playback of the data with judicious selections of filters and automatic gain control

Fig.6.5. Fan-shooting. (From Nettleton, 1940.)

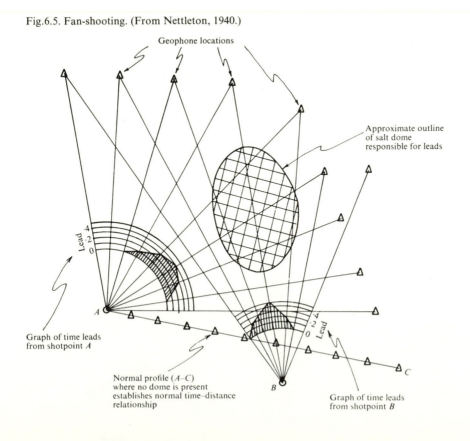

Fig.6.6. Outlining a saltdome using a geophone within the salt. (After Gardner, 1949.) (a) Plan view of the aplanatic surface; (b) isometric view of the raypath and paraboloid of the aplanatic surface.

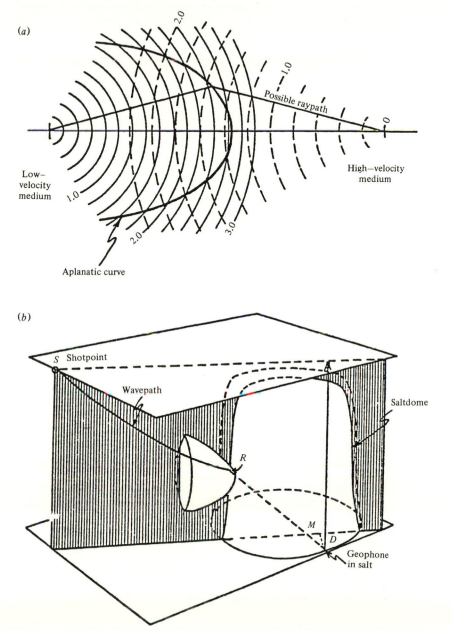

(AGC) may allow one to correlate reflection events with refraction events, thus adding useful information to each type of interpretation. Often the most prominent reflections will not correspond to the most prominent refractions.

Another useful refraction playback technique is to display the data as a *reduced* refraction *section* (fig. 6.8) where arrival times have been shifted by the amount x/V_R, where x is the offset distance and V_R is a value near the refractor's velocity. If V_R were exactly equal to the refractor's velocity the residual times would be the delay times

(which will be discussed in §6.2.2) and relief on the reduced refraction section would correlate with refractor relief (although displaced from the subsurface location of the relief). However, even if V_R is only approximately correct, the use of reduced sections improves considerably the pickability of refraction events, especially secondary refractions.

6.2.2 *Delay-time methods*

(a) *Delay times.* The concept of delay time, introduced by Gardner (1939), is widely used in routine refrac-

Fig.6.7. Sonobuoy operation. (Courtesy Select International Inc.)

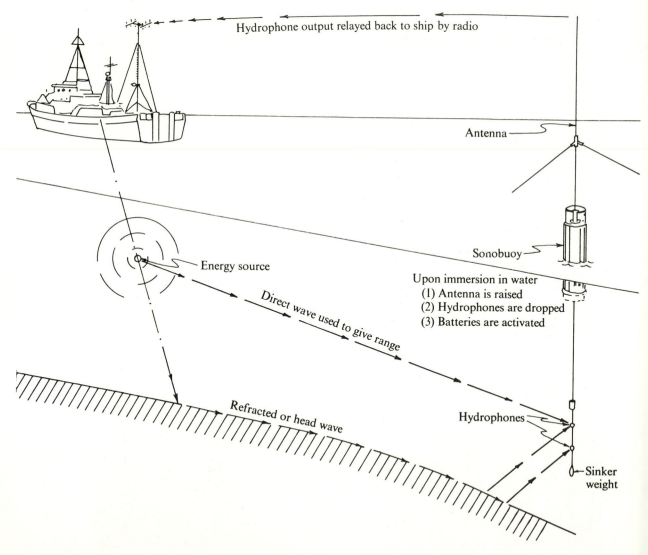

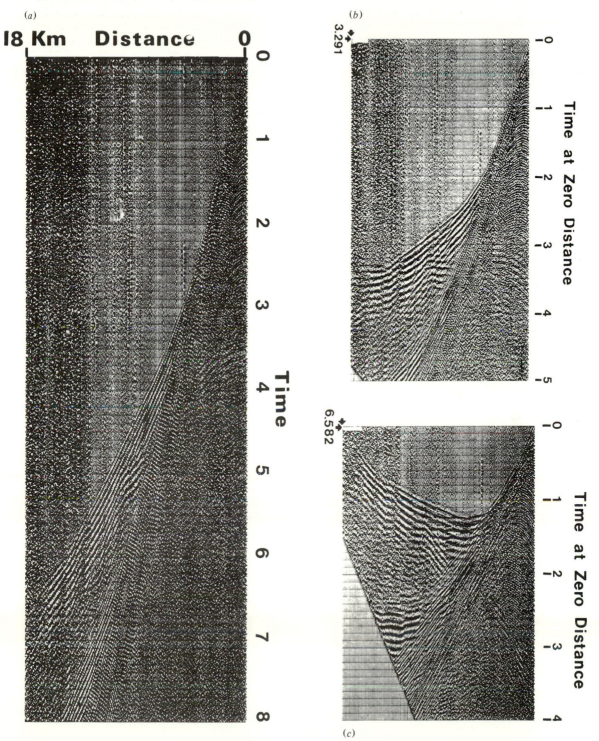

Fig.6.8. Reduced refraction section. (Courtesy Petty-Ray Geophysical.) (a) Conventional refraction section; (b) reduced at 5469 m/s to align highest-velocity events; (c) reduced at 2735 m/s. Subtracting x/V_R makes it easier to separate events and simplifies picking.

tion interpretation, mainly because the various schemes based upon the use of delay times are less susceptible to the difficulties encountered when we attempt to use (3.30)–(3.47) with refractors which are curved or irregular. Assuming that the refraction times have been corrected for elevation and weathering, the *delay time* associated with the path $SMNG$ in fig. 6.9 is the observed refraction time at G, t_g, minus the time required for the wave to travel from P to Q (the projection of the path on the refractor) at the velocity V_2. Writing δ for the delay time, we have

$$
\begin{aligned}
\delta &= t_g - \frac{PQ}{V_2} = \left(\frac{SM + NG}{V_1} + \frac{MN}{V_2}\right) - \frac{PQ}{V_2} \\
&= \left(\frac{SM + NG}{V_1}\right) - \left(\frac{PM + NQ}{V_2}\right) \\
&= \left(\frac{SM}{V_1} - \frac{PM}{V_2}\right) + \left(\frac{NG}{V_1} - \frac{NQ}{V_2}\right) \\
&= \delta_s + \delta_g,
\end{aligned} \tag{6.1}
$$

where δ_s and δ_g are known as the *shotpoint delay time* and the *geophone delay time* since they are associated with the portions of the path down from the shot and up to the geophone.

An approximate value of δ can be found by assuming that the dip is small enough that PQ is approximately equal to the geophone offset x. In this case,

$$
\delta = \delta_s + \delta_g \approx t_g - x/V_2. \tag{6.2}
$$

Provided the dip is less than about 10°, this relation is sufficiently accurate for most purposes. If we substitute the value of t_g obtained from (3.31), (3.41) and (3.42) we see that δ is equal to the intercept time for a horizontal refractor but not for a dipping refractor.

Many interpretation schemes using delay time have been given in the literature, for example, Gardner (1939, 1967), Barthelmes (1946), Tarrant (1956), Wyrobek (1956) and Barry (1967). We shall describe only the latter three.

Fig.6.9. Illustrating delay time.

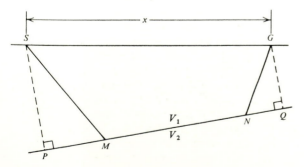

The methods described by Wyrobek and Tarrant are suitable for unreversed profiles while that of Barry works best with reversed profiles.

(b) Barry's method. The scheme described by Barry, like many based on delay times, requires that we resolve the total delay time δ into its component parts, δ_s and δ_g. In fig. 6.10 we show a geophone R for which data are recorded from shots at A and B. The ray BN is reflected at the critical angle, hence Q is the first geophone to record the head wave from B. Let δ_{AM} be the shotpoint delay time for shot A, δ_{NQ} and δ_{PR} the geophone delay times for geophones at Q and R, δ_{AQ} and δ_{AR} the total delay times for the paths, $AMNQ$ and $AMPR$. Then,

$$
\begin{aligned}
\delta_{AQ} &= \delta_{AM} + \delta_{NQ} \\
\delta_{AR} &= \delta_{AM} + \delta_{PR} \\
\Delta\delta &= \delta_{AQ} - \delta_{AR} = \delta_{NQ} - \delta_{PR}.
\end{aligned}
$$

For the shot at B, the shot delay time δ_{BN} is approximately equal to δ_{NQ} provided the dip is small. In this case,

$$
\delta_{BR} = \delta_{BN} + \delta_{PR} \approx \delta_{NQ} + \delta_{PR}.
$$

The geophone delay times are now given by

$$
\left.
\begin{aligned}
\delta_{NQ} &\approx \tfrac{1}{2}(\delta_{BR} + \Delta\delta) \\
\delta_{PR} &\approx \tfrac{1}{2}(\delta_{BR} - \Delta\delta)
\end{aligned}
\right\}. \tag{6.3}
$$

Thus, it is possible to find the geophone delay time at R provided we have data from two shots on the same side and we can find point Q. If we assume that the bed is horizontal at N and is at a depth h_N we have

$$
h_N = V_1 \delta_{BN}/\cos\Theta, \tag{6.4}
$$

$$
BQ = 2h_N \tan\Theta = 2V_1 \delta_{BN}(\tan\Theta/\cos\Theta) = 2V_2\delta_{BN}\tan^2\Theta. \tag{6.5}
$$

The shot delay time δ_{BN} is assumed to be equal to half the intercept time at B; this allows us to calculate an approximate value of BQ and thus determine the delay times for

Fig.6.10. Determining shotpoint and geophone delay times.

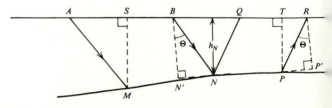

all geophones to the right of Q for which data from A and B were recorded.

The interpretation involves the following steps which are illustrated in fig. 6.11:

(*a*) the corrected traveltimes are plotted,

(*b*) the total delay times are calculated and plotted at the geophone positions,

(*c*) the 'geophone offset distances' (PP' in fig. 6.10) are calculated using the relation $PP' \approx V_2 \delta_{PR} \tan^2 \Theta$ (see problem 6.3) and the delay times in (*b*) are then shifted towards the shotpoint by these amounts,

(*d*) the shifted curves in (*c*) for the reversed profiles should be parallel; any divergence is due to an incorrect value of V_2, hence the value of V_2 is adjusted and steps (*b*) and (*c*) repeated until the curves are parallel (with practice only one adjustment is usually necessary),

(*e*) the total delay times are separated into shotpoint and geophone delay times, the latter being plotted at the points of entry and emergence from the refractor (S and T in fig. 6.10); the delay-time scale can be converted into depth if required using (6.4).

(*c*) *Tarrant's method.* Tarrant (1956) uses delay times to locate the point Q (fig. 6.12*a*) at which the energy arriving at R left the reflector. Denoting the delay time associated with the path QR by δ_g, we have

$$\delta_g = \rho/V_1 - (\rho \cos \phi)/V_2;$$

hence

$$\rho = V_1 \delta_g / (1 - \sin \Theta \cos \phi). \tag{6.6}$$

This is the polar equation of an ellipse. An ellipse is the locus of a point Q (fig. 6.12*b*) which moves so that the ratio (QR/QM) is constant (equal to the eccentricity ε, which is < 1 for an ellipse),

$$\rho/(h + \rho \cos \phi) = \varepsilon,$$

hence

$$\rho = \varepsilon h / (1 - \varepsilon \cos \phi). \tag{6.7}$$

The major axis, $2a = \rho_{\phi=0} + \rho_{\phi=\pi} = 2\varepsilon h/(1 - \varepsilon^2)$. The semi-minor axis, b, can be found by writing $y = \rho \sin \phi$ and finding y_{max}; this gives $b = \varepsilon h (1 - \varepsilon^2)^{-\frac{1}{2}}$. The distance from the focus, R, to the center of the ellipse, O, is equal to $\rho|_{\phi=0} - a = \varepsilon h/(1 - \varepsilon) - \varepsilon h/(1 - \varepsilon^2) = \varepsilon a$. If we take $\varepsilon = \sin \Theta$ and $h = V_2 \delta_g$, (6.7) becomes (6.6).

For a horizontal refractor, we have the ellipse in fig. 6.12*c*, with $a = V_2 \delta_g \tan \Theta \sec \Theta$, $b = V_2 \delta_g \tan \Theta$, and $OR = V_2 \delta_g \tan^2 \Theta$. Also $RQ = b/\cos \Theta = a$ and $\angle OQR = \tan^{-1}(OR/b) = \Theta$, $OQ = OR \cot \Theta = V_2 \delta_g \tan \Theta$.

We can approximate the ellipse in the vicinity of Q with a circle of the same radius of curvature. If we write the equation of the ellipse in the Cartesian form,

Fig.6.11. Illustrating the delay-time method of interpreting reversed profiles. (After Barry, 1967.)

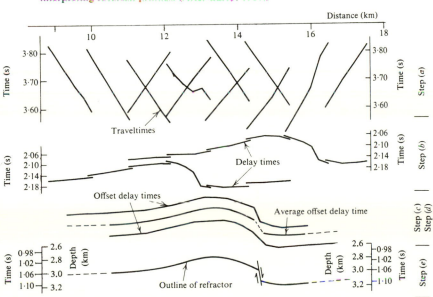

$(x/a)^2 + (y/b)^2 = 1,$

the radius of curvature, r, becomes

$r = \{1 + (y')^2\}^{\frac{3}{2}}/y'',$

where $y' = -(b/a)^2(x/y)$ and $y'' = -(b/a)^2(y - xy')/y^2$; at Q, $y' = 0$ and $y'' = b/a^2$. Hence $r = a^2/b = V_1 \delta_g/\cos^3 \Theta = V_2 \delta_g \tan \Theta \sec^2 \Theta$ and the center C is at the point $(0, r - b)$, that is, $(0, V_2 \delta_g \tan^3 \Theta)$. Also, $\angle CRO = \tan^{-1}(CO/RO) = \Theta$ hence $\angle CRQ$ is a right angle.

To apply the method, we must determine the velocities, V_1 and V_2, and the delay time at the shotpoint, δ_s. Then we compute δ_g from the formula,

$\delta_g = t_R - (x/V_2) - \delta_s.$

We are now able to compute OR, OQ and then locate C by drawing RC perpendicular to RQ. From C we draw an arc of the circle to represent the refracting surface in the vicinity of Q. If the dip is not zero, the point of emergence is Q', the arc QQ' increasing with the dip. Even for moderate dip the elliptical arc QQ' will be close to the circular arc through Q and thus the envelope of circular arcs will outline the refractor closely.

Tarrant's method is useful when the dip is moderate or large and the refractor is curved or irregular. The principal limitation is in the determination of V_2.

(d) *Wyrobek's method.* To illustrate Wyrobek's method we assume a series of unreversed profiles as shown in the upper part of fig. 6.13. The various steps in the interpretation are as follows:

(a) the corrected traveltimes are plotted and the intercept times measured,

(b) the total delay time δ is calculated for each geophone position for each shot and the values plotted at the geophone position (if necessary, a value of V_2 is assumed); by moving the various segments up or down a composite curve similar to a phantom horizon is obtained,

(c) the intercept times divided by 2 are plotted and compared with the composite delay-time curve; divergence between the two curves indicates an incorrect value of V_2 (see below), hence the value used in step (b) is varied until the two curves are 'parallel', after which the half-intercept time curve is completed by interpolation and extrapolation to cover the same range as the composite delay-time curve,

(d) the half-intercept time curve is changed to a depth curve by using (3.32) namely

$h = \frac{1}{2} V_1 t_1/\cos \Theta$

(note that we are ignoring the difference between the vertical depth h and the slant depths h_u, h_d in (3.39) and (3.40)).

Wyrobek's method depends upon the fact that the curve of δ is approximately parallel to the half-intercept time curve. For proof of this result, the reader is referred to problem 6.4. Wyrobek's method does not require reversed profiles because the intercept at a shotpoint does

Fig. 6.12. Illustrating Tarrant's method. (a) Relation between the receiver R and the emergent point Q; (b) showing that the locus of Q is an ellipse with R at one focus; (c) geometry of the ellipse through Q.

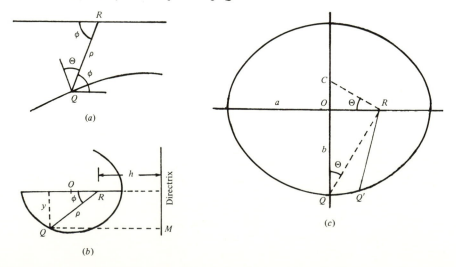

not depend upon the direction in which the cable is laid
out.

Delay-time methods are subject to certain errors
which must be guarded against. As the shotpoint-to-
geophone distance increases, the refraction wavetrain
becomes longer and the energy peak shifts to later cycles.
There is thus the danger that different cycles will be picked
on different profiles and that the error will be interpreted
as an increase in shot delay time. If sufficient data are
available the error is usually obvious. Variations in refractor velocity manifest themselves in local divergences of the
offset total-delay-time curves for pairs of reversed profiles.
However, if some data which do not represent refraction
travel in the refractor under consideration are accidentally
included, the appearance is apt to be the same as if the
refractor velocity were varying. In situations where several
refractors which have nearly the same velocities are
present, unambiguous interpretation may not be possible.

6.2.3 *Wavefront methods*

(*a*) *Thornburgh's method.* Wavefront reconstruction,
usually by graphical means, forms the basis of several
refraction interpretation techniques. The classic paper is
one by Thornburgh (1930); other important articles are
those by Gardner (1949), Baumgarte (1955), Hales (1958),
Hagedoorn (1959), Rockwell (1967) and Schenck (1967).

Fig.6.13. Illustrating Wyrobek's method using
unreversed profiles. (After Wyrobek, 1956.)

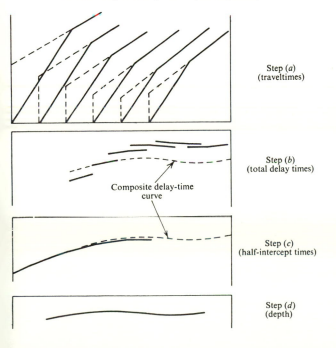

Step (*a*)
(traveltimes)

Step (*b*)
(total delay times)

Composite delay-time
curve

Step (*c*)
(half-intercept times)

Step (*d*)
(depth)

Fig. 6.14 illustrates the basic method of reconstruct-
ing wavefronts. The refraction wavefront which reached
A at $t = 1.600$ s reached $B, C, \ldots$ at the times $1.600 + \Delta t_B$,
$1.600 + \Delta t_C, \ldots$ By drawing arcs with centers $B, C, \ldots$
and radii $V_1 \Delta t_B$, $V_1 \Delta t_C, \ldots$, we can establish the wavefront for $t = 1.600$ s (AZ) as accurately as we wish. Similarly, other refraction wavefronts, such as that shown
for $t = 1.400$ s, can be constructed at any desired traveltime interval. The direct wavefronts from the shot S are
of course the circles shown in the diagram.

In fig. 6.15 we show a series of wavefronts chosen
so that only waves which will be first arrivals are shown
(all secondary arrivals being eliminated in the interests of
simplicity). Between the shotpoint S and the crossover
point C (see (3.34)) the direct wave arrives first. To the
right of C, the wave refracted at the first horizon arrives
first until, to the right of G, the refraction from the deeper
horizon overtakes the shallower refraction.

The two systems of wavefronts representing the
direct wave and the refracted wave from the shallow
horizon intersect along the dashed line ABC; this line,
called the *coincident-time curve* by Thornburgh, passes
through the points where the intersecting wavefronts have
the same traveltimes. The curve $DEFG$ is a coincident-
time curve for the deeper horizon. The coincident-time
curves are tangent to the refractors at A and D where the
incident ray reaches the critical angle (see problem 6.6)
while the points at which the coincident-time curves meet
the surface are marked by abrupt changes in the slopes of
the time–distance plot.

Since the coincident-time curve is tangent to the
refractor, the latter can be found when we have one
profile plus other data, such as the dip, depth, critical
angle – or a second profile (not necessarily reversed) since
we now have two coincident-time curves and the refractor
is the common tangent to the curves.

When reversed profiles are available, the construction of wavefronts provides an elegant method of locating
the refractor. The basic principle is illustrated in fig. 6.16
which shows two wavefronts, MCD and PCE, from shots
at A and B intersecting at an intermediate point C.
Obviously the sum of the two traveltimes from A and B
to C is equal to the reciprocal time between A and B, t_r.
If we had reconstructed the two wavefronts from the time –
distance curve without knowing where the refractor RS
was located, we would draw the wavefronts as MCN and
PCQ, not MCD and PCE. Therefore if we draw pairs of
wavefronts from A and B such that the sum of the traveltimes is t_r, the refractor must pass through the points
of intersection of the appropriate pairs of wavefronts in
fig. 6.16.

Fig.6.14. Reconstruction of wavefronts.

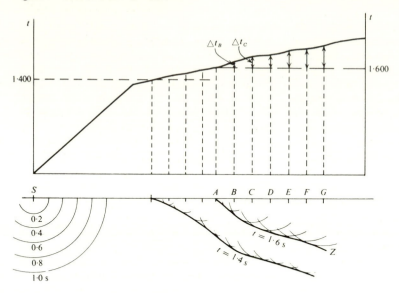

Fig.6.15. Coincident-time curves. (After Thornburgh, 1930.)

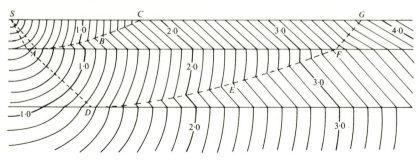

Fig.6.16. Determining refractor position from wavefront intersections.

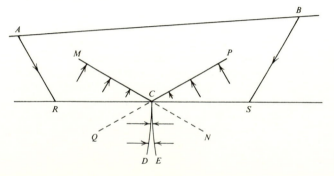

(b) *Hagedoorn's plus-minus method.* Hagedoorn's *plus–minus* method (1959) utilizes a construction similar to that just described. When the refractor is horizontal, the intersecting wavefronts drawn at intervals of Δ milliseconds form diamond-shaped figures (fig. 6.17) whose horizontal and vertical diagonals are equal to $V_2\Delta$ and $V_1\Delta/\cos\theta$ respectively. If we add together the two traveltimes at each intersection and subtract t_r, the resulting 'plus' values equal 0 on the refractor, $+2\Delta$ on the horizontal line through the first set of intersections vertically above those defining the refractor, $+4\Delta$ on the next line up, and so on. Since the distance between each pair of adjacent lines is $V_1\Delta/\cos\theta$, we can use any of the 'plus' lines to plot the refractor. The difference between two traveltimes at an intersection is called the 'minus' value; it is constant along vertical lines passing through the intersections of wavefronts. The distance between successive 'minus' lines as shown in fig. 6.17 is $V_2\Delta$, hence a continuous check on V_2 is possible. Although dip alters the above relations, the changes are small for moderate dip, and the assumption is made that the 'plus' lines are still parallel to the refractor and that the 'minus' lines do not converge or diverge.

(c) *Hales' graphical method.* Graphical methods are well suited to many refraction interpretation problems. When carried out carefully, they often give the requisite accuracy rapidly and they are satisfying to make because the picture unfolds as one carries out the interpretation.

Hales' method (Hales, 1958) is useful where the depth to the refractor varies appreciably, a situation often associated with variation of overburden and refractor velocities. The method requires reversed profiles. The essence of the method is the scheme for locating pairs of points such as A and B (fig. 6.18a) which have a common point of emergence Q, when the dip and depth of the refractor are not initially known. The interpretation procedure will be described first and then the propositions will be proven.

Given reversed refraction profiles as shown in fig. 6.18b, we select an arbitrary point B at which the arrival time is t_{RB}. The point K is located such that $KB = (t_r - t_{RB})$. A line through K at the angle $\alpha = \tan^{-1}(V_1\sin\Theta)$ intersects the reversed profile at time t_{SA} at location A, which is the point on the reversed profile associated with the same point on the refractor (Q on fig. 6.18a) as B. The time t' (fig. 6.18b) and the distance x' can now be read from the reversed profile plot. A line is drawn through A at the critical angle Θ (fig. 6.18c) which intersects the perpendicular bisector of AB at C. An arc is then drawn of radius $\rho = V_1 t'/(2\cos\Theta)$. The refractor is the common tangent to arcs drawn in this way. The angle α given above is not precisely the correct angle α', but the error is negligible, as will be shown.

To establish the soundness of this method, consider the geometry of the triangle AQB (fig. 6.18d) where Q is the refracting point. The refracted waves from R to B and from S to A (fig. 6.18a) leave the refractor at Q. The circle which passes through A, Q, and B is drawn and the values of the several angles can be determined in terms of the critical angle Θ and the dip ξ. The distance $CQ = \rho$

Fig.6.17. Illustrating the plus–minus method.

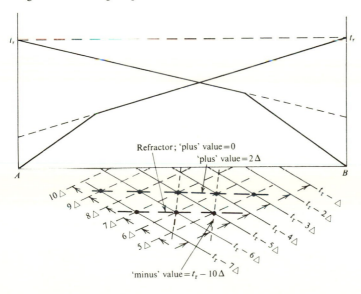

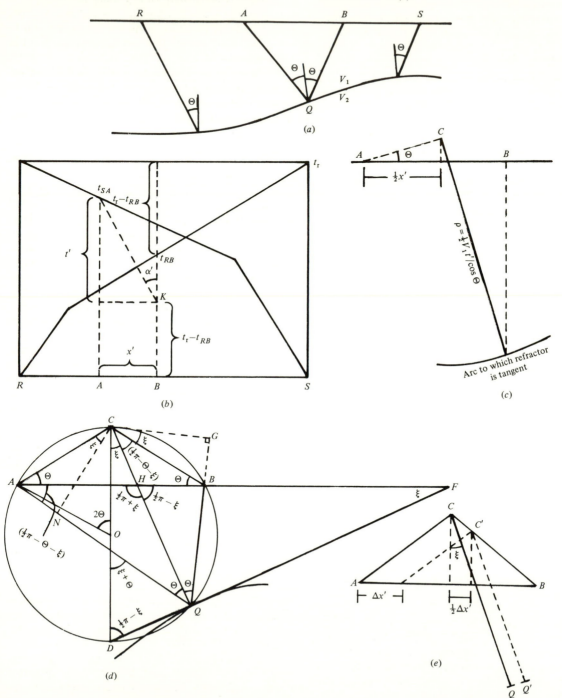

Fig.6.18. Hales' graphical method. (*a*) Relation between two receivers A and B having a common emergent point Q; (*b*) geometrical properties of points on the traveltime curves corresponding to A and B (construction lines are dashed); (*c*) geometrical construction for locating Q; (*d*) geometrical properties of circumscribed circle through A, B and Q; (*e*) effect of errors in x' in (*b*).

can be found by noting that

$$\rho \cos \Theta = QN = AQ - AN$$
$$= QG = BQ + BG.$$

However $AN = CN \tan \xi = CG \tan \xi = BG$; hence adding the two expressions for $\rho \cos \Theta$ gives

$$\rho = (AQ + BQ)/(2 \cos \Theta).$$

From fig. 6.18a we see that

$$t_{RB} + t_{SA} = t_r + (AQ + BQ)/V_1 \,;$$

hence

$$AQ + BQ = V_1 t', \quad \rho = V_1 t'/(2 \cos \Theta).$$

Further,

$$AB = x' = AH + HB$$
$$= \frac{AQ \sin \Theta}{\sin (\tfrac{1}{2}\pi + \xi)} + \frac{BQ \sin \Theta}{\sin (\tfrac{1}{2}\pi - \xi)}$$
$$= (AQ + BQ) \frac{\sin \Theta}{\cos \xi}$$
$$= V_1 t' \sin \Theta / \cos \xi.$$

The angles $\alpha = \tan^{-1}(V_1 \sin \Theta)$ and $\alpha' = \tan^{-1}(x'/t')$ are equal if $\xi = 0$. If $\xi \neq 0$, $\alpha' > \alpha$ so A will be located too close to B by the amount $\Delta x'$, t_{SA} and t' will be slightly too small by the amount $\Delta t'$, and ρ will be too small by $\Delta \rho$. Referring to fig. 6.18e,

$$\Delta t'/\Delta x' = \text{slope of traveltime curve} = \sin (\Theta + \xi)/V_1$$

(for the downdip traveltime curve)

$$\Delta \rho = \frac{V_1 \Delta t'}{2 \cos \Theta} = \frac{\Delta x' \sin (\Theta + \xi)}{2 \cos \Theta}.$$

The point C from which ρ is measured also moves to C' (fig. 6.18e):

$$CC' = \Delta x'/(2 \cos \Theta)$$
$$CQ - C'Q' = CC' \cos (\tfrac{1}{2}\pi - \Theta - \xi)$$
$$= \Delta x' \sin (\Theta + \xi)/(2 \cos \Theta),$$

which is exactly equal to $\Delta \rho$. Hence the only effect of neglecting dip is to displace the refracting point updip by the amount $\tfrac{1}{2}\Delta x'$.

Hales' method requires knowledge of V_1 and V_2 in order to calculate α. Variation of V_2 can be accommodated by calculating V_2 from the slopes of the respective travel-time curves at B and at A (an approximation of the location of A will usually suffice). Variations of V_1 with depth

(usually an increase with increasing depth) can be accommodated by iterating the calculation.

6.3   Interpretation of refraction records

Much of what is termed refraction interpretation, especially the application of equations such as (3.30) to (3.51), should properly be termed 'computation'. The geological interpretation of refraction data, to distinguish it from computation, is much cruder than that of reflection data and usually much more restricted in range of depths involved, detail and precision. Under favorable circumstances refraction data can yield both structural and stratigraphic data but usually only structural information is obtained.

In virgin areas refraction shooting is often done with the twin objectives of determining roughly (1) the shape of the basin, including depth to basement, (2) the nature or rock type of the major lithological units based on their velocities. Velocities in the range 2–3 km/s generally denote sands and shales while velocities of 5–6 km/s usually denote limestone, dolomite or anhydrite. Crystalline basement refractions often have a characteristic envelope and are very strong. Velocities of the various rock types overlap (as shown in fig. 7.1), hence there is generally uncertainty based solely on refraction data. When outcrops and/or well information are available, the interpretation may be more reliable.

The identification of refraction events is usually simpler than reflection events. Traveltimes are usually available for a relatively long range of offsets and hence it is easy to separate reflections and diffractions with their curved alignments from the direct wave, surface waves and refractions with their straight alignments. The direct wave and surface waves are easily distinguished from refractions because of the lower velocities of the former. Usually the only problem is in identifying the different refraction events when several refractors are present.

Record sections, while not as widely used as for reflection interpretation, are very useful, especially in studying second arrivals. The refraction profile in fig. 6.19 shows the direct wave as the first arrival near the shotpoint; refractions from successively deeper refractors become the first arrivals as the offset distance increases. Following the first arrivals the continuations of various events are seen after each has been overtaken by a deeper event. Numerous other events are also seen in the zone of second arrivals; most of these are refractions which never become first arrivals or multiply-reflected refractions (see fig. 4.16).

The simple equations (3.30) through (3.47) can be used when the data are easy to interpret and limited in quantity. Often the chief failure of these equations (and of most refraction interpretation techniques) is in the assumption of V_1, the velocity of the section above the refractor. Most methods assume straight-line raypaths from the refractor upward to the surface. This is usually not true because the overburden velocity is rarely constant. The biggest improvement in the results obtained when using the simple equations to calculate refractor depths often is the result of using a more realistic assumption for V_1 based on information other than that obtainable

from the refraction data themselves (Laski, 1973). Structural interpretation is usually simple provided the data permit accurate matching of up-and-down dip velocities and velocity complications are not present. Faulting is sometimes indicated more clearly on refraction records and the displacement found more accurately than is the case with reflection data. However, the usual paucity of refractors hardly ever permits us to find the variation of displacement with depth, curvature of the fault plane, etc.,

Fig.6.19. Refraction record section. (Courtesy Prakla-Seismos.) (*a*) Record section; (*b*) geologic section.

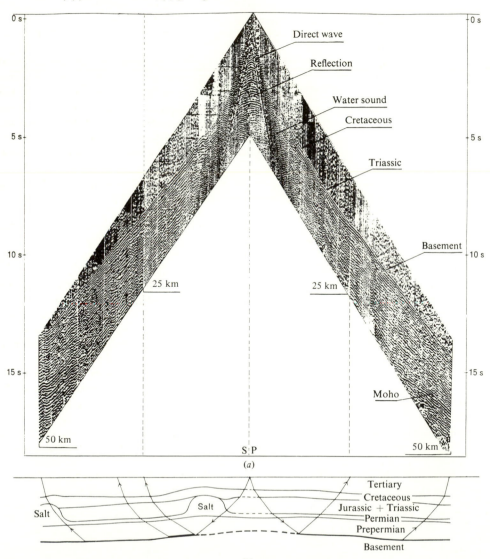

information which under favorable circumstances can be found from reflection sections.

Problems sometimes result from a *hidden zone*, a layer whose velocity is lower than that of the overlying bed so that it never carries a head wave. Energy which would approach it at the critical angle cannot get through the shallower refractors and hence there is no indication of its presence in the refraction arrivals. The low velocity of the hidden layer, however, increases the arrival times of deeper refractors relative to what would be observed if the hidden zone had the same velocity as the overlying bed, hence results in exaggeration of their depths. Another situation, which is also referred to at times as a 'hidden zone', is that of a layer whose velocity is higher than those of the overlying beds but which never produces first arrivals despite this because the layer is too thin and/or its velocity is not sufficiently greater than those of the overlying beds. Such a bed creates a second arrival but the second arrival may not be recognized as a distinct event.

Refraction interpretation often is based solely on first arrivals, primarily because this permits accurate determination of the traveltimes. When we use second arrivals we usually have to pick a later cycle in the wavetrain and estimate traveltime from the measured time. However, velocities based on second arrivals will be accurate and much useful information is available through their study.

Refraction interpretation often involves 'stripping' which is in effect the removal of one layer at a time (Slotnick, 1950). In this method the problem is solved for the first refractor, after which the portions of the time–distance curve for the deeper refractors are adjusted to give the result which would have been obtained if the shotpoint and geophones had been located on the first refracting horizon. The adjustment consists of subtracting the traveltimes along the slant paths from shotpoint down to the refractor and up from the refractor to the geophones, also of decreasing the offsets by the components of the slant paths parallel to the refractor. The new time–distance curve is now solved for the second refracting layer after which this layer can be stripped off and the process continued for deeper refractors.

Several additional considerations affect refraction interpretation. Refraction arrival times are corrected to datum in the same manner as reflection arrival times. While the effect of such corrections on the effective shot-to-geophone distance is usually small for reflection data, this is often not so for refraction travel paths above the refractor, since these may have appreciable horizontal components. Hence the reference datum should be near the surface to minimize such errors.

If enough data are available, interpretational ambi-

guities often can be resolved. However, in an effort to keep survey costs down, only the minimum amount of data may be obtained (or less than the minimum) and some of the checks which increase certainty and remove ambiguities may not be possible.

6.4   Engineering applications

Refraction methods are commonly applied in mineral-exploration and civil-engineering work to measure depth of bedrock and 'rippability'. Refraction velocities in the weathered rock and bedrock underneath soil over-burden indicate rippable rock if they are lower than 2100 to 2400 m/s. The absence of a high-velocity refraction usually indicates that bedrock will not be encountered shallower than about one-third the length of the refraction profile.

While various refraction methods are used with engineering applications (Mooney, 1977), the *ABC method* is the simplest. With the arrangement shown in fig. 6.20 shots are fired from the end points of the spread, *A* and *B*, and the midpoint, *C*. (The 'shot' is usually a hammer blow for shallow overburden or a blasting cap for deeper.) Let t_{AB} be the surface-to-surface traveltime from *A* to *B*, etc.; then (see problem 6.10)

$$h_C = \tfrac{1}{2}(t_{CA} + t_{CB} - t_{AB})\{V_1 V_2/(V_2^2 - V_1^2)^{\frac{1}{2}}\}, \quad (6.8)$$

where V_1 is the overburden and V_2 the bedrock velocity. Frequently $V_2 \gg V_1$ and we can replace the velocity terms by V_1,

$$h_C \approx \tfrac{1}{2}V_1(t_{CA} + t_{CB} - t_{AB}), \quad (6.9)$$

the error in h_C being less than 6% if $V_2 > 3V_1$. This method assumes that the overburden is essentially homogeneous, the depth variation is smooth, the velocity contrast is large and the dip small. Depth calculations by this technique are generally good because they depend on the measurement of only one velocity, V_1, and three traveltimes. While refractor dip can be determined from differences in apparent velocity as seen on reversed profiles, it is more often determined from a series of measurements of depth at different locations of *C*.

Scott *et al.* (1968) describe the use of seismic measurements in the Straight Creek Highway tunnel bore to

Fig.6.20. Refraction profile for determining depth to bedrock.

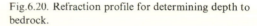

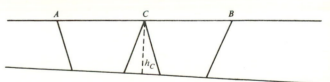

locate fracture zones, the height of the tension arch above the tunnel, the stable rock load and potential weakness factors, as an aid in designing tunnel lining and supports required. He also found a nearly linear correlation between seismic velocity and the rate and cost of construction of the tunnel.

Problems

6.1. Early refraction work searching for saltdomes in the Gulf Coast considered a significant 'lead' to be of the order of 0.250 s. Assuming a range of $3\frac{1}{2}$ miles (5.63 km), a normal sediment velocity at saltdome depth of 2.74 km/s and salt velocity of 4.57 km/s, how much salt travel would this indicate?

6.2. Shotpoint *B* is 2 km east of shotpoint *A*. The following data were obtained with cables extending eastward from *A* and *B* with geophones at 200 m intervals. Interpret the data using Barry's method. Note that *x* is the distance measured from *A*. Take $V_1 = 2.5$ km/s and assume that the delay-time curve for the reversed profiles is sufficiently parallel to yours that step (*d*) can be omitted.

x	t_A	t_B
2.6 km	1.02 s	0.25 s
2.8	1.05	0.34
3.0	1.10	0.43
3.2	1.14	0.52
3.4	1.18	0.61
3.6	1.20	0.70
3.8	1.26	0.78
4.0	1.32	0.87
4.2	1.35	0.96
4.4	1.39	1.05
4.6	1.45	1.10
4.8	1.50	1.14
5.0	1.56	1.20
5.2	1.59	1.22
5.4	1.62	1.28
5.6	1.66	1.31
5.8	1.72	1.36
6.0	1.73	1.42
6.2	1.80	1.47
6.4	1.85	1.53
6.6	1.91	1.56
6.8	1.97	1.59
7.0	2.00	1.63
7.2	2.02	1.67
7.4	2.05	1.70
7.6	2.10	1.73
7.8	2.13	1.78
8.0	2.16	1.81

6.3. Show that PP' in fig. 6.10 is given by

$$PP' = V_2 \delta_{PR} \tan^2 \Theta.$$

6.4. Prove that the half-intercept curve referred to in the discussion of Wyrobek's method in §6.2.2*d* is parallel to the curve of the total delay time δ (see fig. 6.21). Note that the reciprocal time can be written (see (3.41) and (3.42))

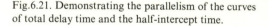

Fig.6.21. Demonstrating the parallelism of the curves of total delay time and the half-intercept time.

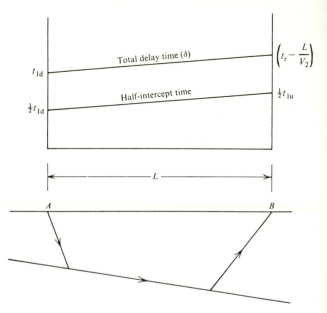

Fig.6.22. Deriving the properties of the coincident-time curve.

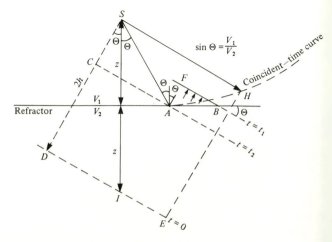

$$t_r = \tfrac{1}{2} \left\{ \left(\frac{L}{V_d} + t_{1d} \right) + \left(\frac{L}{V_u} + t_{1u} \right) \right\}.$$

6.5. Shotpoints *C, D, E, F* and *G* in fig. 6.3 are 5 km apart. The data tabulated below are for three profiles *CE, DF* and *EG* with shotpoints at *C, D,* and *E,* no data being recorded for offsets less than 3 km. For profiles shot from *F* and *G* the intercepts were 1.52 s and 1.60 s respectively. Use Wyrobek's method to interpret the data.

x	t_{CE}	t_{DF}	t_{EG}
3.00 km	1.18 s	1.20 s	1.19 s
3.20	1.22	1.29	1.28
3.40	1.24	1.38	1.35
3.60	1.28	1.45	1.43
3.80	1.35	1.54	1.50
4.00	1.38	1.60	1.58
4.20	1.41	1.70	1.68
4.40	1.47	1.74	1.76
4.60	1.51	1.77	1.82
4.80	1.53	1.80	1.89
5.00	1.58	1.82	2.00
5.20	1.63	1.85	2.06
5.40	1.65	1.91	2.15
5.60	1.69	1.95	2.21
5.80	1.74	1.97	2.29
6.00	1.78	1.99	2.38
6.20	1.82	2.03	2.43
6.40	1.87	2.08	2.46
6.60	1.90	2.12	2.49
6.80	1.94	2.16	2.54
7.00	1.97	2.20	2.57
7.20	2.01	2.25	2.60
7.40	2.06	2.30	2.65
7.60	2.10	2.33	2.68
7.80	2.14	2.37	2.71
8.00	2.17	2.41	2.74
8.20	2.20	2.45	2.77
8.40	2.24	2.47	2.82
8.60	2.30	2.52	2.85
8.80	2.32	2.55	2.89
9.00	2.35	2.61	2.93
9.20	2.38	2.64	2.97
9.40	2.44	2.68	3.00
9.60	2.47	2.73	3.04
9.80	2.50	2.78	3.07
10.00	2.54	2.82	3.10

6.6. Using fig. 6.22, show that: (*a*) *DE*, the 'wavefront for $t = 0$', is at a depth $SD = 2h = 2z \cos \Theta$; (*b*) after *DE* reaches *A*, wavefronts such as *BF* coincide with the head-wave wavefronts; (*c*) the coincident-time curve *AH* is a parabola; (*d*) taking *DE* and *DS* as the *x*- and *y*-axes,

the equation of *AH* is $4hy = x^2 + 4h^2$; (*e*) the coincident-time curve is tangent to the refractor at *A*.

6.7. Interpret the following data using the plus–minus method.

x	t_A	t_B
0.0 km	0.00 s	2.30 s
0.4	0.15	2.23
0.8	0.28	2.15
1.2	0.44	2.09
1.6	0.52	2.04
2.0	0.63	1.98
2.4	0.70	1.92
2.8	0.76	1.85
3.2	0.84	1.80
3.6	0.91	1.72
4.0	0.95	1.64
4.4	1.04	1.60
4.8	1.12	1.55
5.2	1.16	1.47
5.6	1.25	1.40
6.0	1.30	1.32
6.4	1.33	1.28
6.8	1.40	1.24
7.2	1.51	1.18
7.6	1.57	1.10
8.0	1.60	1.04
8.4	1.72	0.96
8.8	1.78	0.90
9.2	1.80	0.83
9.6	1.91	0.76
10.0	1.93	0.66
10.4	2.04	0.52
10.8	2.07	0.39
11.2	2.17	0.25
11.6	2.20	0.12
12.0	2.30	0.00

6.8. The data in the table below show refraction travel-times for geophones spaced 400 m apart between shot-points *A* and *B*, which are separated by 12 km. The columns in the table headed t_A^*, t_B^* give second arrivals. Interpret the data using: (*a*) equations (3.39)–(3.43); (*b*) Tarrant's method; (*c*) the wavefront method illustrated in fig. 6.16; (*d*) Hales' method. On the basis of your results, compare the methods in terms of: (1) time involved; (2) effect of refractor curvature; (3) effect of random errors; and (4) suitability for: (i) routine production or (ii) special effort where high accuracy is essential.

x	t_A	t_B	t_A^*
0.00 km	0.000 s	3.310 s	
0.40	0.182	3.182	
0.80	0.320	3.140	
1.20	0.504	3.063	
1.60	0.680	2.917	
2.00	0.862	2.839	
2.40	0.997	2.714	
2.80	1.170	2.681	1.682 s
3.20	1.342	2.570	1.760
3.60	1.495	2.505	1.858
4.00	1.677	2.442	1.881
4.40	1.821	2.380	1.962
4.80	1.942	2.318	2.053
5.20	2.103	2.220	
5.60	2.150	2.125	
6.00	2.208	2.030	
6.40	2.330	2.003	
6.80	2.422	1.862	
7.20	2.504	1.743	
7.60	2.602	1.622	t_B^*
8.00	2.658	1.610	
8.40	2.720	1.482	1.561
8.80	2.744	1.329	1.440
9.20	2.760	1.140	1.288

x	t_A	t_B	t_B^*
9.60 km	2.855 s	1.018 s	1.202 s
10.00	2.920	0.863	1.177
10.40	2.980	0.660	1.082
10.80	3.065	0.503	
11.20	3.168	0.340	
11.60	3.230	0.198	
12.00	3.310	0.000	

6.9. (*a*) Solve problem 3.17 by stripping off (§6.3) the shallow layer (use the same velocities as in problem 3.17 for the purpose of comparison); (*b*) compare your results with those in problem 3.17; (*c*) what are some of the advantages and disadvantages of stripping?

6.10. Prove (6.8) assuming that the surface is horizontal and the refractor is plane.

6.11. Interpret the data of problem 3.12 using the *ABC* method (see (6.8)). Compare your results with those of problem 3.12. What is an obvious disadvantage of the method based on these data?

6.12. Construct the expected time–distance curve for the Java Sea velocity–depth relation shown in fig. 6.23. Is it feasible to map the top of the 4.25 km/s limestone at a

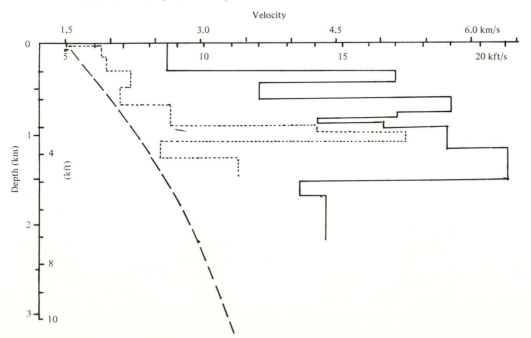

Fig.6.23. Velocity–depth relationship for wells in Illinois Basin (solid curve), Java Sea (dotted curve), and Louisiana Gulf Coast (dashed curve).

depth of about 0.9 km by the use of head waves? What problems are likely to be encountered?

6.13. In refraction mapping of the 5.75 km/s layer at about 0.6 km depth in the Illinois Basin, the overlying shale forms a 'hidden layer.' Using the velocity–depth data from fig. 6.23, determine approximately how much error neglect of the hidden layer will involve.

6.14. The velocity of salt is nearly constant at 4.57 km/s. Calculate the amount of lead time per kilometer of salt diameter as a function of depth assuming the sediments have the Louisiana Gulf Coast velocity distribution shown in fig. 6.23.

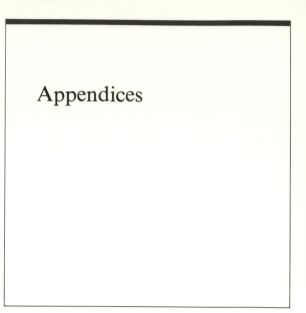

Appendices

A List of abbreviations used

AAPG American Association of Petroleum Geologists
A/D analog-to-digital
AIMME American Institute of Mining and Metallurgical Engineers
AGC automatic gain control
AGI American Geological Institute
API American Petroleum Institute
CDP common-depth-point (method)
CGG Compagnie Générale de Géophysique
D/A digital-to-analog
EAEG European Association of Exploration Geophysicists
GPS Global positioning system
GRC Geophysical Research Corporation
GSA Geological Society of America
GSI Geophysical Service Inc.
IEEE Institute of Electrical and Electronics Engineers
IFP Institute Français du Pétrole
LVL low-velocity layer, also called weathering layer
NMO normal moveout
OPEC Organisation of Petroleum Exporting Countries
OTC Offshore Technology Conference
RDU remote data unit
rms root mean square
SEG Society of Exploration Geophysicists
SEI Seismic Explorations Inc.
SGRM Société Géophysique de Recherches Minières
SI Système International (units)
SIE Southwestern Industrial Electronic Company
SSC Seismograph Service Corporation
TI Texas Instruments
3-D three-dimensional

B   Trademarks and proper names used

Name Whose tradename
ANA *Prakla GMBH*
Aquapulse *Western Geophysical Co. of America*
Aquaseis *Imperial Chemical Industries Ltd*
Argo *Cubic Western Data*
Autotape *Cubic Western Data*
Boomer *EG & G International*
Decca Mainchain *Decca Survey Ltd*
Decca Navigator *Decca Survey Ltd*
Deltapulse *Seiscom Delta*
Dinoseis *ARCO Oil and Gas Co.*
Flexichoc *Institute Français du Pétrole*
Flexotir *Institute Français du Pétrole*
Gassp *Shell Development*
Hi-fix *Decca Survey Ltd*
Hydrodist *Tellurometer*
Hydrosein *Western Geophysical Co. of America*
Lorac *Seismograph Service Corp.*
Marthor *Institute Français du Pétrole*
Maxipulse *Western Geophysical Co. of America*
Miniranger *Motorola Inc.*
Nitramon *E. I. Du Pont de Nemours Co.*
Opseis *Applied Automation Inc.*
Primacord *Ensign Bickford Co.*
Pulse-8 *Decca Survey Ltd*
Raydist *Hastings-Raydist*
RPS *Motorola Inc.*
Seiscrop *Geophysical Service Inc.*
Seisloop *Geophysical Service Inc.*
Sosie *Société Nationale Elf-Aquitaine*
Syslap *Compagnie Générale de Géophysique*
Toran *Sercel S.A.*
Trisponder *Motorola*
Vaporchoc *Compagnie Générale de Géophysique*
Vibroseis *Conoco Inc.*
Wassp *Teledyne Exploration*

Systems developed by governments:
Loran
Navstar
Omega
radar
shoran
sonar
Transit

C   Random numbers

20897 13007 95217 09221 15433 94882 23741 86571 20737 19305 71148
04035 01380 79508 12771 34806 60605 97685 26147 51379 39533 04983
25469 86469 31522 59282 16856 38655 31862 84283 08694 06945 42094
17446 27775 99466 63704 60957 55029 92764 54774 15832 04324 73597
42328 74303 58231 85798 89730 34685 57000 43798 63721 12003 18538
62439 12049 96266 31886 07814

To obtain a random sequence restricted to a given range, choose a rule
and begin applying it at an arbitrary location. For example, to get values
lying between ± 8, we read numbers in pairs and use the first to give the
sign (perhaps making even numbers positive and odd ones negative) and
simply omit any 9 s we may come to. If we wish a different sequence, we
begin at a different place and perhaps omit every other number or every
third number.

D   Units
SI (Système International) units

Prefixes   :

Dimensions	Symbol	Name	Example of use
10^{18}	E	exa	
10^{15}	P	peta	
10^{12}	T	tera	
10^9	G	giga	gigahertz
10^6	M	mega	megawatt
10^3	k	kilo	kilometer
1			
10^{-3}	m	milli	millimeter
10^{-6}	μ	micro	microwatt
10^{-9}	n	nano	nanosecond
10^{-12}	p	pico	picosecond
10^{-15}	f	femto	
10^{-18}	a	atto	

Base units

Dimensions	Symbol	Unit	Equivalencies
Length	m	meter	3.281 feet, (1/0.3048) feet, 39.37 inches, 10^{10} ångström, 0.0006214 statute miles, (1/1609) statute miles, (1/1853.2) nautical miles
Mass	kg	kilogram	2.205 pounds, (1/0.4536) pounds, 0.001102 short tons
Time	s	second	
Current	A	ampere	
Temperature	K	kelvin	293.15 K = 0°C
Intensity	cd	candela	
Plane angle	rad	radian	(57.30°), (1/0.01745) degrees
Solid angle	sr	steradian	

Derived units

Area	m^2	square meter	0.0001 hectares, 0.0002471 acres, 0.3861×10^{-6} square miles
Volume	m^3	cubic meter	0.001 liter, 264.17 US gallons, 6.2898 barrels, 0.0008107 acre feet, 219.97 UK gallons
Density	kg/m^3		0.001 g/cm^3, 0.06243 pounds (mass)/cubic feet
Force	N	newton	kg m/s^2, 0.2248 pounds, 10^5 dynes
Pressure	Pa	pascal	N/m^2, 10^{-5} bars, 0.1450×10^{-3} pounds/square inch, 9.869×10^{-6} atmospheres
Energy, work	J	joule	N m, (1/1055) BTU, (1/4186) kilocalories, 10^7 ergs, 0.73756 foot pounds
Power	W	watt	J/s, 0.001341 horsepower, 3.412 BTU/hour
Frequency	Hz	hertz	cycle/s
Velocity	m/s		1.942 nautical miles/hour, 2.237 miles/hour
Acceleration	m/s^2		10^5 milligals
Charge	C	coulomb	A s

Potential	V	volt	W/A
Resistance	Ω	ohm	V/A
Capacitance	F	farad	A s/V
Magnetic flux	Wb	weber	V s, 10^8 maxwell
Magnetic field strength	T	tesla	Wb/m^2, N/A m, 10^4 gauss, 10^9 gamma
Inductance	H	henry	Wb/A, V s/A

Non-SI units (abbreviations)

ångström Å
foot ft
inch in
milligal mGal
nautical mile/hour knot
pound lb
pound/square inch psi
statute mile st. mile

E Decibel conversion

dB	*amplitude ratio*	*energy ratio*
−120	10^{-6}	10^{-12}
−80	10^{-4}	10^{-8}
−40	0.01	10^{-4}
−20	0.1	0.01
−10	0.316	0.1
−6	0.501	0.251
−3	0.708	0.501
0	1	1
3	1.413	1.997
6	1.995	3.980
10	3.162	10
20	10	100
80	10^4	10^8

F Typical instrument specifications and conventions

(a) Geophones

Geophone transduction constant: 0.25 V/cm/s
Geophone natural frequency tolerance: ±0.5 Hz
Geophone dynamic range: 140 dB
Geophone distortion: < 0.2% at 2 cm/s at 12 Hz
Hydrophone sensitivity:
  6 V/bar = 60 μV/Pa
  10 V/cm/s at 100 Hz, 1 V/cm/s at 10 Hz
Streamer noise: < 15 μV
Ground unrest: 10^{-4}–10^{-6} cm/s

(b) Recording instruments

Frequency response: 3–256 Hz
Time accuracy: 0.005%
Recording range: 90 dB
Dynamic range referred to noise: 115 dB
Linearity: ±0.02%
Distortion: 0.05% (3–256 Hz)
System noise: < 0.2 μV
Crossfeed isolation: 80 dB
Compression/expansion speed: 84 dB/s

(c) Recording conventions (Thigpen *et al.*, 1975)

Channel 1 is towards the north or east; if line is crooked, overall average direction determines (channel sense should not be changed along a line).
Upward kick on a geophone yields a negative number and downswing on monitor record.
Pressure increase on hydrophone yields negative number and downswing.
Recorded Vibroseis sweep leads baseplate velocity by 90°.

References

The numbers in square brackets are the section numbers in which references are cited.

Abbot, H. L. (1878) On the velocity of transmission of earth waves: *Amer. J. Sci. Arts*, Ser. 3, **15**, 178–84. [1.2.1]

Adachi, R. (1954) On a proof of fundamental formula concerning refraction method of geophysical prospecting and some remarks: *Kumamoto J. Sci.*, Ser. A, **2**, 18–23. [3.3.4]

Agnich, F. J. and Dunlap, Jr, R. C. (1959) Standards of performance in petroleum exploration: *Geophysics*, **24**, 916–24. [5.2.1]

Agocs, W. B. (1950) Computation charts for linear increase of velocity with depth: *Geophysics*, **15**, 227–36. [3.2.5]

Aki, K. and Richards, P. G. (1980) *Quantitative Seismology, Theory and Methods*, Vols. I and II: San Francisco, W. H. Freeman. [1.4]

Allen, F. T. (1972) Some characteristics of marine sparker seismic data: *Geophysics*, **37**, 462–70. [5.5.7]

Angona, F. A. (1960) Two-dimensional modeling and its application to seismic problems: *Geophysics*, **25**, 468–82. [4.2.1]

Anstey, N. A. (1970) Signal characteristics and instrument specifications; Vol. 1 of *Seismic Prospecting Instruments*: Berlin, Gebrüder Borntraeger. [1.4, 5.3.5]

Anstey, N. A. (1977) *Seismic Interpretation – the Physical Aspects*: Boston, International Human Resources Development Corp. [1.4]

Attewell, P. B. and Ramana, Y. V. (1966) Wave attenuation and internal friction as functions of frequency in rocks: *Geophysics*, **31**, 1049–56. [2.3.2c]

Barbier, M. G. and Viallix, J. R. (1973) Sosie – A new tool for marine seismology: *Geophysics*, **38**, 673–83. [5.4.9]

Barry, K. M. (1967) Delay time and its application to refraction profile interpretation; in *Seismic Refraction Pros-*

pecting, pp. 348–61 (ed. A. W. Musgrave): Tulsa, SEG. [6.2.2*a, b*]

Barry, K. M., Cavers, D. A. and Kneale, C. W. (1975) Recommended standards for digital tape formats: *Geophysics*, **40**, 344–52. [5.4.8]

Barthelmes, A. J. (1946) Application of continuous profiling to refraction shooting: *Geophysics*, **11**, 24–42. [6.2.2*a*]

Barton, D. C. (1929) The seismic method of mapping geologic structure; in *Geophysical Prospecting*, pp. 572–624: New York, AIMME. [1.2.1, 1.2.2, 1.2.4, 1.2.6, problems c.2, c.3]

Bates, R. L., and Jackson, J. A. (1980) *Glossary of Geology*: Falls Church, Va., AGI. [1.1.1]

Båth, M. (1974) *Spectral Analysis in Geophysics*: Amsterdam, Elsevier. [1.4]

Baumgarte, J. von (1955) Konstruktive Darstellung von seismischen Horizonten unter Berücksichtigung der Strahlenbrechung im Raum: *Geophys. Prosp.*, **3**, 126–62. [6.2.3*a*]

Bedenbender, J. W., Johnston, R. C. and Neitzel, E. B. (1970) Electroacoustic characteristics of marine seismic streamers: *Geophysics*, **35**, 1054–72. [5.5.4]

Birch, F. (1966) Compressibility; elastic constants; in *Handbook of Physical Constants*, pp. 97–173 (ed. S. P. Clark, Jr): GSA Memoir 97. [2.1.4]

Blackman, R. B., and Tukey, J. W. (1958) *The Measurement of Power Spectra*: New York, Dover. [1.4]

Blake, F. C. (1952) Spherical wave propagation in solid media: *J. Acoust. Soc. Am.*, **24**, 211–15. [2.2.9]

Borges, E. (1969) Ein neues seismisches Verfahren san orten von Verwurfen und Auswaschungen in Floz: *Gluckauf Forschft.*, **4**, 201–8. [5.3.9*d*]

Born, W. T. (1960) A review of geophysical instrumentation: *Geophysics*, **25**, 77–91. [1.2.1]

Bortfeld, R. (1962*a*) Exact solution of the reflection and refraction of arbitrary spherical compressional waves at liquid–liquid interfaces and at solid–solid interfaces with equal shear velocities and equal densities: *Geophys. Prosp.*, **10**, 35–67. [2.4.7]

Bortfeld, R. (1962*b*) Reflection and refraction of spherical compressional waves at arbitrary plane interfaces: *Geophys. Prosp.*, **10**, 517–38. [2.4.7]

Braddick, H. J. J. (1965) *Vibrations, Waves and Diffractions*: New York, McGraw-Hill. [2.3.1]

Bradley, J. J. and Fort, A. N. (1966) Internal friction in rocks; in *Handbook of Physical Constants*, pp. 175–93 (ed. S. P. Clark, Jr): GSA Memoir 97. [2.3.2*c*]

Brede, E. C., Johnston, R. C., Sullivan. L. B. and Viger, H. L. (1970) A pneumatic seismic energy source for shallow water/marsh areas: *Geophys. Prosp.*, **18**, 581–99. [5.4.3]

Brillouin, L. (1960) *Wave Propagation and Group Velocity*: New York, Academic Press. [2.3.3]

Brown, R. J. S. (1969) Normal-moveout and velocity relations for flat and dipping beds and for long offsets: *Geophysics*, **34**, 180–95. [3.1.1]

Bullen, K. E. (1965) *An Introduction to the Theory of Seismology*, 3rd ed.: London, Cambridge Univ. Press. [2.2.1]

Burg, J. P. (1964) Three-dimensional filtering with an array of seismometers: *Geophysics*, **29**, 693–713. [5.3.3*e*]

Burnett, C. R., Hirschberg, J. G. and Mack, J. E. (1958) Diffraction and interference; in *Handbook of Physics*, Part 6, chapter 5, pp. 6.81–4 (ed. F. U. Condon and H. Odishaw): New York, McGraw-Hill. [4.3.2*b*]

Cagniard, L. (1962) *Reflection and Refraction of Progressive Seismic Waves*: New York, McGraw-Hill. (Translation by E. A. Flynn and C. H. Dix of L. Cagniard (1939) *Réflexion et réfraction des ondes séismiques progressives*: Paris, Gauthier-Villars.) [2.4.7]

Carlton, D. P. (1946) *The History of the Geophysics Department*: Houston, Humble Oil and Refining Co. [1.2.4]

Cassand, J., Damotte, B, Fontanel, A., Grau, G., Hemon, C. and Lavergne, M. (1971) *Seismic Filtering*: Tulsa, SEG. (Translated by N. Rothenburg from *Le Filtrage en Sismique*, 1966: Paris, Editions Technip.) [1.4]

Cheng, D. K. (1959) *Analysis of Linear Systems*: Reading, Mass., Addison-Wesley. [1.4]

Claerbout, J. F. (1976) *Fundamentals of Geophysical Data Processing*: New York, McGraw-Hill. [1.4]

Clay, C. S., and Medwin, H. (1977) *Acoustical Oceanography*: New York, John Wiley. [2.4.8]

Coffeen, J. A. (1978) *Seismic Exploration Fundamentals*: Tulsa, Petroleum Publishing Co. [1.4]

Daly, J. W. (1948) An instrument for plotting reflection data on the assumption of a linear increase of velocity: *Geophysics*, **13**, 153–7. [3.2.5]

DeGolyer, E. (1935) Notes on the early history of applied geophysics in the petroleum industry: *Trans. Soc. Pet. Geophysicists*, **6**, 1–10. (Reprinted in *Early Geophysical Papers of the Society of Exploration Geophysicists* (1947), pp. 245–54: Tulsa, SEG.) [1.2.1]

Denham, L. R. (1981) Extending the resolution of seismic reflection exploration: *J. Canadian Soc. Exp. Geophysicists* (in press). [5.3.8]

Dennison, A. T. (1953) The design of electromagnetic geophones: *Geophys. Prosp.*, **1**, 3–28. [5.4.4*b, d*]

Dix, C. H. (1952) *Seismic Prospecting for Oil*: New York, Harper. [1.4]

Dix, C. H. (1954) The method of Cagniard in seismic pulse problems: *Geophysics*, **19**, 722–38. [2.4.7]

Dix, C. H. (1955) Seismic velocities from surface measurements: *Geophysics*, **20**, 68–86. [3.2.3, problems c.3]

Dobrin, M. B. (1951) Dispersion in seismic surface waves: *Geophysics*, **16**, 63–80. [2.2.10*a, b*]

Dobrin, M. B. (1976) *Introduction to Geophysical Prospecting*, 3rd ed.: New York, McGraw-Hill. [1.4]

Elkins, T. A. (1970) *A Brief History of Gulf's Geophysical Prospecting*: Pittsburgh, Gulf Research and Development Co. [1.2.1]

Ergin, K. (1952) Energy ratios of seismic waves reflected and refracted at a rock–water boundary: *Bull. Seis. Soc. Am.*, **42**, 349–72. [2.4.2]

Eve, A. S. and Keys, D. A. (1928) *Applied Geophysics*: Cambridge, Cambridge Univ. Press. [1.4]

Evenden, B. S. and Stone, D. R. (1971) Instrument performance and testing. Vol. 2 of *Seismic Prospecting Instruments*: Berlin, Gebrüder Borntraeger. [1.4, 5.4.5]

Ewing, W. M., Jardetzky, W. S. and Press, F. (1957) *Elastic Waves in Layered Media*: New York, McGraw-Hill. [2.2.1, 2.2.10c, 2.4.7, 2.4.8]

Farriol, R., Michon, D., Muniz, R. and Staron, P. (1970) Study and comparison of marine seismic source signatures: Paper at SEG 1970 annual meeting. [5.5.3f]

Fitch, A. A. (1976) *Seismic Reflection Interpretation*: Berlin, Gebrüder Borntraeger. [1.4]

Futterman, W. I. (1962) Dispersive body waves: *J. Geophys. Res.*, **67**, 5279–91. [2.3.3]

Gal'perin, E. L. (1974) *Vertical Seismic Profiling* (Translated by A. J. Hermont): Tulsa, SEG. [5.3.9a]

Gardner, L. W. (1939) An areal plan of mapping subsurface structure by refraction shooting: *Geophysics*, **4**, 247–59 [6.2.2a]

Gardner, L. W. (1947) Vertical velocities from reflection shooting: *Geophysics*, **12**, 221–8. [problems c.3]

Gardner, L. W. (1949) Seismograph determination of saltdome boundary using well detector deep on dome flank: *Geophysics*, **14**, 29–38. [6.1.3, 6.2.3a]

Gardner, L. W. (1967) Refraction seismograph profile interpretation; in *Seismic Refraction Prospecting*, pp. 338–47 (ed. A. W. Musgrave): Tulsa, SEG. [6.2.2a]

Giles, B. F. (1968) Pneumatic acoustic energy source: *Geophys. Prosp.*, **16**, 21–53. [5.5.3b]

Godfrey, L. M., Stewart, J. D. and Schweiger, F. (1968) Application of dinoseis in Canada: *Geophysics*, **33**, 65–77. [5.4.3]

Goupillaud, P. L. (1976) Signal design in the 'Vibroseis' technique: *Geophysics*, **41**, 1291–1304. [5.4.3]

Grant, F. S. and West, G. F. (1965) *Interpretation Theory in Applied Geophysics*: New York, McGraw-Hill. [1.4, 2.2.11, 2.4.7, 2.4.8]

Green, C. H. (1979) John Clarence Karcher, 1894–1978, Father of the reflection seismograph: *Geophysics*, **44**, 1018–21. [1.2.1]

Hagedoorn, J. G. (1954) A process of seismic reflection interpretation: *Geophys. Prosp.*, **2**, 85–127. [4.3.2b]

Hagedoorn, J. G. (1959) The plus–minus method of interpreting seismic refraction sections: *Geophys. Prosp.*, **7**, 158–82. [6.2.3a, b]

Halbouty, M. T. (1970) *Geology of Giant Petroleum Fields*: Tulsa, AAPG Memoir 14. [1.3.3]

Hales, F. W. (1958) An accurate graphical method for interpreting seismic refraction lines: *Geophys. Prosp.*, **6**, 285–94. [6.2.3a, c]

Hecker, O. (1900) Ergebnisse de Messung von Bodenbewegungen bei einer Sprengung: *Gerland's Beiträge zur Geophysik*, **4**, 98–104. [1.2.1]

Heiland, C. A. (1929a) Modern instruments and methods of seismic prospecting; in *Geophysical Prospecting*, pp. 625–53: New York, AIMME. [1.2.1]

Heiland, C. A. (1929b) Geophysical methods of prospecting – principles and recent successes: *Quarterly of Colorado School of Mines*, **24**, no. 1. [1.2.1]

Heiland, C. A. (1940) *Geophysical Exploration*: New York, Prentice-Hall. [1.4]

Hilterman, F. J. (1970) Three-dimensional seismic modeling: *Geophysics*, **35**, 1020–37. [4.3.3]

Howell, B. Jr (1959) *Introduction to Geophysics*: New York, McGraw-Hill. [2.2.10a]

Jaeger, J. C. (1958) *Elasticity, fracture and flow*: London, Methuen. [Overview, chapter 2]

Jakosky, J. J. (1950) *Exploration Geophysics*: 2nd ed.: Newport Beach, Calif., Trija Publishing. [1.4]

Jeffreys, H. (1926) On compressional waves in two superposed layers: *Proc. Camb. Phil. Soc.*, **22**, 472–81. [2.4.7]

Jeffreys, H. (1952) *The Earth*, 3rd ed.: Cambridge, Cambridge Univ. Press. [1.4]

Jenkins, F. A. and White, H. F. (1957) *Fundamentals of Optics*: New York, McGraw-Hill. [4.3.2a]

Johnson, S. H. (1976) Interpretation of split-spread refraction data in terms of plane dipping layers: *Geophysics*, **41**, 418–24. [3.3.4]

Johnston, R. R. (1980) North American drilling activity in 1979: *AAPG, Bull.*, **64**, 1295–1330. [1.3.3]

Kanasewich, E. R. (1973) *Time Sequence Analysis in Geophysics*: Edmonton, Univ. of Alberta Press. [1.4]

Karcher, J. C. (1974), The reflection seismograph: its invention and use in the discovery of oil and gas fields: unpublished manuscript. [1.2.2, 1.2.5]

Kennett, P. and Ireson, R. L. (1977) Vertical seismic profiling: recent advances in techniques for data acquisition, processing and interpretation: Paper presented at 47th annual meeting of SEG, Calgary. [5.3.9a]

King, V. L. (1973) Sea bed geology from sparker profiles, Vermillion Block 321, Offshore Louisiana: *1973 Offshore Technology Conference Preprints*, paper 1802; Dallas, OTC. [5.5.7]

Knott, C. G. (1899) Reflexion and refraction of elastic waves, with seismological applications: *Phil. Mag.*, **48**, 64–97. [1.2.1, 2.4.2]

Knudsen, W. C. (1961) Elimination of secondary pressure pulses in offshore exploration: *Geophysics*, **26**, 425–36. [5.5.3c]

Koefoed, O. (1962) Reflection and transmission coefficients for plane longitudinal incident waves: *Geophys. Prosp.*, **10**, 304–51. [2.4.6]

Kramer, F. S., Peterson, R. A. and Walter, W. C., eds. (1968) *Seismic Energy Sources – 1968 Handbook*: Pasadena, Bendix United Geophysical. [1.4, 5.5.2, 5.5.3f]

Krey, T. C. (1963) Channel waves as a tool of applied geophysics in coal mining: *Geophysics*, **28**, 701–14. [5.3.9d]

Kronberger, F. P. and Frye, D. W. (1971) Positioning of marine surveys with an integrated satellite navigation system: *Geophys. Prosp.*, **19**, 487–500. [5.5.5e]

Kulhánek, O. (1976) *Introduction to Digital Filtering in Geophysics*: Amsterdam, Elsevier. [1.4]

Laing, W. E. and Searcy, F. (1975) *Geophysics – the First Fifty Years*: Houston, Conoco. [1.2.1]

Lamb, H. (1960) *Statics*: New York, Cambridge Univ. Press. [2.2.10*d*]

Lamer, A. (1970) Couplage sol-géophone: *Geophys. Prosp.*, **18**, 300–19. [5.3.3*g*]

Laski, J. D. (1973) Computation of the time–distance curve for a dipping refractor and velocity increasing with depth in the overburden: *Geophys. Prosp.*, **21**, 366–78. [6.3]

Laster, S. J. and Linville, A. F. (1968) Preferential excitation of refractive interfaces by use of a source array: *Geophysics*, **33**, 49–64. [5.3.3*g*]

Lavergne, M. (1970) Emission by underwater explosions: *Geophysics*, **35**, 419–35. [5.5.3*c*]

Lee, Y. W. (1960) *Statistical Theory of Communication*: New York, Wiley. [1.4]

Leet, L. D. (1938) *Practical Seismology and Seismic Prospecting*: New York, Appleton-Century. [1.2.1, 1.4]

Love, A. E. H. (1927) *Some Problems of Geodynamics*: London, Cambridge Univ. Press. [1.2.1]

Love, A. E. H. (1944) *A Treatise on the Mathematical Theory of Elasticity*: New York, Dover. [2.1.4]

Malamphy, M. C. (1929) Factors in design of portable field seismographs: *Oil Weekly*, 22 March, 1929. [1.2.2]

Mallet, R. (1848) On the dynamics of earthquakes; being an attempt to reduce their observed phenomena to the known laws of wave motion in solids and fluids: *Trans. Roy. Irish Acad.*, **21**, 50–106. [1.2.1]

Mallet, R. (1851) Second report on the facts of earthquake phenomena: *BAAS*, **21**, 272–320. [1.2.1]

Mayne, W. H. (1962) Common-reflection-point horizontal data-stacking techniques: *Geophysics*, **27**, 927–38. [5.3.1]

Mayne, W. H. (1967) Practical considerations in the use of common reflection point techniques: *Geophysics*, **32**, 225–9 [5.3.1]

Mayne, W. H. and Quay, R. G. (1971) Seismic signatures of large air guns: *Geophysics*, **36**, 1162–73. [5.5.3*b*]

McDonal, F. J., Angona, F. A., Mills, R. L., Sengbush, R. L., Van Nostrand, R. G. and White, J. E. (1958) Attenuation of shear and compressional waves in Pierre Shale: *Geophysics*, **23**, 421–39. [2.3.2*c*]

McGee, J. E. and Palmer, R. L. (1967) Early refraction practices; in *Seismic Refraction Prospecting*, pp. 3–11 (ed. A. W. Musgrave): Tulsa, SEG. [1.2.1]

McKay, A. E. (1954) Review of pattern shooting: *Geophysics*, **19**, 420–37. [5.3.3*g*]

McQuillin, R., Bacon, M. and Barclay, W. (1979) *An Introduction to Seismic Interpretation*: Houston, Gulf Publishing. Co. [1.4, 5.5.3*f*]

Meidav, T. (1969) Hammer reflection seismics in engineering geophysics: *Geophysics*, **34**, 383–95. [6.1.4]

Meiners, E. P., Lenz, L. L., Dalby, A. E. and Hornsby, J. M. (1972) Recommended standards for digital tape formats: *Geophysics*, **37**, 36–44. [5.4.8]

Meissner, R. (1967) Exploring deep interfaces by seismic wide-angle measurements: *Geophys. Prosp.*, **15**, 598–617. [2.4.6]

Millahn, K. O. (1980) In-seam seismics: position and development: *Prakla-Seismos Report*, **80**, no. 2+3, 19–30. [5.3.9*c, d*]

Milne, J. (1885) Seismic experiments: *Trans. Seis. Soc. Japan* **8**, 1–82. [1.2.1]

Mintrop, L. (1931) *On the History of the Seismic Method for the Investigation of Underground Formations and Mineral Deposits*: Hannover, Germany, Seismos. [1.2.1, 1.2.2]

Mooney, H. M. (1977) *Handbook of Engineering Geophysics*: Minneapolis, Bison Instruments. [1.4, 6.4]

Morgan, N. A. (1970) Wavelet maps – a new analysis tool for reflection seismograms: *Geophysics*, **35**, 447–60. [5.3.1]

Mossman, R. W., Heim, G. E. and Dalton, F. E. (1973) Vibroseis applications to engineering work in an urban area: *Geophysics*, **38**, 489–99. [5.4.3]

Musgrave, A. W., ed. (1967) *Seismic Refraction Prospecting*: Tulsa, SEG. [1.4]

Muskat, M. and Meres, M. W. (1940*a*) Reflection and transmission coefficients for plane waves in elastic media: *Geophysics*, **5**, 115–48. [2.4.6]

Muskat, M. and Meres, M. W. (1940*b*) The seismic wave energy reflected from various types of stratified horizons: *Geophysics*, **5**, 149–55. [2.4.6]

Neidell, N. S. and Poggiagliolmi, F. (1977) Stratigraphic modeling and interpretation; in *Seismic Stratigraphy – Applications to Hydrocarbon Exploration*, pp. 389–416 (ed. C. E. Payton): Tulsa, AAPG Memoir 26. [4.3.2*b*]

Nettleton, L. L. (1940) *Geophysical Prospecting for Oil*: New York, McGraw-Hill. [1.4, 6.1.2]

Newman, P. (1973) Divergence effects in a layered earth: *Geophysics*, **38**, 481–8. [2.3.1]

Newman, P. and Mahoney, J. T. (1973) Patterns – with a pinch of salt: *Geophys. Prosp.*, **21**, 197–219. [5.3.3*g*]

Northwood, E. J., Weisinger, R. C. and Bradley, J. J. (1967) Recommended standards for digital tape formats: *Geophysics*, **32**, 1073–84. [5.4.8]

O'Brien. P. N. S. (1965) Geophone distortion of seismic pulses and its compensation: *Geophys. Prosp.*, **13**, 283–305. [5.4.4*e*]

O'Doherty, R. F. and Anstey, N. A. (1971) Reflections on amplitudes: *Geophys. Prosp.*, **19**, 430–58. [4.2.2*b*]

Officer, C. B., Jr (1958) *Introduction to the Theory of Sound Transmission*: New York, McGraw-Hill. [2.4.8]

Olhovich, V. A. (1964) The causes of noise in seismic reflection and refraction work: *Geophysics*, **29**, 1015–30. [4.4.1]

Owen, E. W. (1975) *Trek of the Oil Finders: A History of Exploration for Petroleum*: Tulsa, AAPG Memoir 6. [1.2.1, 1.2.2]

Parr Jr, J. O. and Mayne, W. H. (1955) A new method of pattern shooting: *Geophysics*, **20**, 539–64. [5.3.3*e*]

Pautsch, E. (1927) *Methods of Applied Geophysics*: Houston, Minor Printing Co. [problems c.4]

Payton, C. E., ed. (1977) *Seismic Stratigraphy – Applications to Hydrocarbon Exploration*: Tulsa, AAPG Memoir 26. [1.4]

Peterson. R. A. and Dobrin, M. B. (1966) *A Pictorial Digital Atlas*: Pasadena, Calif., United Geophysical. [1.4]

Peterson, R. A., and Walter, W. C. (1976, 1977, 1978) *Seismic Imaging Atlas*, Vols. I, II and III: Pasadena, Calif., United Geophysical. [1.4]

Petty, O. S. (1976) *Seismic Reflections*: Houston, Geosource. [1.2.1, 1.2.4, 1.2.6]

Postma, G. W. (1955) Wave propagation in a stratified medium: *Geophysics*, **20**, 780–806. [2.1.4]

Poulter, T. C. (1950) The Poulter seismic method of geophysical exploration: *Geophysics*, **15**, 181–207. [5.4.2]

Rayleigh, Lord (1885) On waves propagated along the plane surface of an elastic solid: *Proc. London Math. Soc.*, **17**, 4–11. [1.2.1]

Rayleigh, Lord (1917) On the pressure developed in a liquid during the collapse of a spherical cavity: *Phil. Mag.*, **34**, 94–8. [5.5.3*f*]

Ricker, N. (1940) The form and nature of seismic waves and the structure of seismograms: *Geophysics*, **5**, 348–66. [4.3.4]

Ricker, N. (1944) Wavelet functions and their polynomials: *Geophysics*, **9**, 314–23. [4.3.4]

Ricker, N. (1953*a*) The form and laws of propagation of seismic wavelets: *Geophysics*, **18**, 10–40. [4.3.4]

Ricker, N. (1953*b*) Wavelet contraction, wavelet expansion, and the control of seismic resolution: *Geophysics*, **18**, 769–92. [4.3.2*a*]

Rieber, F. (1936) A new reflection system with controlled directional sensitivity. *Geophysics*, **1**, 97–106. [1.2.7]

Roark, R. L. (1976) Versatile energy source control system for seismic exploration applications: *1976 Offshore Technology Conference Preprints*, paper 2514; Dallas, OTC. [5.5.3*f*]

Robinson, E. A. (1967) *Multichannel Time Series Analysis with Digital Computer Programs*: San Francisco, Holden-Day. [1.4]

Robinson, E. A. and Treitel, S. (1973) *The Robinson–Treitel Reader*: Tulsa, Seismograph Service. [1.4]

Robinson, E. A. and Treitel, S. (1980) *Geophysical Signal Analysis*: Englewood Cliffs, N. J., Prentice-Hall. [1.4]

Rockwell, D. W. (1967) A general wavefront method; in *Seismic Refraction Prospecting*, pp. 363–415 (ed. A. W. Musgrave): Tulsa, SEG. [5.6.3, 6.2.3*a*]

Rosaire, E. E. (1935) On the strategy and tactics of exploration for petroleum: *J. Soc. Pet. Geophysicists*, **6**, 11–26. (Reprinted in *Early Geophysical Papers of the Society of Exploration Geophysicists* (1947), pp. 255–70: Tulsa, SEG.) [1.2.1]

Rosaire, E. E. and Adler, J. L. (1934) Applications and limitations of dip shooting: *Bull. AAPG*, **18**, 119–32. [1.2.6]

Rosaire, E. E. and Lester, O. C. Jr (1932) Seismological discovery and partial detail of Vermillion Bay salt dome, *Bull. AAPG*, **16**, 51–9. (Reprinted in *Early*

Geophysical Papers of the Society of Exploration Geophysicists (1947), pp. 381–9: Tulsa, SEG.) [1.2.1, 1.2.6]

Saul, T. and Higson, G. R. (1971) The detection of faults in coal panels by a seismic transmission method: *Int. J. Rock Mech. Min. Sci.*, **8**, 483–99. [5.3.9*d*]

Savarensky, A. (1975) *Seismic Waves*: Moscow, MIR. [2.2.1, 2.2.9]

Savit, C. H., and Siems, L. E. (1977) A 500-channel streamer system: *1977 Offshore Technology Conference Preprints*, paper 2833; Dallas, OTC. [problems c.5]

Schenck, F. L. (1967) Refraction solutions and wavefront targeting: in *Seismic Refraction Prospecting*, pp. 416–25 (ed. A. W. Musgrave): Tulsa, SEG. [6.2.3*a*]

Scherbatskoy, S. A. and Neufeld, J. (1937) Fundamental relations in seismometry: *Geophysics*, **2**, 188–212. [5.4.4*b*]

Schneider, W. A. (1978) Integral formulation for migration in two and three dimensions: *Geophysics*, **43**, 49–76. [4.3.2*c*]

Schoenberger, M. (1970) Optimization and implementation of marine seismic arrays: *Geophysics*, **35**, 1038–53. [5.3.3*g*]

Schoenberger, M. and Levin, F. K. (1978) Apparent attenuation due to intrabed multiples: *Geophysics*, **43**, 730–7. [4.2.2*b*]

Scholte, J. G. (1947) The range of existence of Rayleigh and Stoneley waves: *Monthly Notices, Roy. Astron. Soc., Geophys. Supp.*, **5**, 120–6. [2.2.10*c*]

Schriever, W. (1952) Reflection seismograph prospecting – how it started: *Geophysics*, **17**, 936–42. [1.2.1, 1.2.2]

Schulze-Gattermann, R. (1972) Physical aspects of the 'airpulser' as a seismic energy source: *Geophys. Prosp.*, **20**, 155–92. [5.5.3*b*]

Scott, J. H., Lee, F. T., Carroll, R. D. and Robinson, C. S. (1968) The relationship of geophysical measurements to engineering and construction parameters in the Straight Creek Tunnel Pilot Boring, Colorado: *Int. J. Rock Mech. Min. Sci.*, **5**, 1–30. [6.4]

SEG (1948, 1956) *Geophysical Case Histories*; Vols. I and II: Tulsa, SEG. [1.4]

Ségonzac, Ph. D. de and Laherrére, J. (1959) Application of the continuous velocity log to anisotropy measurements in Northern Sahara; results and consequences: *Geophys. Prosp.*, **7**, 202–17. [2.2.11]

Senti, R. J. (1981) Geophysical activity in 1980: *Geophysics*, **46**, 1316–33. [1.3.2]

Shah, P. M. and Levin, F. K. (1973) Gross properties of time–distance curves: *Geophysics*, **38**, 643–56. [3.2.3]

Shaw, H., Bruckshaw, J. M. and Newing, S. T. (1931) *Applied Geophysics*: London, His Majesty's Stationery Office. [1.2.1]

Sheriff, R. E. (1973) *Encyclopedic Dictionary of Exploration Geophysics*: Tulsa, SEG. [1.1.1, 1.4, 5.3.4, 5.4.1, 5.5.4, 5.5.5*c*, *e*]

Sheriff, R. E. (1974) Navigation requirements for geophysical exploration: *Geophys. Prosp.*, **22**, 526–33. [5.5.5*a*]

Sheriff, R. E. (1976) Inferring stratigraphy from seismic data: *AAPG Bull.*, **60**, 528–42. [4.3.2*a*]

Sheriff, R. E. (1977) Limitations on resolution of seismic reflections and geologic detail derivable from them: in *Seismic Stratigraphy – Applications to Hydrocarbon Exploration*, pp. 3–14 (ed. C. E. Payton): Tulsa, AAPG Memoir 26. [4.3.2*a*]

Sheriff, R. E. (1978) *A First Course in Geophysical Exploration and Interpretation*: Boston, International Human Resources Development Corp. [1.4, 5.3.5]

Sheriff, R. E. (1980) *Seismic Stratigraphy*: Boston, International Human Resources Development Corp. [1.4]

Sheriff, R. E. and Lauhoff, T. A. (1977) Marine geophysical exploration – the state of the art: *I.E.E.E. Trans. on Geoscience Electronics*, **GE–15**, 67–73. [5.5.6]

Sherwood, J. W. C. and Trorey, A. W. (1965) Minimum-phase and related properties of the response of a horizontally-stratified absorptive earth to plane acoustic waves; *Geophysics*, **30**, 191–97. [4.3.4]

Shortley, G. and Williams, D. (1950) *Physics*: New York, Prentice Hall. [2.2.12]

Sieck, H. C. and Self, G. W. (1977) Analysis of high resolution seismic data; in *Seismic Stratigraphy – Applications to Hydrocarbon Exploration*, pp. 353–85 (ed. C. E. Payton): Tulsa, AAPG Memoir 26. [problems c.5]

Silvia, M. T. and Robinson, E. A. (1979) *Deconvolution of Geophysical Time Series in the Exploration for Oil and Natural Gas*: Amsterdam, Elsevier. [1.4]

Sittig, M., ed. (1980) *Geophysical and Geochemical Techniques for Exploration of Hydrocarbons and Minerals*: Park Ridge, N.J., Noyes Data Corp. [1.4]

Slotnick, M. M. (1950) A graphical method for the interpretation of refraction profile data: *Geophysics*, **15**, 163–80. [6.3]

Slotnick, M. M. (1959) *Lessons in Seismic Computing*: Tulsa, SEG. [1.4]

Sokolnikoff. Y. (1958) *Mathematical Theory of Elasticity*: New York, McGraw-Hill. [Overview, chapter 2]

Spradley, L. H. (1976) Analysis of position accuracies from Satellite systems – a 1976 update: *1976 Offshore Technology Conference Preprints*, paper 2462; Dallas, OTC. [5.5.5*e*]

Stoep, P. M. V. (1966) Velocity anisotropy measurements in wells: *Geophysics* **31**, 900–16. [2.2.11]

Stoneley, R. (1924) Elastic waves at the surface of separation of two solids: *Proc. Roy. Soc. (London)*, **A–106**, 416–28. [1.2.1, 2.2.10*c*]

Stoneley, R. (1949) The seismological implications of aeolotropy in continental structures: *Monthly Notices, Roy. Astron. Soc. Geophys. Supp.* **5**, 343–53. [2.2.11]

Sweet, G. E. (1978) *History of Geophysical Prospecting*: Sudbury, Suffolk, England, Spearman. [1.2.1, 1.2.2, 1.2.3]

Tarrant, L. H. (1956) A rapid method of determining the form of a seismic refractor from line profile results: *Geophys. Prosp.*, **4**, 131–9. [6.2.2*a, c*]

Tatham, R. H. and Stoffa, P. L. (1976) V_P/V_S: a potential hydrocarbon indicator: *Geophysics*, **41**, 837–49. [5.7]

Telford, W. M., Geldart, L. P., Sheriff, R. E. and Keys, D. A. (1976) *Applied Geophysics*: Cambridge, England, Cambridge Univ. Press [1.4]

Thigpen, B. B., Dalby, A. E. and Landrum, R. (1975) Special report of the subcommittee on polarity standards: *Geophysics*, **40**, 694–99. [Appendix *F*]

Thornburgh, H. R. (1930) Wavefront diagrams in seismic interpretation: *Bull AAPG* **14**, 185–200. [6.2.3*a*]

Toksoz, M. N. and Johnston, D. H. (1981), Seismic Wave Attenuation: Tulsa, SEG (Geophysical reprint series No. 2) [2.3.2]

Tooley, R. D., Spencer, T. W. and Sagoci, H. F. (1965), Reflection and transmission of plane compressional waves: *Geophysics*, **30**, 552–70. [2.4.6]

Trorey, A. W. (1970) A simple theory for seismic diffractions: *Geophysics*, **35**, 762–84. [2.3.5*b, c, e*]

Trorey, A. W. (1977), Diffractions for arbitrary source–receiver locations: *Geophysics*, **42**, 1177–82. [2.3.5*b*]

Tullos, F. N. and Reid, A. C. (1969) Seismic attenuation of Gulf Coast sediments: *Geophysics*, **34**, 516–28. [2.3.2*c*]

Udden, J. A. (1920) Suggestions of a new method of making underground observations: *Bull. AAPG*, **4**, 83–5. (Reprinted in *Geophysics*, **16**, 715–6.) [1.2.2]

Uhrig, L. F. and van Melle, F. A. (1955) Velocity anisotropy in stratified media: *Geophysics*, **20**, 774–9. [2.1.4, 2.2.11]

Walton, G. G. (1972) Three-dimensional seismic method: *Geophysics*, **37**, 417–30. [5.3.7*c*]

Ward, R. W. and Hewitt, M. R. (1977) Monofrequency borehole traveltime survey: *Geophysics*, **42**, 1137–45. [2.3.3]

Washburn, H. W. (1937) Experimental determination of the transient characteristics of seismograph apparatus: *Geophysics*, **2**, 243–52. [5.4.4*b*]

Waters, K. H. (1978) *Reflection Seismology*: New York, Wiley. [1.4, 2.3.2*c*, 5.4.3, problems c.4]

Weatherby, B. B. (1940) The history and development of seismic prospecting: *Geophysics*, **5**, 215–30. [1.2.1]

White, J. E. (1965) *Seismic Waves – Radiation, Transmission and Attenuation*: New York, McGraw-Hill. [1.4, 2.2.10*d*, 2.2.11, 2.3.2*d*]

White, J. E. (1966) Static friction as a source of seismic attenuation: *Geophysics*, **31**, 333–9. [2.3.2*d*]

Whitfill, W. A. (1970) The seismic streamer in the marine seismic system: *1970 Offshore Technology Conference Preprints*, paper 1238; Dallas, OTC. [5.5.4]

Whittlesey, J. R. B., Neidell, N. S. and Arrington, G. R. (1980) Marine cross-dip seismic surveys: three-dimensional recording and mapping: *1980 Offshore Technology Conference Preprints*, paper 3847; Dallas, OTC. [5.3.7*b*]

Widess, M. B. (1973) How thin is a thin bed? *Geophysics*, **38**, 1176–80. [4.3.2*a*]

Wiechert, E. and Zoeppritz, K. (1907) Uber Erdbebenwellen: *Nachrichten von der Königlichen Gesellschaft der Wissenschaften zur Göttingen*, 415–549. Berlin. [1.2.1]

Willis, H. F. (1941) Underwater explosions – time interval between successive explosions: *British Report* WA-47-21. [5.5.3*f*]

Wood, L. C., Heiser, R. C., Treitel, S. and Riley, P. L. (1978) The debubbling of marine source signatures: *Geophysics*, 43, 715–29. [5.5.3*f*]

Wylie, Jr, C. R. (1966) *Advanced Engineering Mathematics*, 3rd ed: New York, McGraw-Hill. [5.4.4*b*]

Wyrobek, S. M. (1956) Application of delay and intercept times in the interpretation of multilayer refraction time–distance curves: *Geophys. Prosp.*, **4**, 112–30. [6.2.2*a, d*]

Zoeppritz, K. (1919) Über reflexion und durchgang seismischer Wellen durch Unstetigkerlsfläschen: Berlin, *Über Erdbebenwellen VII B, Nachrichten der Königlichen Gesellschaft der Wissenschaften zu Göttingen, Math-Phys.*, **K1**, 57–84. [2.4.4]

Index